Understanding Nature

Understanding Nature is a new kind of ecology textbook: a straightforward resource that teaches natural history and ecological content, and a way to instruct students that will nurture both Earth and self. While meeting the textbook guidelines set forth by the Ecological Society of America, *Understanding Nature* has a unique ecotherapy theme, using a historical framework to teach ecological theory to undergraduates.

This textbook presents all the core information without being unnecessarily wordy or lengthy, using simple, relatable language and discussing ecology in ways that any student can apply in real life. Uniquely, it is also a manual on how to improve one's relationship with the Earth. This is accomplished through coverage of natural history, ecology, and applications, together with suggested field activities that start each chapter and thinking questions that end each chapter. The book includes traditional ecological knowledge as well as the history of scientific ecological knowledge.

Understanding Nature teaches theory and applications that will heal the Earth. It also teaches long-term sustainability practices for one's psyche. Professor Louise Weber is both an ecologist and a certified ecopsychologist, challenging ecology instructors to rethink what and how they teach about nature. Her book bridges the gap between students taking ecology to become ecologists and those taking ecology as a requirement, who will use the knowledge to become informed citizens.

Understanding Nature
Ecology for a New Generation

Louise M. Weber

Department of Biology and Environmental Science, University of Saint Francis -
Fort Wayne, USA

CRC Press
Taylor & Francis Group
Boca Raton London New York

CRC Press is an imprint of the
Taylor & Francis Group, an **informa** business

Designed cover image: Diane Dickens

First edition published 2023
by CRC Press
4 Park Square, Milton Park, Abingdon, Oxon, OX14 4RN

and by CRC Press
6000 Broken Sound Parkway NW, Suite 300, Boca Raton, FL 33487-2742

Library of Congress Cataloging-in-Publication Data
Names: Weber, Louise M., author.
Title: Understanding nature : ecology for a new generation / Louise M. Weber,
Department of Biology and Environmental Science, University of Saint Francis, USA.
Description: First edition. | Boca Raton, FL : CRC Press, 2023. |
Includes bibliographical references and index.
Identifiers: LCCN 2022047833 (print) | LCCN 2022047834 (ebook) |
ISBN 9781032222615 (hbk) | ISBN 9781032222608 (pbk) | ISBN 9781003271833 (ebk)
Subjects: LCSH: Ecology–Textbook. | National history–Textbook.
Classification: LCC QH541 .W353 2023 (print) | LCC QH541 (ebook) |
DDC 577–dc23/eng/20221031
LC record available at https://lccn.loc.gov/2022047833
LC ebook record available at https://lccn.loc.gov/2022047834

ISBN: 9781032222615 (hbk)
ISBN: 9781032222608 (pbk)
ISBN: 9781003271833 (ebk)

DOI: 10.1201/9781003271833

Typeset in Minion Pro
by codeMantra

Access the companion website: www.Routledge.com/9781032222608

To the Metzger family, a great gang, who engendered my love for nature at dear old Beer Lake.

Contents

Acknowledgments

Every book has a story, and this story began largely at my family's annual reunions when I fell in love with nature. I wanted to know everything about nature, so I read books, made observations, and began at some point to write down what I learned. All of that became this book. Further inspiration came in 1985 during my first year of graduate school at Notre Dame. Dr. Robert McIntosh, ecology historian and editor of the natural history journal *American Midland Naturalist*, admonished us students to someday, somebody, write an ecology textbook using a historic framework. I was the least experienced student in the room and thought surely one of the other students would raise a hand to volunteer. This lightning bolt of an idea was meant for me and remained on my conscience for four decades.

Because it has been a career project, I thank my professors, my graduate school advisors, Dr. David Lodge at Notre Dame, and Dr. Susan Haig at Clemson, along with the other students in my research groups. I thank the hundreds of students in my ecology courses over the decades at Warren Wilson College in North Carolina and the University of Saint Francis in Indiana for enduring multiple editions of the unpublished manuscript. Many students provided corrections and pointed out jokes used twice during lecture. Others worked on illustrations and organizational aspects of this book. Thank you, Kyle Ibholm, Annie Bussierre, Alison Larrocca, Andy Driscoll, Pat Arrico, Elizabeth Martin, Little Bear Byrne, Amelia Snyder, Hannah McMerriman, and all of those who wrote helpful comments on course evals. Warren Wilson students critiqued the book from the perspective of some of the finest environmental studies majors in the U.S. Saint Francis students were mainly health pre-professionals and dared me to make them interested. I struck gold once I related ecology to mental health.

The major artists who contributed were Diane Dickens for the cover, students at USF, Grant Giacomin, Ellen Robbins (who also provided technical advice, proofreading, and designed my website), Lauren Gill, and friend, Emily Tauber-Steffen. I thank Helen Weber-McReynolds for writing the activity information on wilderness skills and for several semesters of teaching these attributes to my students during snowstorms. I thank other members of my immediate family for their ongoing support.

Reviewers of draft chapters over the years include Dr. Amy Boyd, Dr. Mark Brenner, and several anonymous ecologists. They have my utmost thanks. I thank TAA (Textbook and Academic Authors) for several workshops on how to write a textbook. Specifically, I thank Michael D. Spiegler for his counsel and encouragement. I thank the University of Saint Francis for a semester-long sabbatical and time spent at Holy Wisdom Monastery where I pulled together much of the remaining work. I thank friends at both Warren Wilson College and University of Saint Francis for ongoing support, including John Brock, Vicki Garlock, Laura Lengnick, Mallory McDuff, Alan Nauts, Naomi and Phillip Otterness, Catherine Reid, and Gretchen and Bennett Whipple. I thank the Ladies Chainsaw Club and Prairie Team at the Southwest Conservation Club.

Finally, I thank Mother Nature for inspiring this book and all it entails. It is still a beautiful world, even though She doesn't deserve how we, her most intelligent species, have treated Her.

More background for the book, review questions, and ongoing announcements can be found at *louiseweber.net*.

About the Author

Louise M. Weber, PhD, is Full-time Professor in the Department of Biology and Environmental Science at the University of Saint Francis, Fort Wayne, Indiana, USA, where she is also Division Director of Sciences and Program Director of Biology. She teaches or has taught Ecology and Diversity, Ecology, Evolution, Conservation Biology, Wildlife Biology, Introduction to Environmental Studies, Ornithology, Vertebrate Zoology, Invertebrate Zoology, Entomology, Research Design, Principles of Biology, General Biology, Human Biology, Human Anatomy, and Physiology I. For more information, visit her website at https://www.louiseweber.net

Introduction

In 1911, Anna Comstock wrote:

> Out-of-door life takes the child afield and keeps him in the open air, which not only helps him physically and occupies his mind with sane subjects, but keeps him out of mischief... This is the age of nerve tension, and the relaxation which comes from the comforting companionship found in woods and fields is, without doubt, the best remedy for this condition.

Anna Comstock, through her *Handbook of Nature Study*, was a leader in the nature study movement of the early 1900s (Armitage 2009). The movement was so successful that nature study was a mainstay in U.S. elementary schools until the 1950s. Comstock and others were known to directly influence Aldo Leopold and Rachel Carson, instrumental in bringing about the environmental movement of the 1960s (Figure 1.1).

Despite advancements in the environmental movement, other things have not changed. Almost a century after Comstock, Richard Louv (2005) wrote about the lack of familiarity children have with the outdoors, a condition Louv called "nature deficit disorder." In Louv's words:

> Stress reduction, greater physical health, a deeper sense of spirit, more creativity, a sense of play, even a safer life – these are the rewards that await a family when it invites more nature into children's lives.

Louv's book *Last Child in the Woods* is a plea for experiencing direct exposure to nature. The culprits bringing about nature deficit disorder, according to Louv, include obsession with screen electronics, stranger danger (fearful parents keeping children indoors), and overly structured childhoods, including participation in organized sports.

The problems are not solely with young people. Regarding adults, Comstock wrote:

> During many years, I have been watching teachers in our public schools in their conscientious and ceaseless work: and so far as I can foretell, the fate that awaits them finally is either nerve exhaustion or nerve atrophy.

Comstock's remedy:

> Did you ever try a vigorous walk in the open air in the open country every Saturday or every Sunday of your teaching year?

A 21ST-CENTURY RESPONSE

Ecopsychology together with its applied counterpart ecotherapy focuses on the synthesis of ecology and psychology. A basic assumption of these fields is that humans are deeply stressed by living in what Buzzell and Chalquist (2009) call "a collective narcissistic over-scheduled, runaway train that is modern society." Our disconnect with nature has been a major contributor to the epidemics of

DOI: 10.1201/9781003271833-1

Figure 1.1 Anna Comstock, leader of the nature study movement of the early 1900s (*Wiki Commons*).

our time: obesity, diabetes, suicide, depression, anxiety, chronic illness, and eye disease. Even the current epidemic of loneliness is at least partly a result of less direct interaction with people and more private time with technology (Brown 2019).

Yet, despite the vastness of society's ills, a major assumption of ecopsychology is that each of us has a readily available potential for healing. We access it simply by going outside and engaging with nature – with few side effects and near instant results (Nabhen et al. 2020, Bratman et al. 2019). When outdoors, our minds enter a state of what ecopsychologists call "awe" (Ballew and Omoto 2018). Our physiology immediately changes (lower blood pressure, lower stress hormones, more relaxed eyesight, improved focus, better immunity). It resets

our standards, resets our brains, and creates defining moments, especially in children.

THE GOALS OF THIS BOOK

The main goal is to help readers build a personal relationship with the Earth, under the premise that those who love nature will walk more softly and work to heal Her. For students who intend to raise children, own property, travel to distant places, take visits to wilderness areas, or be leaders when environmental problems arise, this book can lead to a deeper relationship with the Earth. Whether the readers are health pre-professionals who may eventually prescribe outdoor time for patients, or those seeking a nature-related occupation, this book can provide a source of therapy for Earth and self.

Natural history is included because it is the parent discipline of ecology. It draws us to nature as children, leads professionals to the study of ecology, and gives us joy. If we are to discover our truest eco-self, natural history can become a practice as well as a set of facts. It is available to anyone, even in urban settings.

Joy is essential. Ecopsychologists have observed that in the magnitude of the environmental crisis, people withdraw with a sense of helplessness (Roszak 2009). The environmental movement has been so successful that it has invented a problem too large to solve, in many people's mind. The response can be regression to what feels good, creating further epidemics such as shopping, gambling, substance abuse, and screen addictions (Brown 2019). The answer is not to scold people for their addictions and lack of concern, as environmentalists sometimes do, but to reconnect humans with nature in an expression of love rather than guilt and fear (Roszak 2009). The key remedy is to have people engage in **reciprocity** – healing self by healing Earth.

Beyond these goals, this book is fully an ecology textbook, meeting the guidelines set forth in the Four-Dimensional Ecology Education framework (4DEE) outlined by the Ecological Society of America. As such it covers ecological theory as well as natural history and applications so that students can live what they learn. *Understanding Nature* suggests field activities at the start of chapters and lists thinking questions and further activities at the ends of chapters.

ECOLOGY'S MAGIC GLASSES

An understanding of ecology begins by realizing there are many ways to see nature. Chapter 2 lays out the history of ecological thought. It introduces traditional ecological knowledge (indigenous knowledge about nature) and the history of scientific ecological knowledge (in the tradition of Western science). Even within scientific ecological knowledge ecologists look at nature and see different elements, including populations, communities, ecosystems, and landscapes.

Blink once and a mountainside is one or more **populations** (the members of one species living together in one place). Blink again and the same mountainside becomes several species interacting to form an **ecological community** (an assemblage of species that occurs in one place). Blink three times and the mountainside is an **ecosystem** (characterized by energy flows and nutrient cycles). The mountainside could also be a mosaic in **landscape ecology** (with its interactions between spatial patterns and ecological processes). Other ecologists may focus on trade-offs as in ecophysiology, evolutionary ecology, or behavioral ecology.

These ecological views are different from what natural historians (naturalists) see on the mountain. Naturalists focus on the names of species and the major characteristics of each. Besides biology, **natural history** includes geology, astronomy, and other natural sciences (Chapters 3–9). The ecological views are also different from the environmental problems and solutions that environmentalists see on the mountain (Chapter 32 on wicked problems). Environmentalists address issues with human dimensions like energy use or water quality. **Environmentalism** is characterized by **advocacy**, citizens reporting problems and seeking solutions. In contrast, ecologists have focused more on theory, determining why species are where they are and do what they do. In the relationship between the two disciplines, ecology is to environmental studies as physics is to engineering (Krebs 2001).

Finally, for the many ecologists who are conservation biologists, wildlife managers, fisheries biologists, range managers, foresters, or others in applied fields, they will find their specialty on the mountainside. **Conservation biologists** protect rare and endangered species and conserve native biodiversity (Chapter 28). **Wildlife managers** have traditionally focused on conservation of game species (Chapters 25–27). Other specialties are defined by their names.

WHAT IS ECOLOGY?

A common definition of **ecology** is the scientific study of interactions among organisms and with their environment (Figure 1.2). The organisms include those in all kingdoms and domains. The interactions are:

- *intra*specific (within species): competition, cannibalism, parasitism, mutualism, facilitation, adaptation, and evolution.

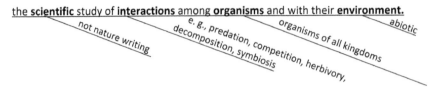

Figure 1.2 A popular definition of ecology depicted in a fishbone diagram.

- *interspecific* (between species): competition, predation, herbivory, symbiosis, mutualism, parasitism, decomposition, succession, facilitation, adaptation, and evolution.
- **between organisms and abiotic factors**: the flow of energy, the cycling of matter, and the interactions between species and their habitats.

A second common definition is the study of the distribution and abundance of organisms (Krebs 2001). This theme goes back at least as far as **Charles Darwin**, often considered the **founder of ecology** (McIntosh 1985). At the time Darwin finished writing *Origin of Species* in 1859, natural historians were describing patterns of abundance and distribution for organisms across the globe. In fact, by Darwin's time, some biologists had moved beyond mere description to providing explanations for distributions. This developed into ecology with the addition of Darwin's theory of natural selection. Explaining why organisms were where they were became a central goal that is still with us. Table 1.1 lists other important goals of ecology.

Beyond definitions, the attempt to develop theory within the four major subdisciplines has characterized ecology since 1859 (Figure 1.3). Darwin provided a theory to explain distribution and abundance of species. Building on Darwin's success, ecologists after him made it their life work to find more theory that would elucidate the principles that govern the field biology part of nature.

Community ecologists were first to begin the search for theory (late 1800s) (McIntosh 1985). These were mainly botanists who tend to study more than one species at a time because plants intertwine with other plants in their roots and layers.

Table 1.1 Summary of the goals, objectives, and approaches in ecology

Major goal of ecology	Major objectives within the goals of ecology	Four major approaches within theoretical ecology
To elucidate the principles that govern the field biology part of nature, accomplished by understanding: 1. relationships among organisms and relationships with the environment. 2. abundance and distribution of organisms at different spatial and temporal scales. 3. the flow of energy and the cycling of matter in ecosystems. 4. evolution as it applies to ecology. 5. the worldwide effort to better appreciate, respect, and protect the Earth.	1. develop theory to foster prediction, not just observation of nature. 2. develop techniques to apply theory. 3. apply theory to protect and restore nature, battle environmental problems, manage the environment, and sustainably obtain resources. 4. teach ecology at ever deeper levels and to ever wider audiences by cataloging the major vocabulary, principles, techniques, and accomplishments within ecology.	1. population ecology 2. community ecology 3. ecosystem ecology 4. landscape ecology

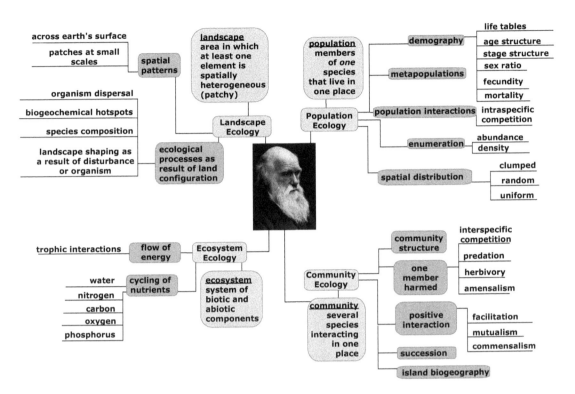

Figure 1.3 The evolution of an idea: the major ecology disciplines and their topics depicted in a spider diagram.

Population ecologists were next (early 1900s). These were mainly zoologists who tend to study one species at a time because animal movement makes it hard to concentrate on more than one individual at a time, much less more than one species.

Ecosystem ecology did not come to full recognition until the 1960s, simultaneously with the environmental movement, examining pollutants and other issues. Ecosystem ecology had a different focus than population or community ecology. **Ecosystem ecology** is "machine theory applied to nature" (Golley 1993). Its approach is abiotic, specifically focused on matter as it recycles within a system, and flow of energy as it moves through the system. In its classic sense, organisms are just bags of chemicals that use energy.

Landscape ecology came to recognition in the 1970s when satellite photos and early forms of layered maps were used to categorize large blocks of the Earth's surface. Each of the applied fields like conservation biology, wildlife management, and ecological restoration had their own histories. They were less focused on theory and more directed toward technique development.

Community and population ecology in particular were focused on forming theory for the first 100 years of ecology. Theory is important, but often underappreciated by students. Theory can do two things. It creates an organizing scheme, simplifying complex ideas into categories for better understanding. It also allows prediction, important in any discipline with applications. Just as it is one thing to report the winners of a horse race and another to predict them, so too ecological theory does not merely report an accumulation of facts. It explains the underlying causes and predicts outcomes.

In particular, theorists in ecology have investigated the **economy of nature** idea, a balance of nature, living things operating within self-correcting cycles, an idea suggested by Darwin and others before his time (Figure 1.4). The balance assumption led early ecologists to believe that a set of laws, theories, and principles governed nature. Once defined, ecologists hoped ecological laws could be used to predict outcomes, obtain resources, manage nature, protect humans from nature, and protect nature from humans.

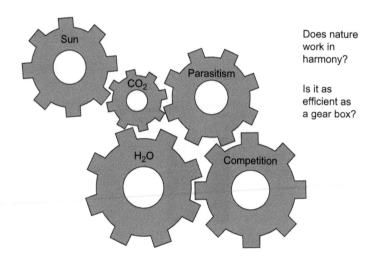

Does nature work in harmony?

Is it as efficient as a gear box?

Figure 1.4 The balance of nature idea is controversial. Is it really as efficient as a machine? (*G. Giacomin*).

Was the 100-year search for theory successful? I tempt you to look ahead at the final chapter to see a complete list of theories, laws, and principles thought to be true today in ecology. Until then, there can be two conclusions. The balance of nature idea is currently not held in high regard by ecologists after 100 years of searching (Kricher 2009). "Equilibria" are hard to find. Chance and abiotic (nonliving) factors are thought to play a greater role than previously appreciated.

Secondly, the list of theories in ecology is not as tightly categorized as the periodic table in chemistry or the laws of thermodynamics in physics. Ecology has fewer and looser theories than chemistry and physics, but has theory, nonetheless. The major ecological theories to arise since Darwin are:

- optimal foraging, trophic-level interactions, keystone species, equilibrium theory, nonequilibrium theory;
- patch dynamics, shifting mosaics, metapopulations, alternate stable states, stochastic models of population growth.

The fewer and looser theory does not mean we do not understand anything. Ecologists understand a great deal, and you will too after reading this book. The ecology theory established to date has been critically important in dealing with pandemics as well as endangered species management; pollution control; restoration of lakes, rivers, and grasslands; control of pest species; and many other applications. The search for theory continues in emerging topics such as:

- tropical ecology,
- open systems, human biomes,
- remote sensing, microbial ecology,
- deep marine and deep soil ecology,
- biogeography of freshwater communities,
- mutualisms and facilitation,
- biology of species survival in zoos, botanical gardens, fragmented areas, lawns, and nature reserves.

More than anything, the world is desperate for answers on climate change, sustainable harvest of resources, and preservation of its own humanity. This brings us back to Anna Comstock and the nature study movement of the early 1900s.

WHAT BECAME OF THE EARLY NATURE STUDY MOVEMENT?

Required natural history courses fell out of the curriculum as world wars turned education toward resource exploitation (Armitage 2009). During the 1960s, the environmental movement turned nature study toward the recognition of environmental problems and the study of systems. These and other factors led educators away from the Comstock-era nature study movement.

In colleges, science courses in the -ologies, such as entomology, mammalogy, and paleontology dropped out of the curriculum as genetics, molecular science, and health care became more immediate. Fewer students trained in natural history, and thus, fewer teachers taught nature study in schools. A circle of decline ensued. How do we reinvigorate?

TODAY'S NATURE STUDY

As important as it is to identify and name organisms and other natural entities, today's nature study must move beyond just that. If the nature study movement died once in the shadow of laboratory sciences, a more all-encompassing coverage of nature is required today. If nature really is remedy, bringing about the wholeness we seek, natural history combined with the major subdisciplines would be beneficial in understanding nature.

The best nature study would include elements of population, community, ecosystem, and landscape ecology, neobiogeography, post-modern evolution, conservation biology, wildlife biology, restoration ecology, traditional ecological knowledge, scientific ecological knowledge, wilderness skills, art, writing, and drawing to inform and inspire. Most importantly, nature study would foster an intimate personal relationship with the Earth that includes reciprocity. How do we get there?

THE STRATEGY FOR THIS BOOK

Nature study requires experience in nature, not just reading a book. Toward that end, field exercises and thinking questions are included in each chapter. Readers do not have to be in wilderness to complete the activities. A local park with a natural area helps, but if one can get to the outdoors, or even see the sky while stuck in traffic, nature study can happen. To benefit fully, however, nature study becomes a lifelong practice.

CLOSING ARGUMENTS

Ultimately, we must each forge our own relationship with the Earth and decide why such a relationship matters. In the words of Wangari Maathai (1940–2011), Greenbelt Movement Founder (Figure 1.5):

> We are called to assist the Earth, to heal her wounds, and in the process to heal our own.

Figure 1.5 Wangari Maathai, first African woman to win a Nobel Peace Prize. ("Wangari Maathai" by Oregon State University is licensed under CC-BY-SA.2.)

THINKING QUESTIONS

1. What is meant by "place-based study" or "a sense of place"?
2. What if humans were not in the place where you live? What would nature look like?
3. How could we live without using up the resources for ourselves or the people who come after us?
4. Rural areas in the U.S. and elsewhere often have no broadband and limited cell service. Beyond a mere inconvenience, people in these regions cannot fill out applications, and businesses cannot offer services via the internet. Make a list of other effects, positive and negative, that might be the result of this situation.
5. Aldo Leopold argues that to effect meaningful change in the world, we need to focus not just on people's behavior but on their "intellectual emphasis, loyalties, affections, and convictions." How do you reach people in your world in this way?

REFERENCES

Armitage, K.C. 2009. *The Nature Study Movement: The Forgotten Popularizer of America's Conservation Ethic*. University of Kansas Press, Lawrence, KS.

Ballew, M.T., and A.M. Omoto. 2018. Absorption: how nature experiences promote awe and other positive emotions. *Ecopsychology* 10:26–35. http://doi.org/10.1089/eco.2017.0044

Bratman, G.N., C.B. Anderson, M.G. Berman, and 24 other authors. 2019. Nature and mental health: an ecosystem service perspective. *Science Advances* 5. http://doi.org/10.1126/sciadv.aax0903

Brown, T. 2019. Does technology make us more alone? https://www.itchronicles.com/technology/does-technology-make-us-more-alone/

Buzzell, L., and C. Chalquist. 2009. *Ecotherapy: Healing with Nature in Mind*. Counterpoint, Berkely, CA.

Comstock, A.B. 1911/1986. *Handbook of Nature Study*. Cornell University Press, Cornell, New York.

Golley, F.B. 1993. *A History of the Ecosystem Concept in Ecology*. Yale University Press, New Haven, CT.

Krebs, C.J. 2001. *Ecology: The Experimental Analysis of Distribution and Abundance*. 5e. Benjamin Cummings, San Francisco, CA.

Kritcher, J. 2009. *The Balance of Nature: Ecology's Enduring Myth*. Princeton University Press, Princeton, NJ.

Louv, R. 2005. *Last Child in the Woods: Saving Our Children from Nature-Deficit Disorder*. Algonquin Books of Chapel Hill, Chapel Hill, NC.

McIntosh, R. 1985. *The Background of Ecology*. Cambridge University Press, Cambridge, UK.

Nabhan, G.P., L. Orlando, L. Smith Monti, and J. Aronson. 2020. Hands-on ecological restoration as a nature-based health intervention: reciprocal restoration for people and ecosystems. *Ecopsychology* 12:195–202. http://doi.org/10.1089/eco.2020.0003

Roszak, T. 2009. A Psyche as Big as the Earth. In L. Buzzell and C. Chalquist (eds.) *Ecotherapy: Healing with Nature in Mind*. Counterpoint, Berkely, CA.

2

We stand on their shoulders

DOI: 10.1201/9781003271833-2

ACTIVITY: Understanding SEK, TEK, and the Lay of the Land

Goal: Understand the primary land features that merge with cultural history.
Objective: Map and visit prominent sites of indigenous culture in your geographic region.

Step 1: Answer the following: for the Native Peoples who live/lived on the land where you are now, how do/did they refer to themselves? When did they first come to the land? What is the history of Western settlers and their influence on the Native Peoples? Do names of towns, cities, and places of interest reflect the presence of Native Peoples in your location? What do these names mean?

Step 2: Historically, native cultures used waterways extensively, especially rivers. Take the opportunity to look at topographic maps and become acquainted with waterways and contours (elevations) of your location. If topo maps of the local area are not available, Google Earth can be used.

Step 3: While examining your map, what is the elevation for where you are now? Go to a high building and look out the windows. What is the highest ground on campus or in your neighborhood? What is the highest elevation of your city or local area? Does elevation play a part in the cultural history of your area?

Step 4: Examine the contours on the map. When rain falls where you are, where does it go? Follow the flow all the way to the ocean on your continent. Where is the nearest continental divide from one major watershed to another?

Step 5: If you have not already done so, bring up Google Earth and examine your geographic location and its surroundings. Search the area using satellite images. Notice that some sections of rivers look like they may have been straightened. Satellite images may reveal old riverbeds.

Step 6: Make a copy of the relevant parts of the topo map together with any copies of historical maps that you may have, showing life just after European settlement. Traverse the pertinent areas by auto, bike, or foot. Follow the path of river travel, portage, or trading routes. Draw on your map the path you take and circle places that have significance. Streets may have names like Portage Avenue, Indian Creek, or Tecumseh Boulevard that serve as markers for important activities. Indicate the presence of old riverbeds, now with straightened rivers.

INTRODUCTION

Do scientists search for meaning or truth? Is there a difference? Does it matter? This chapter is meant to help deepen your response to these questions. The questions will be asked again at the chapter's end when perhaps you will have a different perspective. This chapter is about different ways of understanding nature and the history of this understanding in Western science.

NATURAL HISTORY MEETS TRADITIONAL ECOLOGICAL KNOWLEDGE

Traditional ecological knowledge (TEK) is the cumulative body of knowledge, practice, and belief about the relationship of living beings with one another and with their environment (Berkes 2012). TEK refers to indigenous or local knowledge and includes "an intimate and detailed knowledge of plants, animals, natural phenomena, and use of appropriate technologies for hunting, fishing, trapping, agriculture and forestry..." (Inglis 1993). TEK is handed down through generations by cultural transmission.

Dr. Robin Wall Kimmerer, of Pottawatomie ancestry and now a professor at State University of New York, invites us to learn about the traditional way of sustaining life (Kimmerer 2012, 2019). She laments, "it is not the land which is broken, but our relationship to land. Restoring the human relationship with the Earth will require physical involvement and the realization that the four-leggeds and basket makers might have something to teach us." Kimmerer (2012) writes, "cultivating a relationship with the living Earth should be an essential component of higher education."

According to Kimmerer (2008, 2019), traditional cultures already have ways of thinking about climate, classification, taxonomy, changing landscapes, monitoring, assessment, manipulation of landscape, and sustainable agriculture. Familiarity with population dynamics, pharmacology, and an evolving sense of this is included. There is great seriousness about these topics because the one who gets it wrong will not eat well. Kimmerer contrasts **traditional ecological knowledge** with scientific ecological knowledge (SEK) in Table 2.1.

As an example of TEK, consider the worldview of the Cree near James Bay in Canada as described by anthropologist F. Berkes and the Cree themselves (Berkes 2012). In this view, it is the animals, not people, who control the success of a hunt. If an animal decides to make itself available, the hunter is successful. The hunters have obligations to show respect to the animals. Lack of respect affects hunting because animals can retaliate by returning the discourtesy. To show respect, the killing should be done quickly and simply, without mess, using a gun appropriate for the size of the animal. Campsites are to be left tidy and clean. There are proper ways of carrying game. Once in camp, offerings may be made to the animal as a show of respect, perhaps with tobacco or pieces of meat or skin thrown into the fire.

Unlike the understanding in SEK for fish and wildlife management, in the Cree view, the cyclic disappearance and reappearance of game animals is related to the willingness of animals to be hunted (Berkes 2012). The disappearing animals go underwater or under the land. Continued use of the resources is important to achieve a sustainable, productive harvest. For this reason, some of the Cree trappers divide their territory into three or four sectors and trap heavily in one area only, resting the others. This keeps the system from reaching the critical point at which vegetation would be depleted by beaver through underuse by hunters.

As opposed to the popular current view that the best conservation would mean not hunting the animals at all, the Cree do not consider killing of game as an act of violence (Berkes 2012). The hunter loves the animals because they sustain the hunter's family. To the Cree, if the game want to be left alone, they would let the hunters know. In Cree conservation, "When you don't use a resource, you lose respect for it."

Table 2.1 Comparing the characteristics of TEK to SEK according to Robin Wall Kimmerer (2008)

Traditional ecological knowledge (TEK)	Scientific ecological knowledge (SEK)
Qualitative	Quantitative
Holistic and primarily characterized by respect and reciprocity	Mainly reductionist
Long term	Synchronic, short term, 4-year grant cycles
Resource users have knowledge (for example, basket makers)	Elite group of scientists hold the knowledge
Embedded in social, ethical, and spiritual	Mechanistic, purely rational
Explanations are spiritual and instinctual	Explanations consider nature as a machine
Nature as subject, considering even "non-human person"	Nature as object

TEK is not about content, but about process, writes Berkes (2012). TEK is a way of life. It is the actual living of life, with knowledge embedded in culture. It is knowing the waterholes, the songs that go with the trails, and the names of places.

For Berkes, sharing information about TEK is in the interest of developing a new ecological ethic. In the words of Berkes, TEK has the power to address shortcomings of contemporary Western thought and practice. Possibilities include restoring the unity of mind and nature, providing intuitive wisdom for developing a self-identity distinct from the world around us, and restoring a cosmology based on morality toward nature. In short, TEK could bring nature and culture together to re-integrate humans back into the ecosystem, unifying the mind and nature.

THE WOLF, THE RAVEN

To Pierotti (2010), the relationship between TEK and SEK is like that between wolves and ravens. The two species have coexisted for millennia, with ravens depending on wolves to provide food, and wolves depending on ravens to guide them to sources of food. As a pair they converge within a relationship that is sometimes friendly and sometimes hostile. We will return to this theme at the end of the chapter.

THE HISTORY OF SEK

Scientific ecological knowledge (SEK) refers to the Western understanding of nature. It had its early history in Europe. Specifically, it was the Greeks and their invention of the alphabet that made SEK possible (Kritcher 2009). As early as 2,500 years ago, the ancient Greek philosopher Anaximander taught that life arose in water. Simpler forms came before complex, and humans came from fish that evolved to live on dry land.

Later, a more prominent Greek philosopher, **Aristotle** (384–322 B.C.), did not believe in evolution, steering human civilization away from this concept for most of the last two millennia (Worster 1994). For Aristotle and his teacher Plato, form referred to genus and species of an organism, but these were immutable. One did not become another, even though the living forms could be ranked on a scale from highest to lowest in degrees of perfection (Minkoff 1983).

This ranking was called the scale of nature – *scala naturae*. Plants resided low on the scale and humans represented a higher degree of perfection (Figure 2.1). Quadruped animals that gave birth to live young were higher than flying mammals. Aristotle believed that no mobility occurred along the ladder.

For 18 centuries, Aristotle's hierarchy provided the fundamental understanding of nature. The ideas were not questioned or developed for at least the first ten centuries. The Christian-Jewish tradition discouraged scientific speculation, believing that science was a distraction from the sacred writings of the Bible. Species were created by God or developed by spontaneous generation.

During medieval times through the 18th century, even Christian intellectuals supported the *scala naturae*. They embellished it into the idea of the **Great Chain of Being**, adding the Trinity,

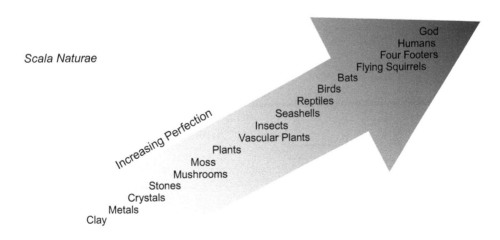

Figure 2.1 Aristotle's *scala naturae* (G. Giacomin).

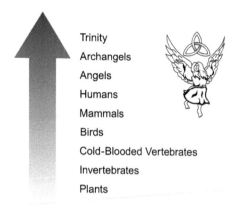

Trinity

Archangels

Angels

Humans

Mammals

Birds

Cold-Blooded Vertebrates

Invertebrates

Plants

Figure 2.2 The great chain of being, as embellished by theologians (*E. Robbins, G. Giacomin*).

archangels, and angels (Figure 2.2). Humans were the pinnacle of Earthly life (Minkoff 1983). Every organism had a place. Questioning this order went against God's purposes. The overarching sin was to aspire to a position higher than the one appointed in the hierarchy by questioning authority.

The reigning **paradigm** (set of beliefs) toward nature for two millennia in Western thought can be summarized as:

1. God created nature and all species one by one,
2. species, climate, and the Earth were immutable (unchangeable),

3. species stayed the same as God created them and these species were global,
4. a balance of nature kept everything in an orderly self-regulated system,
5. nature was controlled by a benevolent deity.

It was as if God were sitting at mission controls directing nature. God stood over and apart as the creator but was not in nature and could not be questioned.

THE RENAISSANCE AND NATURAL HISTORY

The paradigm began breaking down during the **Renaissance** (from 1500 to 1800) with the discovery of new continents and the realization that the Earth was round (Minkoff 1983). René Descartes introduced the experimental approach and **reductionism** (explanations derived from understanding the component parts), which eventually launched biology as a science. The discovery of blood circulation and the dissection of cadavers brought about an understanding of physiology. The issue of whether species could change over time (evolution) arose in the 1700s.

Further advancement came from the **Protestant Reformation** in the 1500s. Besides attacking papal authority, this was a time when intellectuals questioned everything (Figure 2.3). Galileo (1564–1542)

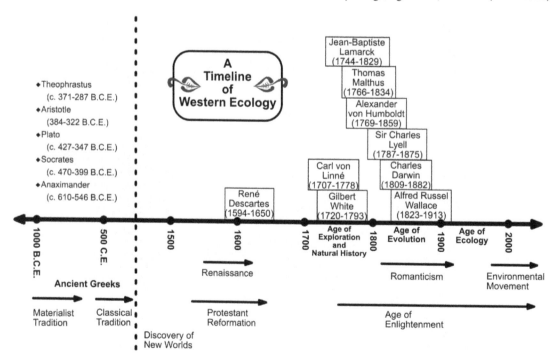

Figure 2.3 Timeline for the origins of Western ecology (*E. Robbins*).

taught that the universe operated by mathematical laws. Francis Bacon (1561–1626) argued that old, primitive systems of understanding should be abandoned, calling them idols (Pierotti 2010).

The French Revolution beginning in 1789 brought further changes within the **Age of Enlightenment**. This was the advent of progressive, modern ideas still in place today for science, philosophy, music, and political theories (Worster 1994). As part of the Enlightenment, European classical composers wrote very orderly music. Thomas Jefferson and other founders had great influence in ordering society and promoting democracy.

The Age of Enlightenment eventually brought about the **Industrial Revolution**. In a mere 100 years, Europeans emerged from relative ignorance and poverty. It came with a Western attitude that Native Peoples were ignorant and superstitious and that humans were meant to be dominant over nature, with a utilitarian philosophy that saw nature as a commodity (Pierotti 2010). Europeans regarded themselves as dominating the world.

1700s – AGE OF EXPLORATION AND NATURAL HISTORY

The 1700s could be called the age of exploration and natural history. This was the golden era of what we now call **biogeography**, the branch of biology that attempts to document and understand geographic patterns of biological diversity (Lomolino et al. 2016). Few plant and animal species were classified by the early 1700s, but this changed as biologists rode along on ships to explore and map the world's biota (Worster 1994). Johann Reinhold Forester (1729–1798) traveled with Captain Cook in 1778. Karl Ludwig Willdenow (1765–1812) and his Prussian student, **Friedrich Heinrich Alexander von Humboldt** (1769–1859), were other contributors.

LINNAEUS

As natural historians returned from their explorations, they needed a standard classification scheme for describing their collections. Carl von Linné (1707–1778), a Swedish botanist, wrote *Systema Naturae* in which he developed a species classification strategy still used as the basis for the Latin binomial nomenclature (genus-species) system. Carl von Linné named 12,000 species himself and was such a proponent that he became known as "Carolus Linnaeus," matching the style of genus, species endings in his scheme. Even with this development, Linnaeus argued that all interactions between organisms and the environment were directly controlled by God.

1800s – AGE OF EVOLUTION

By the 1800s, natural historians moved from documenting world patterns of biota to ascribing reasons for these patterns (Lomolino et al. 2016). Interpretations began to deemphasize the role of the Creator and increase emphasis on natural forces. The science began to teach that climate controlled nature rather than the direct hand of God.

Humboldt and Comte de Buffon (1707–1788) before him discovered, through a technique using isothermal lines, that different altitudinal areas on mountain ranges had the same vegetative zones as different latitudinal areas (Figures 2.4 and 2.5). In other words, the vegetation zones from the bottom of a mountain to the top corresponded to the vegetation zones from the equator to the poles.

Th relationship between altitude and latitude explained why tropical rain forests in separate places had similar life forms even though they had different species. The father of geology, Charles Lyell, concluded that the Earth's climate was mutable over long periods. Sea levels had changed. Tropical climates had once occurred in temperate areas. Species had gone extinct. These conclusions laid the groundwork for further thought on evolution.

GILBERT WHITE

Gilbert White was not a trained scientist, but had great influence on science and on Charles Darwin, Henry David Thoreau, and other notable figures. White lived in Selbourne, 50 miles southwest of London, serving as a pastor and living close to the land in his country estate. He wrote one of the best-loved books in the English language, *The Natural History of Selbourne*. The book took 18 years to write and was published in 1789. Its popularity in the 1800s led to over a hundred editions, making it one of the most reprinted books of all time. It is available

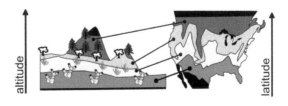

Figure 2.4 Mountain vegetative regions and their corresponding life zones. (*Adapted from C.H. Merriam 1890 by E. Robbins.*)

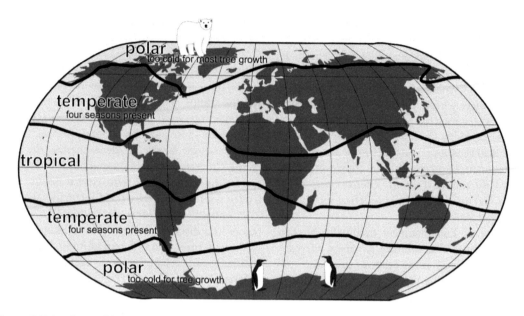

Figure 2.5 Isothermal lines not parallel to the equator. Each line represents the same average temperatures across the globe. (*Adapted from Bailey 2014 by E. Robbins.*)

in most libraries or free online as an open-source ebook.

The book consists of letters and journal entries about the natural history of Selbourne. Its content and style contrasted with other biology books of its time that emphasized physiology and corpses. Some of the most amusing sections are about a turtle named Timothy who for nearly 40 years lived in the garden of White's Aunt Rebecca Snookes, then later with Gilbert White. Regarding Timothy's occasional escapes from the garden, White wrote,

> The motives that impel him to undertake these rambles seem to be of the amorous kind: his fancy then becomes intent on sexual attachments, which transport him beyond his usual gravity, and induced him to forget for a time his ordinary solemn deportment.

White's book became popular because it contrasted dramatically with the destruction occurring in Great Britain during the 1800s (Worster 1994). England was changing drastically because of the Industrial Revolution, with classic English countryside transforming into factories and towns. White's book lived on because of nostalgia

for a time lost. Keeping a naturalist journal became a fad. Thoreau kept White's book at Walden and made his own journal into an American classic, *Walden*. Charles Darwin, with White's influence, kept a naturalist journal for years in which he reworked ideas that eventually developed into the theory of natural selection. Darwin even made a pilgrimage to Selbourne in 1850 as a tribute to White's book.

CHARLES DARWIN AND THE FOUNDING OF ECOLOGY

Charles Darwin was born on the same day as Abraham Lincoln in 1809. He was the grandchild of Erasmus Darwin (1731–1802), who had written published works about evolution (E. Darwin 1794). As a young adult, Darwin was invited to go on a 5-year voyage around the world on **His Majesty's Ship (H.M.S.) Beagle** (Figure 2.6). The ship's captain employed Darwin as his dinner companion, then later renamed Darwin as naturalist (Quammen 2009). The *Beagle* set out in 1831 to chart the coast of South America and several islands. In 1835, it arrived in the Galápagos, 800 miles off the Ecuadorian coast.

Darwin visited the Galápagos for 5 weeks, observing tropical penguins, lizards in the surf, brightly

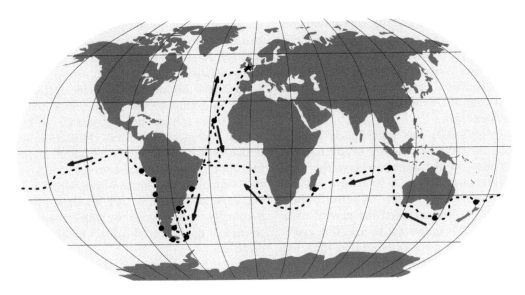

Figure 2.6 The voyage of the H.M.S. *Beagle*. (*Adapted from various sources by E. Robbins.*)

colored seabirds, and giant tortoises (Worster 1994). A dozen or more species of finch-like birds were similar except for beaks modified to eat different foods. The Galápagos observations, as well as fossils collected from the South American mainland began to influence Darwin's ideas about evolution. In Uruguay, while on the *Beagle*, Darwin received a mail package containing the second volume of Lyell's *Principles of Geology*. Lyell's book described life on islands as a matter of competition. This idea was the nugget for Darwin's thoughts on the struggle for existence within evolution.

Back in England, Darwin lived in London for the first 4 years after his voyage, preparing a book about his travels, which was published in 1839. In London, his ideas on evolution were further developed by reading **Thomas Malthus'** *Essay on Population*. Published in 1798, Malthus' book described the struggle for existence among social classes of people. It was while reading Malthus that the idea of **natural selection** became clear to Darwin. "Favorable variations would tend to be preserved and unfavorable ones to be destroyed" (Darwin 1887).

Worster (1994) speculates that perhaps it was both the influence of Malthus and the crowded, unpleasant life of "dirty, odious" London that explains the full title of Darwin's book, *On the Origin of Species by Means of Natural Selection, or The Preservation of Favored Races in the Struggle for Life*.

The stress and busy life of a well-known natural historian in London did not match Darwin's temperament. He moved to the countryside 50 miles away once he married into the Wedgwood family, known for their fine porcelain. In the country he had more privacy and took up family life as well as more writing. With the help of his family, he found abundant time for simple experiments, on earthworms, pigeons, plants, and other organisms in his backyard (Costa 2017).

Origin of Species was not published until 1859, which was 21 years after Darwin left London. During the interim, Darwin gathered a great deal of supporting data for the idea of natural selection, from embryology, comparative anatomy, paleontology, and animal breeding. He spent much of this time writing other documents, mainly with a natural history focus. In all he wrote over 100 articles and essays on different subjects in addition to more than 20 books, making him a mature writer by the time *Origin* was published.

Alfred Russel Wallace unwittingly pushed Darwin to publication. Wallace had come to the idea of natural selection while in a delirious fever state during his own biogeographic voyage in the Pacific. He sent his ideas to Darwin, so Darwin could pass them to Lyell – if he thought they were worthy. Wallace himself argued that it should be Darwin who developed the idea of natural selection and presented it to the world. Darwin had higher social status in England, and Darwin's grandfather had already published on some aspects of evolution. Darwin graciously arranged for papers to be read by both Wallace and him at the Linnaean Society of London. Darwin also explained on the very first page of *Origin* that Wallace had developed the same idea.

Origin of Species was immediately popular and controversial all around the Western world. Getting most of the attention were the ideas that adaptations could arise by chance and that humans had ape ancestors. Within 15 years, most scientists had accepted the idea of natural selection. It is important to note that evolution, the idea that species change over time, had already been conceived before Darwin. Even Charles' grandfather had written about evolution. Biologists were familiar with the idea from several avenues. Fossils had been discovered that seemed to be precursors to species alive today. Selective breeding for farm animals was familiar. Vestigial organs had been discovered such as a tail in the human fetus and a lateral toe on pigs and dogs.

What was new was the idea of natural selection as a mechanism for evolution. In 1809, **Jean Baptiste Lamarck** described the conventional understanding of the mechanism up to that time. Lamarck argued that evolution happens because of the **inheritance of acquired characteristics**. For instance, the son of a blacksmith could inherit the arm strength that his father developed over time. Children could inherit scars from relatives. Body parts that go unused, like the little toe, would eventually disappear.

After publication, vocal allies of Darwin promoted the idea of natural selection and were prepared to defend it. This included anatomist **Thomas Huxley**, who wrote to Darwin, "I am sharpening up my claws and beak in readiness." One outcome of Darwin's work was **social Darwinism**, natural selection applied to human societies. It originated with **Herbert Spencer** (1820–1903), an English philosopher who had coined the term "**survival of the fittest**" to describe natural selection. Spencer described natural selection as a violent struggle, "**red in tooth and claw.**" Darwin argued that it was usually a more peaceful matter, but the more dramatic rendition became popular. Social

Darwinism came to be used in some social and business entities to justify domination. Adherents included Nazis, Marxists, Andrew Carnegie, and other capitalists.

In the end, evolution, and specifically Darwin's books on evolution, had an enormous influence on Western culture, not just science. A list of the ten best ideas in Western civilization would undoubtedly include natural selection. Darwin's idea took on prominence because it questioned the assumption that humans were separate from and superior to nature, such a radical idea that eventually ecology would be called **a subversive science**. It questioned the direct hand of God in creation.

The theory of natural selection led to the founding of ecology and the search for more ecological theory. Much of the rest of this book tells that story.

THE WOLF AND THE RAVEN

Ravens depend on wolves to provide food, and wolves depend on ravens to guide them to sources of food. In the relationship between TEK and SEK, which is the wolf, and which is the raven? According to Berkes (2012), Western thought assumes there are immutable laws based on the search for universal truths. The goal of its science is to discover these truths with the aim of predicting and controlling nature. Western scientists are expected to describe events in nature objectively and with detachment. Berkes further characterizes Western thought by the presence of "high priests," the experts who know best, with a utilitarian attitude toward nature. Western culture sees itself as the beacon of knowledge, pushing back the sea of ignorance. With enough push, all people could live in one correct way of thinking according to rational processes. Has this worked, Berkes asks, or do we need a revisioning?

To Berkes Western thought is symbolized by man's dominion over nature. In contrast, TEK is a picture of life and spirit, incorporating people who belong to the land and have a relationship of peaceful coexistence with other beings. In TEK, there is explicit human-nature reciprocity in which animals have obligations to nourish humans in return for respect and other proper behavior. Berkes concedes that Western thought has more power in our current world. He writes that SEK and TEK can be pursued separately but in parallel, two canoes traveling side by side down the river of life. The options going forward are not just to either abandon traditional belief or resist the dominant society. There is a third option: to combine old and new. Initiatives may include conservation projects, ecological rehabilitation, ecotourism, cultivation of medicinal plants, and use of valuable crop varieties. Berkes sees certain environmental ethics as filling the gaps for SEK. Leopold's land ethic, a sense of place, biophilia, and adaptive management are themes that have congruence.

TO WHAT ENDS? NEXT STEPS

Do Western scientists search for meaning or truth? Is there a difference? Does it matter? For now, and as a way of summarizing Chapter 2 challenge yourself to the Thinking Questions below. From here, the next seven chapters cover natural history and what are commonly known as biomes. The root of their study began in the 1700s. Study continues to this day, especially regarding the conservation problems that plague biomes of the 21st century.

THINKING QUESTIONS

1. Examine a world map and find the Galápagos Islands. Are they to the east or west of Ecuador? How do you think Ecuador got its name?
2. In working toward his theory why do you think it was helpful for Darwin to join a pigeon-breeding club in England after returning from Galápagos?
3. Do you think we still live under the legacy of ten centuries of anti-evolution teaching? Explain.
4. Make your own Great Chain of Being by listing 15 common plant or animal species from where you live and attempt to rank them according to their relevance.
5. Aboriginal people were able to survive and thrive on a diet of wild plants and animals for food, medicine, and fiber. Basic foods included fruits (typically berries), green vegetables including water plants and seaweeds, underground roots, tubers, fungi, and rhizomes. For your local geographic area, what were the major native plants and animals consumed (Turner 1995)?
6. New research shows that some fungus species have substantial protein. Find out which species groups have the highest protein concentrations and determine whether these species are native in your location.
7. Via internet find the Gitxsan Classification system for plants used by some indigenous peoples. Describe how it works.
8. Research the phenomenon of mounds and effigies in Native American culture. Design on paper a series of effigies for your campus, community, or family that would demonstrate appropriate modern-day reflections of your culture.
9. If you live in a place with multiple mountain peaks or islands within waterways, research the names given to them by Native Peoples, as well as English names on old maps. Alternatively, create your own names for islands in a waterway such as a river and make a map of their exact location. Draw the profile, features, and vegetation.
10. Reach out to Native Peoples' groups in your area. Find out about public events, local museums, memorial statues, or workshops that can be a first step. When working with traditional knowledge, it is important to consider the generations of understanding and the spiritual significance that underlies this knowledge. While some traditional knowledge is general and openly shared, some information is sensitive and can only be shared when permission is given.

REFERENCES

Berkes, F. 2012. *Sacred Ecology*. 2e. Routledge, Oxfordshire, UK.

Costa, J.T. 2017. *Darwin's Backyard: How Small Experiments Led to a Big Idea*. W.W. Norton, New York.

Darwin, E. 1794–1796. *Zoonomia; or the Laws of Organic Life*. Thomas and Andrews, Boston, MA.

Darwin, C. 1887. *The Life and Letters of Charles Darwin, Including an Autobiographical Chapter*. John Murray Publishers, London.

Inglis, J.T. 1993. *Traditional Ecological Knowledge: Concepts and Cases*. International Program on Traditional Ecological Knowledge International Development Research Centre, Ottawa, ON.

Kimmerer, R.W. 2008. Keynote Address at the Environmental Summit Conference, June 2008, Syracuse, NY.

Kimmerer, R.W. 2012. Searching for synergy: integrating traditional and scientific ecological knowledge in environmental science education. *Journal of Environmental Studies and Sciences* 2:317–323, https://doi.org/10.1007/s13412-012-0091-y.

Kimmerer, R.W. 2019. Reciprocal Healing Conference, Natural History Institute, Sedona, AZ.

Kritcher, J. 2009. *The Balance of Nature: Ecology's Enduring Myth*. Princeton University Press, Princeton, NJ.

Lomolino, M.V., B.R. Riddle, and R.J. Whittaker. 2016. *Biogeography*. 5e. Sinauer, Sunderland, MA.

Minkoff, E. 1983. *Evolutionary Biology*. Addison-Wesley, Reading, MA.

Pierotti, R. 2010. *Indigenous Knowledge, Ecology, and Evolutionary Biology*. Routledge, Oxfordshire, UK.

Quammen, D. February 2009. Darwin's first clues. National Geographic.

Turner, N.J. 1995. *Food Plants of Coastal First Peoples*. Royal British Columbia Museum, Victoria, British Columbia and University of British Columbia Press, Vancouver, BC.

Worster, D. 1994. *Nature's Economy: A History of Ecological Ideas*. 2e. Cambridge University Press, Cambridge, UK.

3

Biomes, life forms, and ecoregions

ACTIVITY: Warming Up Your Outdoor Skills

Goal: Slow down. Experience nature as therapy. Give back to nature and heal yourself in return. Get in touch with your senses and concentrate on the present. This is the essence of many meditation practices. **Objective**: Gain up to 20 points by finding items in a treasure hunt. Place the appropriate evidence for each item in a brown paper lunch sack, including drawings, and turn in the bag with a scorecard like the one below, but tailored for your biome. On the scorecard circle the points you think you have achieved.

SCORECARD

1. Collect a twig from a tree species with opposite branching, correctly ID species .... 1 pt
   Do the same for alternate branching .... 1 pt
   And whorled branching .... 1 pt
2. Collect a pine twig with needles. Correctly identify the tree's species and location on a campus map.. 1 pt
3. Correctly draw the profile of the pine tree .... 0.5 pt
4. Collect a spruce twig with needles. Correctly identify the tree's species and location on a campus map .... 0.5 pt
5. Correctly draw the profile of the spruce tree you find .... 0.5 pt
6. Crush and smell the needle of a spruce on campus. Wet your rhinarium and smell again. Describe the smell as it seems to you. Did you get the same aroma when the rhinarium was wet? .... 0.5 pt
7. Get an estimate of the height of a tree using the thumb method. Have someone stand at the base of a tall tree. Walk a distance away until the person at the base is one thumb high when viewing your thumb at arm's length. Determine how many thumbs high the tree is in feet or meters based on the height of the person.
   How tall is the tree (signify units)? _____ 1 pt
8. Using the thumb method compare the height of a tall deciduous tree to the height of a mature coniferous tree. Which is taller? _____ **Signify** units. Height of deciduous tree _____
   Height of coniferous tree _____ 1.5 pt
9. Collect part of a plant that grows only in water. Identify the plant species ....2 pt
10. Find the tracks of three different animal species. Draw them. Mark their location on a campus map.. 3 pt

DOI: 10.1201/9781003271833-3

INTRODUCTION

Suppose you flew around the world examining land life and recording the dominant type of plant in each area. You would find that areas with similar climates had similar plant life forms. **Life form** of a plant refers to its **morphology** (shape, structure). For example, succulents occur in deserts, and broad-leafed plants occur in moist forests. Examples of other plant life forms are shrubs, **deciduous** trees (drop their leaves seasonally), **conifers** (cone bearing), and **herbaceous** (nonwoody) plants.

By examining geographic patterns of biological diversity, you would be engaging in **biogeography** (Lomolino et al. 2016). In other words, biogeographers ask the questions, where are organisms distributed and why are they distributed there. Biogeography underlies all ecology. Through his voyage on the H.M.S. *Beagle*, Charles Darwin himself was a biogeographer, although the term was not used at the time.

Even before Darwin, plant geographers of the 1700s and 1800s discovered the relationship between climate and life forms. They used the words "association" or "formation" to refer to blocks of similar vegetation with similar climate. This idea developed into zoological realms, biomes, and now ecoregions.

ARE THERE ANIMAL LIFE FORMS?

Once plant life forms were described in associations, animals were classified into geographic blocks in 1878 when Philip Schlater (1829–1913), a friend of Charles Darwin, defined regions for birds. Alfred Russel Wallace then divided the world into six **zoological realms** for all animals, designations still used to represent six major continental areas evolving in relative isolation (Figure 3.1). Biogeographical regions for animals took longer to establish than plants because most animals are small, difficult to collect, and have fewer direct relationships with climate.

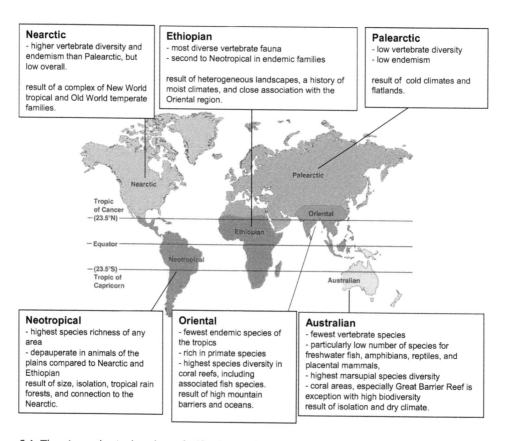

Nearctic
- higher vertebrate diversity and endemism than Palearctic, but low overall.

result of a complex of New World tropical and Old World temperate families.

Ethiopian
- most diverse vertebrate fauna
- second to Neotropical in endemic families

result of heterogeneous landscapes, a history of moist climates, and close association with the Oriental region.

Palearctic
- low vertebrate diversity
- low endemism

result of cold climates and flatlands.

Neotropical
- highest species richness of any area
- depauperate in animals of the plains compared to Nearctic and Ethiopian
result of size, isolation, tropical rain forests, and connection to the Nearctic.

Oriental
- fewest endemic species of the tropics
- rich in primate species
- highest species diversity in coral reefs, including associated fish species.
result of high mountain barriers and oceans.

Australian
- fewest vertebrate species
- particularly low number of species for freshwater fish, amphibians, reptiles, and placental mammals,
- highest marsupial species diversity
- coral areas, especially Great Barrier Reef is exception with high biodiversity
result of isolation and dry climate.

Figure 3.1 The six zoological realms of Alfred Russel Wallace. Major biodiversity characteristics are listed for each realm (Lomolino et al. 2016). In addition to the realm names, note that Holarctic is sometimes used to refer to the totality of the Nearctic and Palearctic Regions, and Wallacea refers to the complex of Pacific Islands between the Oriental and Australian realms (*Weber*).

Further understanding of the biogeography of animals began with recognition of several **name rules,** also known as **ecogeographic rules** (Table 3.1). Many name rules identify differences within or among animal species by **latitude** (distance north or south of the equator).

Table 3.1 Name (ecographic) rules and the biogeography of animals. Although named as rules, these are patterns at best, fraught with exceptions.

Gloger's rule	Allen's rule	Bergmann's rule
For related forms of endothermic animals, darker colors occur in more humid environments. C.W.L. Gloger (1883)	For warm-blooded animals, those in hotter environments have longer limbs and appendages. J.A. Allen (1877)	From C. Bergmann (1847) rewritten to: Tendency for average body mass of the geographic population of an animal species to increase with latitude (Lomolino et al. 2016). Appears valid for mammals, salamanders, turtles, parasitic flatworms, and ants.
Examples from the north include the white fur of Polar Bear, Arctic Fox, and Arctic Hare.	Supported by a limited number of isolated cases with many exceptions.	
Explanation: the pattern is the opposite of what you would expect if light-colored animals reflect sun and dark colors absorb it. Coloration is mainly a result of crypsis to avoid visual detection. Humid climates usually have dark shadows and dense vegetation. Dry areas have light soils or snow.	**Explanation**: thermoregulatory but needs more research.	**Explanation**: rather than a result of lower surface:volume, larger animals are more able to withstand environmentally harsh conditions. A similar pattern is deep-sea gigantism, with a positive correlation between depth and body size, probably for same reasons as in Bergmann's Rule.
Cope's rule	**Buffon's law**	**Vermeij's rule**
Over evolutionary time, the fossil record shows many lines of organisms continually increasing in body size until they go extinct. E.D. Cope (1887)	Environmentally similar, but isolated regions have distinct assemblages of mammals and birds. Comte de Buffon (1749)	Antipredator defenses (such as thicker, more sculpted shells) are developed to a greater degree in low latitudes. G. Vermeij (1978) Appears valid for marine gastropods, bivalves, barnacles, sponges, and terrestrial plant seeds (Lomolino et al. 2016).
	Explanation: species evolve when isolation occurs. Profound idea in Buffon's time because it opposed Linnaeus who ascribed the direct hand of God for distribution patterns.	**Explanation**: predation pressure higher in tropics.

(Continued)

Table 3.1 (*Continued*) Name (ecographic) rules and the biogeography of animals. Although named as rules, these are patterns at best, fraught with exceptions.

Thorson's rule	Latitudinal biotic interaction hypothesis	Rapoport's rule
Invertebrates with direct development are more likely to become isolated and have higher speciation rates. Thorson (1930s)	Biotic interactions are more intense at lower rather than higher latitudes. Dobzhansky (1950)	Geographic range size tends to increase with latitude. E. Rapoport (1975)
Appears valid for mollusks, gastropods, and crustaceans, but needs further research. For marine invertebrates, more diversity tends to occur with increasing latitude.	While true for some species groups for some interactions, should not be assumed as a blanket rule for all species and all interactions.	Exhibited by mammals, birds, mollusks, amphibians, reptiles, trees, fish, crayfish, amphipods, and beetles. Many exceptions.
Explanation: when dispersal is limited because of direct development, it produces more speciation than for indirect development, which disperses eggs or planktonic larvae far and wide, thus preventing isolation and speciation.	**Explanation**: higher species diversity at low latitudes, a long time since glaciation, and lack of freezing winters lead to more species packing and thus stronger interactions.	**Explanation**: may be that only organisms with a broad range of tolerance for variable climates can survive there, and as a side effect occupy broader ranges.
Rensch's rule	Foster's island rule	Hanski's rule
Mammals and birds in colder climates have larger litter/clutch sizes. Additional parts of the rule address size dimorphism of male versus female birds and mammals. Gaston (2009)	When living on islands, smaller-bodied species become larger-bodied, and larger-bodied species smaller-bodied compared with mainland areas. E.g., Pygmy Hippo and Key Deer. Gaston (2009)	More widespread species tend to be more abundant than species with small geographic ranges (positive correlation between distribution and abundance). Krebs (2001)

Source: Summarized from Lomolino et al. (2016).

HOW DID THE CONCEPT OF BIOMES ARISE?

The biome idea was introduced in a 1939 book by Victor Shelford (1877–1968) and Fredric Clements (1874–1945). **Biomes** are geographic blocks with plants and animals that are characteristic for that climate (Figure 3.2). They are identified by the dominant life form of the plants that are in the late stage of succession for that region. Physical factors such as soils and slopes are largely ignored.

WHAT DETERMINES WHERE BIOMES ARE LOCATED?

At the broadest level three factors determine why plant life forms occur where they do: evapotranspiration, frost line, and fire. **Evapotranspiration**

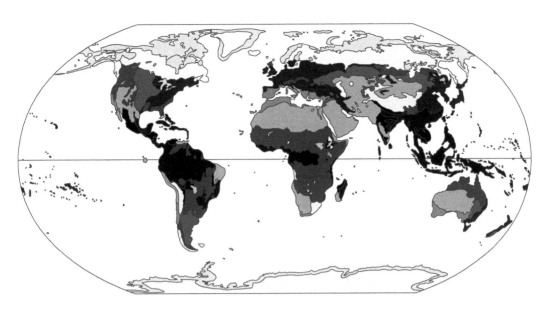

Figure 3.2 Biomes of the world. (*Biomes of the World.svg 2012 from World Wildlife Fund modified by G. Giacomin.*)

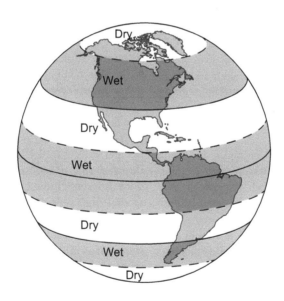

Figure 3.3 Zones of wet and dry alternating from the equator to the poles (*G. Giocomin*).

is a combination of evaporation and **transpiration** (the water evaporating from plants through open stomata). **Frost line** is the freezing depth within the soil from the surface. Beyond these, wind speed and soil type provide limiting factors for plants.

Because plant life form is based largely on climate, biomes occur in bands approximately parallel to the equator (Figure 3.3). Deviations to these parallel bands appear when mountain ranges and coastlines run in a north/south orientation perpendicular to prevailing winds. These alter the climate for hundreds of kilometers, for example, in the Americas where the Andes and Rocky Mountains create a rain shadow across much of the continents.

WHAT ARE THE LIMITATIONS OF BIOMES?

The number and naming of biomes has varied among textbooks since 1939. Without standard designations it is difficult for conservation agencies to communicate globally, a problem for tracking climate change, biodiversity losses, and other environmental problems. Secondly, biomes have traditionally covered divisions on land only, not water because biomes are identified by plant life form, and not all waterways have plants.

Additionally, biome delineation has always been a problem for map makers. Boundaries between biomes cannot be distinguished at fine scales, and distinct pockets of one may occur deep within sections of another. Note, too, that boundaries change with climate and human influence. Likewise, characteristics of one biome may blend with another. For instance, parts of the Great Basin in the western U.S. host grass species growing between shrubs, with temperature and rainfall equivalent to nearby regions dominated by cactus. Is this grassland, shrubland, or desert? Depending on the observer, it could be any of the three.

Besides these limitations, ecologists and conservationists need a better understanding of biome formation, not just a map and description. For example, shrubland covers an enormous percentage of the Earth's land. It sometimes occurs naturally with high species diversity. It also occurs inadvertently because humans create a shrubby wasteland from a more valuable ecological community. Land managers need to understand what subtle climatic and management practices tip the biological community toward shrub.

WHAT ARE ECOREGIONS AND ARE THEY BETTER THAN BIOMES?

The ecoregion system was devised by Robert Bailey in the 1970s to create a standard classification for describing geographic areas of the Earth (Bailey 2014). Bailey's system created a common language and standardization for the benefit of climate change monitoring and inventories in general. It encompassed waterways including oceans, recognizing that natural resources interact with their surroundings. Bailey focused on controlling factors that distinguish each area to provide an understanding of the mechanisms responsible for patterns in landform.

Bailey assigned a three-digit number for each area of the Earth, with numbers reflecting commonalities at a hierarchy of three scales. The first digit reflects the climatic zone at the widest scale within four **domains**:

- polar regions are 100s,
- humid temperate regions are 200s,
- dry regions are 300s,
- humid tropical regions are 400s,
- an M in front of the number designates a mountain ecoregion.

The domains run in bands that are roughly parallel from the equator to the poles. These bands are not perfectly parallel because dryness occurs on the west side of continents, and mountains produce different climatic regions than flatlands. The second digit reflects the climatic category within the domain. Up to 30 **divisions** have been defined, such as hot continental, warm continental, or subtropical. The third digit reflects the category within the division. Up to 98 **provinces** have been described and are based on life form modified by climate. Each unique three-numbered area is called an **ecoregion.**

For oceans, water temperature provides the primary means of classification and follows a similar pattern as for land:

- polar seas are 500s,
- temperate seas are 600s,
- tropical seas are 700s,
- an S stands for shelf and designates near-shore areas at depths less than 200 m.

Marine shelf regions have been further categorized through a system called Marine Ecoregions of the World (MEOW) by Spalding et al. (2007) who recognized 232 unique shelf ecoregions. A website is available with maps. Freshwater areas have been further categorized in a system called Freshwater Ecoregions of the World (FEOW) by Abell et al. (2008). A website is available with maps.

The ecoregion system has immense relevance for our current global reality because of its standardization. It has been adopted by the U.S. Forest Service and other U.S. government agencies as well as nongovernmental agencies including The Nature Conservancy, Sierra Club, and others. The limitation of ecoregions for ecologists is that Bailey is a geographer, not an ecologist. He defines ecosystems by climatic regions and thus primarily by abiotic factors (temperature, moisture, soil, topography) – then life form. Life form provides just one way of recognizing differences between climatic regions.

Thus, ecoregions are not exactly equivalent to biomes. For many purposes, the difference is not important, but sometimes ecologists want a stricter focus on life forms. For example, when distinguishing the reason why shrubland versus grassland occurs in a particular place, the causes can be subtle and completely biotic.

WORLD WILDLIFE FUND DESIGNATIONS

Because ecoregions are not exactly biomes, the World Wildlife Fund (WWF) devised a classification based on ecoregions, but with more focus on distinct biota. In the WWF system, 14 major land ecoregions, seven major freshwater ecoregions, and five major marine ecoregions are recognized. Each major type is further subdivided into seven biogeographic realms (Afrotropical, Australasia, Indo-Malayan, Nearctic, Neotropical, Oceania, Palearctic). The WWF provides distinct examples of each major type with a description.

THE 14 LAND BIOMES RECOGNIZED BY WWF

1. Tundra
2. Boreal forests/taiga
3. Temperate grasslands, savannas, and shrublands
4. Tropical and subtropical grasslands, savannas, and shrublands
5. Montane grassland and shrubland
6. Desert and xeric shrublands
7. Flooded grasslands and savannas
8. Temperate broadleaf and mixed forests
9. Temperate coniferous forests
10. Mediterranean forests, woodlands, and scrubs
11. Tropical and subtropical moist broadleaf forests
12. Tropical and subtropical coniferous forests
13. Tropical and subtropical dry broadleaf forests
14. Mangrove

NEXT STEPS

Figures 3.4–3.8 show the ecoregions by continents as designated by the WWF. Names may be shortened on the maps for convenience. Chapters 4–7 provide more detail about each major biome.

Figure 3.4 Biomes of N. and S. America. (*Adapted from worldwildlife.org by G. Giacomin.*)

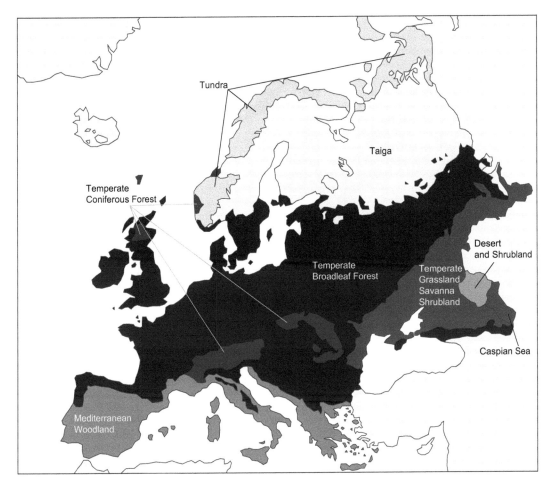

Figure 3.5 Biomes of Europe. (*Adapted from worldwildlife.org by G. Giacomin.*)

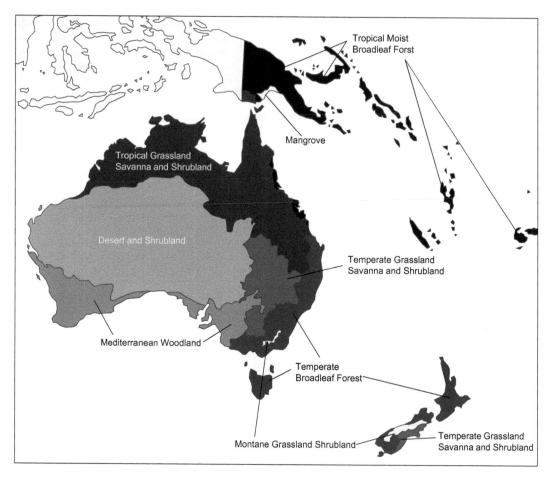

Figure 3.6 Biomes of Australia, New Zealand, and Pacific Islands. (*Adapted from worldwildlife.org by G. Giacomin.*)

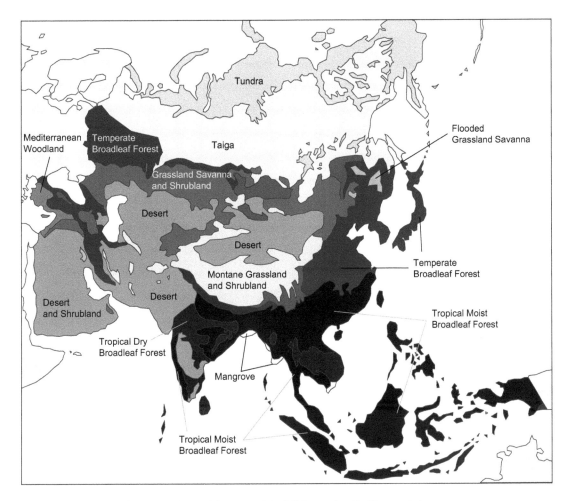

Figure 3.7 Biomes of Asia. (*Adapted from worldwildlife.org by G. Giacomin.*)

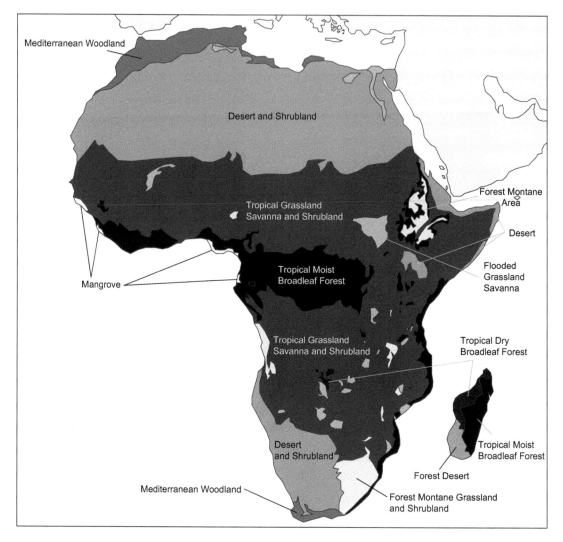

Figure 3.8 Biomes of Africa. "Montane" refers to mountains. (*Adapted from worldwildlife.org by G. Giacomin.*)

THINKING QUESTIONS

1. In which biome is your school situated?
2. In which ecoregion is your school situated?
3. Which is the largest continent? Which is the second? Rank all seven continents by size before searching the internet.
4. Are there other name (ecographic) rules besides those listed in this chapter? Search the internet and list.
5. Investigate the Japanese concept of "forest bathing" and describe it.
6. How is "ecotherapy" different from forest bathing? Describe the origin of ecotherapy and its main tenets.
7. Within peer-reviewed literature find two articles on the subject of ecoregions.

REFERENCES

Abell, R., M. Thieme, C. Revanga, and 25 other authors. 2008. Freshwater ecoregions of the world: a new map of biogeographic regions for freshwater biodiversity conservation. *BioScience* 58:403–414. https://doi.org/10.1641/B580507

Bailey, R.G. 2014. *Ecosystem Geography from Ecoregions to Sites*. 2e. Springer, New York.

Gaston, K.J. 2009. Geographic range. In: S.A. Levin (ed.), *The Princeton Guide to Ecology*. Princeton University Press, Princeton, NJ.

Krebs, C.J. 2001. *Ecology: The Experimental Analysis of Distribution and Abundance*. 5e. Benjamin Cummings, San Francisco, CA.

Lomolino, M.V., B.R. Riddle, and R.J. Whittaker. 2016. Biogeography. 5e. Sinauer, Sunderland, MA.

Spalding, M.D., H.E. Fox, G.R. Allen, and 12 other authors. 2007. Marine ecoregions of the world: a bioregionalization of coastal and shelf areas. *Bioscience* 57:573–583. https://doi.org/10.1641/B570707

4

Biomes: Tundra and taiga

Goal: To restore a personal relationship with the Moon in its wondrous yet common-place existence.

Step 1 (Definitions): Define Blue Moon, Super Moon, Wolf Moon, Blood Moon, and eclipses. How often do Blue Moons occur? When will the next Super Moon occur? What are the names given to Full Moons throughout the year? When will the next lunar eclipse be visible in your geographic area? Why are Full Moons sometimes blood red? What accounts for the redness?

Step 2 (Orbits and Moon Phases):

a. Make a sketch and indicate the direction of Earth's rotation as if you were viewing Earth from above looking down at the North Pole.

b. Make a sketch with the Sun on the right, the Earth in the middle, and the Moon in the following places in orbit around the Earth:
   i. between the Earth and Sun
   ii. directly opposite the Earth from the Sun
   iii. the angle between the Sun, Earth, and Moon is 90°, with Moon below Earth. Place another Moon at the same angle above.
   iv. indicate the lighted portion for each Moon in your sketch to show the major Moon phases.

c. Label the Moons in your sketch with the following:
   i. New Moon
   ii. Full Moon
   iii. First Quarter
   iv. Third Quarter

d. Add more Moons that are gibbous and crescent.

e. Show with arrows the direction the Moon is orbiting. (The Moon orbits the Earth in the same direction as the Earth spins.)

f. Indicate on your sketch which Moons are waxing and which are waning.

g. Be able to correctly identify the names of Moon phases from pictures provided by your instructor or quizzes found on the internet. Be able to correctly identify waxing and waning Moons from pictures.

Step 3 (waxing, waning, and finding direction): Examine the Moon on a clear night. Notice that in the northern hemisphere a waxing Moon is instantly recognizable because the right

---

DOI: 10.1201/9781003271833-4

side is illuminated, i.e., **the Moon waxes from right to left**. A waning Moon is illuminated on the left. These rules change in the southern hemisphere where the Moon waxes from left to right.

a. Notice that the crescent does not sit straight up. It is cradled at an angle. To tell direction, the straight line connecting the two points of the crescent point south when viewed in the northern hemisphere. The higher the Moon in the sky, the more precise the directionality.

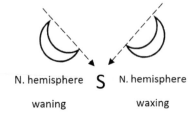

N. hemisphere   **S**   N. hemisphere

waning   waxing

b. For finding direction during summer in the northern hemisphere use the Big Dipper to find the location of the North Star (Polaris).

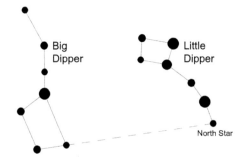

Big Dipper

Little Dipper

North Star

c. For finding direction during winter in the northern hemisphere use Orion. Specifically, find the middle star in Orion's belt in line with the main star in his sword. This line points south (Gooley 2014).

Orion

S

Step 4 (celebrating celestial holidays):

a. In two columns write the month and day of the solstices, equinoxes, and cross quarters (half-way between solstice and equinox).
b. In a third column write corresponding holidays near the celestial dates (e.g., Christmas, May Day, Ground Hog Day, religious holidays).
c. What explains why the holidays we celebrate are near the same times as celestial events?

WHAT IS MEANT BY "TUNDRA?"

Arctic tundra is characterized by short vegetation occurring in the far northern parts of the Old and New Worlds (Figure 4.1). Ice caps at the North Pole are not included because they have no land and no vegetation. Alaska, Canada, Greenland, Scandinavia, and Russia have arctic tundra. **Alpine tundra** occurs on high mountain tops throughout the world.

DOES ANTARCTICA HAVE TUNDRA?

In Antarctica, a landmass is present, but no tundra exists. Antarctica presently supports little vegetation because it is too cold, covered with ice, has little topsoil, or is too mountainous as on the Antarctic Peninsula (Figure 4.2). As icecaps melt, more rocky areas will be exposed, but currently Antarctica has no trees or terrestrial large mammals. Vertebrate life is abundant in some places and consists of seabirds and seals, with a few introduced rabbits, cats, and humans.

WHAT IS THE VEGETATIVE LIFE FORM FOR TUNDRA?

Most biomes are characterized by a typical plant life form, such as grass or shrub. **Tundra has many life forms** (lichens, mosses, sedges, grasses, forbs, shrubs, and dwarf trees) but they are not tall. Some species are miniatures of plant species that grow tall in other biomes.

WHAT ARE OTHER CHARACTERISTICS OF ARCTIC TUNDRA?

Arctic tundra is Earth's newest biome at 10,000 years old. Precipitation is less than 25 cm per year, equivalent to desert, but unlike desert, tundra has little evaporation. Thus, during summer, when surface sediments thaw, the tundra is characterized by boggy, wet pools, teaming with life. Below the immediate surface, the soil stays frozen as **permafrost**, which is impenetrable to roots. Trees of the tundra, when they occur, are most plentiful near rivers and lakes.

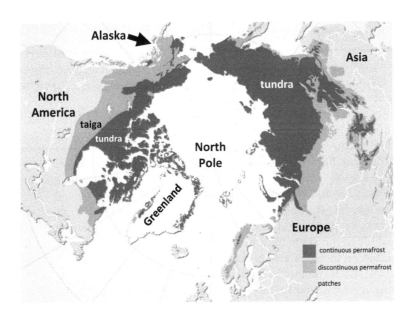

Figure 4.1 The North Pole, with no land mass, but continents surround it. (*Modified from UNEP.org.*)

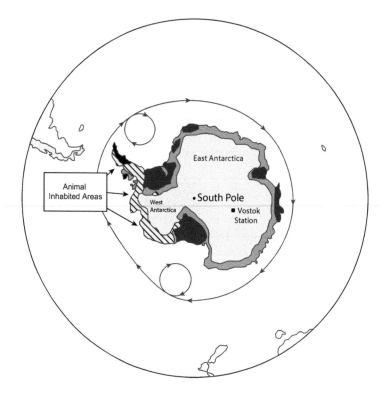

Figure 4.2 Antarctica's eastern and western ice sheets of the 20th century, landmass, and mountains. The Antarctic Peninsula points toward S. America and has little plant life, but abundant animal life in places (*G. Giacomin*).

ANIMAL LIFE IN ARCTIC TUNDRA

A profusion of lakes and bogs allows an abundance of Black Fly, Deer Fly, and mosquitoes. The larval insects of these species dwell in the sediment below water and are prey for waterfowl (ducks, geese, and swans) and shorebirds (sandpipers, plovers, dunlin, and others). Shorebird species have migrated great distances to nest in the arctic. Baird's Sandpiper, one of the longest distance migratory animals on Earth, spends the non-breeding season at the southern tip of S. America and the breeding season in the arctic.

Low species diversity but high abundance characterizes the animal life in the arctic tundra. The major vertebrate herbivores include Caribou (reindeer), Musk Oxen, Arctic Hare, Arctic Ground Squirrel, and ptarmigan (a type of bird). Lemmings eat vegetation year-round, even under the snow. Caribou migrate, spending summers in the tundra and winters in the taiga. Musk Oxen spend summers in valleys and winters on windswept ridges. Predators of the tundra include Gray Wolf, Arctic Fox, and Snowy Owl. Major fish species include cod, flatfish, salmon, and trout. Decomposition and nutrient cycling are slow within the soil food chain. Dung may take 5–10 years to decompose. Trees may never decompose. Thus, the tundra is a carbon sink as long as the permafrost stays frozen.

CONSERVATION PROBLEMS OF ARCTIC TUNDRA

An extremely fragile biome, life in the tundra is already stressed with extremes. Slight disruptions can have major repercussions.

- Climate change from greenhouse gases is expected to heat polar regions inordinately (7°C–8°C compared to 1°C–2°C in tropical areas).
- Permafrost melting is already dramatic, making large cliff-like depressions and release of gases.
- Oilfield and pipeline disruptions have permanently disturbed fragile soil and plant life and created barriers for migrating herbivores. Tire ruts can remain for decades.
- Oil spills at times have devastated marine life.

- Mining waste and industry have contaminated air and lichens (a chief food of many herbivores).
- Radioactive waste has been dumped in the Arctic Ocean, and plastic waste tends to accumulate there.
- On the hopeful side, conservation reserves have been established in many regions including much of an entire territory in Canada called Nunavut (Figure 4.3).

WHAT ARE CHARACTERISTICS OF ALPINE TUNDRA?

Alpine tundra occurs not in high latitude, but in high elevation. Vegetative growth forms are like those in the arctic. Other differences emerge (Table 4.1).

Few animals use alpine tundra for breeding unlike the arctic, but those that do are now facing climate change problems.

WHAT IS MEANT BY TAIGA?

Taiga is a Russian word meaning land of little sticks. Coniferous forest forms a circumpolar band south of the arctic tundra (Figure 4.4). Alaska, Canada, Scandinavia, and Siberia are the major sites. Taiga also occurs as fingers down the Appalachian, Rocky, and Cascade mountains of N. America. In Canada, taiga is known as **boreal forest**. Taiga is the largest biome, the most extensive forest in the world, even exceeding the tropical forests. Taiga is a young biome. Only 100 generations of conifers have passed since the glaciers (Askins 2002).

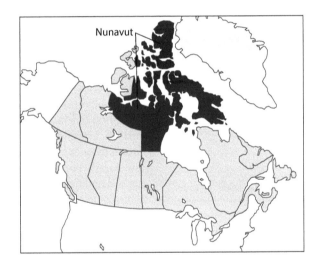

Figure 4.3 Nunavut is the largest and northernmost territory in Canada. Established in 1999, it has governance strongly influenced by the Inuit. Total land area is just smaller than Mexico.

Table 4.1 Arctic and alpine tundra compared

	Arctic	Alpine
Permafrost	Yes	No – except in highest
Dark winters	Yes	No
Precipitation	Scant	Heavy
Drainage	Poor	Good
Fragmentation	Low	High
Animal breeding	Abundant	Rare

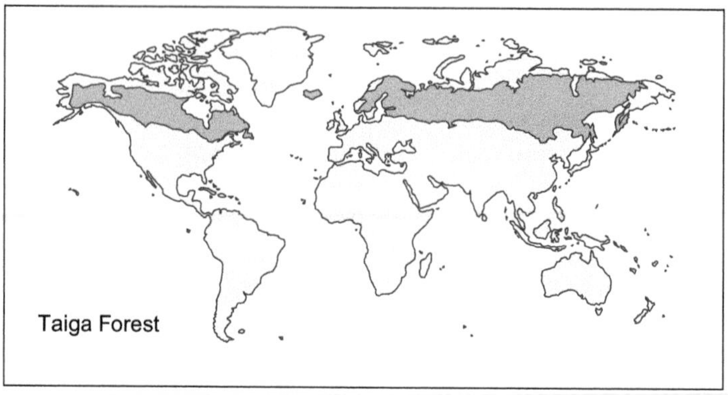

Figure 4.4 The taiga, a circumpolar band south of the arctic, mainly found in Alaska, Canada, Scandinavia, and Siberia (*G. Giacomin*).

WHAT IS THE PLANT LIFE FORM FOR TAIGA?

Coniferous trees are the dominant life form, consisting mostly of spruces, but pines, firs, and larches are also present. Trees with a cone shape shed snow easily, sometimes forming a closed canopy not in the treetops, but two-thirds of the way down from the top. The needle shape of the leaves allows photosynthesis from all directions in dim sunlight. The dark tree color encourages absorption of light. Thick bark protects against fire. Other vegetative life forms include mosses, grasses, lichens, and fungi.

WHAT ARE OTHER CHARACTERISTICS OF TAIGA?

The boundary between taiga and tundra transitions rather than forms a sharp line, but in general, the taiga has more precipitation than tundra (30–85 cm/year). Precipitation falls in the form of snow, rain, and dew, but levels are still low compared to most biomes. Air temperature is more extreme than tundra (both higher and lower) and more variable and stormy than in tundra owing to less maritime influence.

Fires can be extensive (250,000 ac., 100,000 ha) started by lightning. Dead branches, resin in wood, and the litter layer of needles make the forest flammable. Insect infestation (Spruce Budworm) and blowdowns create patches of dead trees, which make the forest more fire prone.

CONSERVATION PROBLEMS OF TAIGA

Traditionally, the taiga was not greatly modified in Canada and Russia, but the situation is changing rapidly.

- Wood is cut for lumber, paper, and cardboard.
- Poor regeneration in thin soils, flooding, erosion in clearcuts, climate change, and melting of permafrost have resulted in the southward expansion of tundra and the creation of more muskeg. Large blocks of impenetrable dead falling trees in bogs are known as **drunken forests**, where tree roots are weakened by melting soil and dead trees fall into each other.
- Mining for iron ore, gold, coal, natural gas, tar sands, and oil has increased, and is producing air pollution and acid precipitation that continually circulates in the circumpolar atmosphere.
- Massive hydroelectric schemes drain some areas and flood others.
- Prevalence of forest fire has increased as has mass insect damage to trees. Smoke from fires creates persistent haze.
- In Scandinavia and Finland, intensive forestry has been part of the culture for decades. Although sustainable, it has created a young homogeneous forest lacking the biological diversity of natural forests and the loss of distinctive species.

THINKING QUESTIONS

1. Although coniferous trees in the taiga have dark bark and needles, Gooley (2017) has noted that deciduous species in the taiga tend to have white bark, including species of birch, alder, aspen, and others. What accounts for the prevalence of white-barked trees in the northern forests?

2. What is the difference in appearance of spruces, pines, firs, and larches? What makes something a gymnosperm? How is it different from an angiosperm?

3. What are the "northern lights" and what accounts for their images?

4. Take the temperature of the ground surface below deep snow and in nearby areas free of snow. Compare and provide an explanation for consistent differences.

5. Catch common insects of a few species and place them in a large glass jar with a thermometer. Place the jar in the refrigerator overnight. The next day remove the jar and record the temperature at which each species becomes active. Make conjectures about the adaptive environment of each species based on the activity temperature (from Lederer 1984).

6. Gooley (2014) has written chapters of ideas for finding direction without the aid of GPS or compass. Verify or refute the following suggestions for finding your direction during snow:

   a. Snow leaves white lines on the windy side of trees. If you know the wind direction you know the direction.
   b. Tree bark is duller and darker on the poleward side of a tree opposite the sunlight.
   c. Trees tend to have more leaves and branches on their equator facing side.
   d. Solar panels on houses face toward the equator.
   e. Satellite dishes face the southwest in the Midwest and Eastern U.S. and southeast in the Western U.S.

7. Look up the taxonomic classification for lichens. Shake different types of lichen within a jar that has a little ethanol in it. Examine the sediment in the alcohol with a dissecting microscope and look for small animals, microbes, and bits of lichen.

8. Collembola (springtails; snow fleas) hop on the snow surface, usually in large groups. Investigate the taxonomy and lineage of these organisms. Are they insects? Are they fleas? How can they move in frigid temperatures? Where were they just before forming groups on the snow's surface? What mechanism and anatomy provides their jumping ability? Are they dangerous to humans or pets? Do they bite?

9. Jonna Jinton is a northern Sweden artist who, among other things, records the sound and images of ice forming on lakes in early winter. Listen for these sounds at a lake near you as cracks develop on ice. Record the patterns of images within ice. Like Joanna, put your ear on the ice of a lake and listen (wear hat and hood).

10. Within peer-review literature find two articles on the subject of tundra and taiga.

11. For study of the natural history of conifers Needham (1916) suggests a field exercise in winter and completion of a table in a notebook with several columns. For each conifer species write names along with a drawing of overall tree shape, color and shape of bark, canopy closure, length and color of needles, and whether needles are cylindric, flat, keeled, grooved, etc., bundled or solitary, opposite or alternate, form of cones, and diagrams of distinctive features.

REFERENCES

Askins, R.A. 2002. *Restoring North America's Birds*. Yale University Press, New Haven, CT.

Gooley, T. 2014. *The Lost Art of Using Nature's Signs*. The Experiment. LLC, New York.

Gooley, T. 2017. *How to Read Nature*. The Experiment. LLC, New York.

Lederer, R.J. 1984. *Ecology and Field Biology*. Benjamin Cummings, Menlo Park, CA.

Needham, J.G. 1916. *The Natural History of the Farm*. Comstock Publishing, Ithaca, NY.

5

Biomes: Grassland

ACTIVITY: Grassland Ecology in Three Dimensions

Goal: Gain appreciation for grassland by exploring its layers, species composition, animal qualities, and potential for grazing.
Time and location: Best done in late summer or early fall within a native plant grassland. A
weedy field or unmowed lawn could be used. Pasture is fine for some activities.

EXERCISE 1: IS THE VEGETATION STRATIFIED (LAYERED)?

a. Use pin flags or other markers to delineate a 1 m² square plot in a knee-high or taller grassland
   that is growing and not mowed, burned, or grazed.
b. When viewed from the side look for distinct layers (e.g., ground layer, subcanopy). On graph
   paper draw what is seen, labeling each layer, true to scale.

EXERCISE 2: WHAT SPECIES EXIST IN THE PLOT?

Use plant field guides to identify at least seven plant species within the plot. Classify each species as grass or forb, if grass (warm or cool and turf or bunch), web rooted or taproot, what place
within stratification.

EXERCISE 3: HOW GOOD IS THE HABITAT FOR ANIMALS?

Vertical canopy cover measures percent space covered by vegetation near the ground and at
increasing heights. It provides an assessment of habitat for aboveground animals because they
receive thermal protection, camouflage, nesting cover, and food from dense vegetation.

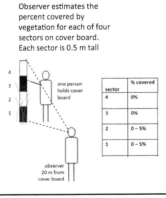

Observer estimates the
percent covered by
vegetation for each of four
sectors on cover board.
Each sector is 0.5 m tall

one person
holds cover
board

sector	% covered
4	0%
3	0%
2	0 – 5%
1	0 – 5%

observer
20 m from
cover board

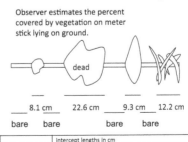

Observer estimates the percent
covered by vegetation on meter
stick lying on ground.

dead

___ 8.1 cm   22.6 cm   9.3 cm   12.2 cm

bare   bare   bare   bare

	intercept lengths in cm	
live vegetation	8.1 + 9.3 + 12.2	= 29.6%
dead vegetation	22.6	= 22.6%
bare ground		= 100-52.2 = 47.8%
dung patches	0	

DOI: 10.1201/9781003271833-5

EXERCISE 4: WHAT IS THE FORAGING QUALITY OF A PASTURE?

Crown cover measures percent live vegetation when viewed as if hovering over a pasture. It is a partial way of assessing the foraging quality of pasture. Place a meter stick near the ground in pasture grass. Look down from above and measure the summed length (and thus percent) of the stick covered by live versus dead vegetation versus bare space.

WHAT IS MEANT BY "GRASSLAND?"

Grassland is characterized by the dominant presence of grasses in vast flat areas too wet to be desert but too dry to support a closed canopy of trees. Grassland can be further divided into temperate and tropical regions. In temperate areas grassland tends to be a treeless plain except near waterways where trees are common. In tropical areas, grassland tends to be **savanna** with trees that do not form a closed canopy, and grass dominates the understory.

WHAT ARE CHARACTERISTICS OF TEMPERATE GRASSLAND?

Temperate grassland occurs in areas outside the tropics and polar regions. Other words for temperate grassland include **prairie** and **plain** (N. America), **steppe** (Asia), **puszta** (Hungary), **veld** (Africa), and **pampas** (S. America).

In temperate grassland summers are hot, and winters are cold with precipitation mainly in the form of summer storms and late spring snows. This is a pattern often the result of rain shadow from mountains (see Chapter 8). Multiyear periodic droughts occur roughly in 11-year cycles. Shrubland tends to occur instead of grassland if rain is erratic and comes during winter.

The fibrous roots of grasses hold water like a sponge, resisting drought and holding soil in place. Grassland is disturbance dependent. Its plant species evolved with frequent grazing, fire, or flooding. Dead plants build up above ground until disturbance occurs, but deep roots remain below. In some areas, fire is necessary to keep shrubs out and monocots (grasses) dominating over dicots.

Grazing or mowing as a disturbance tends to favor dicots over monocots. Most grassland plant species have evolved protections from disturbance. For instance, the apical meristems of grasses (the growing portion) are close to the ground and adapted to having their tops eaten by grazers or burned by fires.

Old growth non-disturbed grassland has no ecological value. Dead vegetation that builds up on the surface can shade out germinating seeds and become vulnerable to fire. Thus, the best conservation practice is to mimic the natural cycles of disturbance.

WHAT IS THE PLANT LIFE FORM FOR TEMPERATE GRASSLAND?

Dominant are **grasses** (nonwoody angiosperm monocots in the family Poaceae) mixed with **forbs** (nonwoody angiosperm dicots like Daisy and Goldenrod). Trees may persist (less than 10%), usually near waterways.

Grasses are classified into two growth forms (bunch or turf-forming), and warm or cool season. **Bunch** grasses grow in small tufts with deep roots, each bunch enlarging in diameter over time. N. American examples include Little Bluestem and Prairie Dropseed. Bare spaces naturally occur between bunches and provide nesting sites for ground-nesting bird species such as Northern Bobwhite.

Turf-forming grasses (also called **sod**) create a dense matrix with extensive shallow shoots that grow laterally and form new individuals (Figure 5.1a). Examples include several non-native species in the U.S. like the fescues and Kentucky Bluegrass. Some native species are turf-formers. Non-native turf-formers have been planted widely for lawns and cattle grazing. They require extensive measures to uproot and restore to natives because of their mat-like roots.

Warm season grasses (like Big Bluestem, Needle Grass, Blue Grama, and Indian Grass in the U.S.) are greenest in summer when they grow the most (sometimes 5–10 cm per day) and produce flowers (Figure 5.1b). They become brown and senescent in winter. **Cool season** grasses (like fescues and Kentucky Bluegrass) grow primarily in spring and

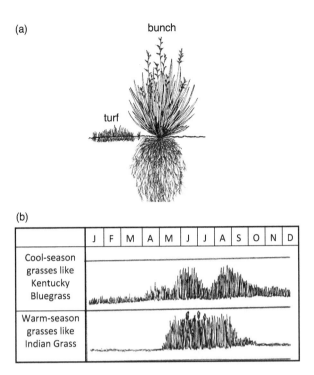

(a) bunch

turf

(b)

	J	F	M	A	M	J	J	A	S	O	N	D
Cool-season grasses like Kentucky Bluegrass												
Warm-season grasses like Indian Grass												

Figure 5.1 Categories of grass: (a) bunch versus turf-forming and (b) warm versus cool season (*E. Robbins*).

late summer. They remain green all year if sufficient water and sun are present and snow cover is scant. Their water needs are high as natives of cool, wet areas. Because they stay green for many months they have been widely planted for cattle but are drought-prone compared to warm season grasses.

THE DESERT-FOREST DIAGONAL IN TEMPERATE GRASSLAND

Temperate regions of continents display a consistent climatic pattern (Figures 5.2 and 5.3). In the U.S., this is demonstrated by rainfall increasing as one moves east, and temperature decreasing as one moves north. The result is a desert-forest diagonal from southwest to northeast, with shrubland and grassland in the mid-continent. Grass height increases moving east.

ANIMAL LIFE IN TEMPERATE GRASSLAND

Herbivores and their carnivorous predators are abundant in grassland. Most herbivores are invertebrates under the soil especially in the form of nematodes (roundworms). Above ground, herbivorous insects like grasshoppers are numerous. Herds of large mammals sometimes occur, although even large herds make up only a small fraction of the total biomass of animals below and above ground.

Before 12,000 years ago, horses, rhinos, camels, and mammoths were native grazers of the Great Plains in N. America (Askins 2002). Most of these large-bodied species disappeared probably because of overhunting by the first human inhabitants. Bison, elk, deer, and pronghorn remained perhaps because of descending from European mammals, better adapted at defending themselves from humans.

At the time European settlers arrived in N. America, Bison by the tens of millions roamed across the plains in large herds. Five species of prairie dogs, plus jackrabbits, ground squirrels, gophers, and voles grazed (Apfelbaum and Haney 2010). Prairie dogs played the role of keystone species, providing burrows for owls and snakes, attracting carnivores that fed on the guild.

Relatively few bird species are native to N. American temperate grassland compared to forests (Askins 2001). In the U.S. tallgrass prairies have no unique bird species. Shortgrass prairies have five species that appear to have evolved alongside grazers such as Bison and prairie dogs. Overall, a patchwork of disturbance is necessary

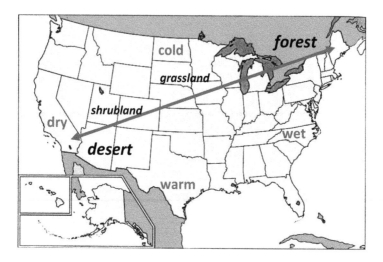

Figure 5.2 Biomes forming a desert-forest diagonal in the U.S. Temperate grassland, savanna, and shrubland occur in areas too wet to be desert, but too dry for a closed canopy of trees.

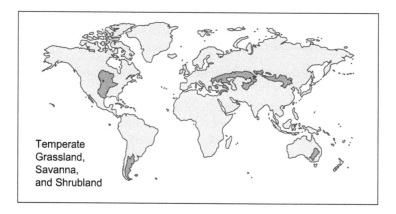

Figure 5.3 World Wildlife Fund biome categories are "Temperate Grassland, Savanna, and Shrubland" and "Tropical Grassland, Savanna, and Shrubland." This chapter focuses on temperate and tropical grassland. Chapter 6 describes shrubland and Chapter 7 describes savanna (*G. Giacomin*).

for maintaining prairie animal diversity, which has evolved with grazing in different stages of succession.

GRASSLAND CONSERVATION PROBLEMS

Worldwide, grassland is one of the largest biomes and the human favorite, recreated as savanna on lawns and golf courses. Grassland is the basis for much of the human diet, producing grains and sites of grazing. However, grassland has been the easiest biome to alter. Little is currently left in a natural state. Once changed it is difficult to achieve

widescale restoration. In N. America, prospects for conserving shortgrass and mixed-grass areas of the Great Plains are the best hope for providing large expanses of native prairie overall. One victory has been saving the American Bison from the brink of extinction and now successfully reintroduced to many areas throughout the Great Plains and raised on farms.

THE GREAT PLAINS OF NORTH AMERICA

The best-known grassland region in N. America is in the southern midsection of the continent (Figure 5.4).

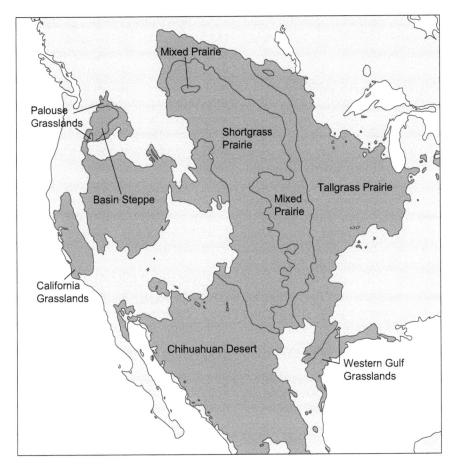

Figure 5.4 Grassland regions of the western two-thirds of the U.S. and neighboring regions (*G. Giacomin*).

This is sometimes called the "little Serengeti" because of its vast size. Shortgrass, mixed grass, and tallgrass sections lie in a moisture gradient from dry to wet, west to east.

Easternmost in N. America, **tallgrass prairie** makes up the wettest of the Great Plains. Patches occur as far east as Ohio. Of all types in the Great Plains, tallgrass incurred the greatest losses and the most soil broken by plow and development, as much as 98% (Wilsey et al. 2019).

Mixed-grass prairie occurs as a transition between shortgrass and tallgrass supporting intermediate grass height. Approximately 76% has been converted to cropland and development (Wilsey et al. 2019). Much of what is left is grazed.

Shortgrass prairie grades into desert and shrubland to its west and mixed grass to the east. It is adapted for sparse rainfall by having dense sod with little forb growth and shallow roots that soak

up water. Drought and grazing rather than fire are the main disturbances. It is too dry for most crops.

NORTH AMERICAN GRASSLAND NOT IN THE GREAT PLAINS

The **Chihuahuan Desert** covers parts of Texas, Arizona, New Mexico, and northern Mexico and is highly diverse in species. Grassland in a mosaic with shrubland covers 20%. In other areas bunchgrass intermixed with oak, creosote, and mesquite is adapted to frequent fire but is being destroyed by grazing and fire suppression. The **Central Valley of California** occurs in a rain shadow from mountains. It once hosted a highly diverse grassland, but has largely been replaced by development, non-natives, and agriculture. Less than 1% of the native area remains. **The Palouse Prairie of the Great Basin** between the Rocky Mountains and Cascade

Mountains of the U.S. northwest occurs in a rain shadow. Virtually all has been converted to shrub, non-native grasses, and crops.

OPEN CANOPY AREAS AND GRASSLAND IN EASTERN N. AMERICA

Small grasslands and brambly thickets are a natural feature within the forest matrix across the northeast U.S. (Askins 2002). In the U.S. southeast, even larger areas of savanna, grassland, wet grassland, mountain balds, and forest with grass understory were once extensive (Noss 2012). Remnants remain as a result of weather events like hurricanes and lightening-strike fires, insect infestations in trees, and beaver activity (once extensive). Native Americans used fire to enhance abundance of game, clear areas for villages, and provide resources. Even meteor activity in the past is thought to have produced openings in the form of small lakes and bogs.

It was not rain shadows that created these eastern U.S. grasslands. It was probably animals. Beginning 15,000 years ago vertebrate herbivores included caribou, mastodons, Giant Ground Sloth, Long-nosed Peccary, and Camel in the U.S. (Noss 2012, Lomolino et al. 2016). They kept areas open until humans crossed the Bering land bridge approximately 12,000 years ago or before. Early human inhabitants likely followed herds of large mammals such as mammoths and mastodons, hunting them for sustenance. Most of the N. American megafauna went extinct by 10,000 B.C., probably the victims of human overhunting, although Woodland Bison and Elk remained until the 1700s. After the megafauna were gone, the people of eastern N. America switched to hunting deer, turkey, and small game, often using fire to keep areas open. Balds on mountaintops remain to this day but are closing unless intentionally lit fire keeps them open.

Conservation practices for providing grassland/thicket vegetation in the eastern U.S. include the creation of permanent thickets under powerlines, maintenance of unmowed grass between runways at airports, and management of the Conservation Reserve Program (which pays farmers to take cropland out of production for a time and keep it in perennial grasses).

TROPICAL GRASSLAND, SAVANNA, AND SHRUBLAND

Tropical and subtropical grassland occurs on the poleward side of tropical forests that lie on the equator (Figure 5.5). Typically, grasslands are adjacent to mountains, oceans, deserts, or forests that in some way cause them to have dramatic and predictable wet and dry seasons. In the tropics, grassland is rarely a treeless plain. Instead, savanna (sparse trees not forming a closed canopy) occurs with an understory of grass. Savanna typically has tall bunchgrass that can sustain poor soils with little humus. The grasses lie dormant during long, hot droughts, then come to life during short rainy seasons.

Earth's most spectacular herds of large mammal herbivores and their predators inhabit the savannas, but the herds can only be sustained where large

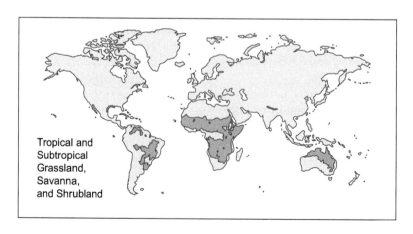

Tropical and Subtropical Grassland, Savanna, and Shrubland

Figure 5.5 Grassland, savanna, and shrubland on the poleward side of tropical forests (*G. Giacomin*).

contiguous blocks remain. The herds must follow seasonal rainfall and escape their usual predators during at least part of the year, thus needing large migratory routes. Vast insect swarms and mound-building termites are characteristic.

The migrations make it possible for many species to flourish in one place. This species packing has fostered the evolution of specializations among herbivores and among predators. Altogether, this bountiful animal life is only possible because of the combination of migration, seasonality, and the quality of grass, which enables the system to withstand fire, drought, and herbivore specialists.

CONSERVATION OF TROPICAL GRASSLAND, SAVANNA, AND SHRUBLAND

For generations many grasslands were sparsely populated by humans because of dangerous predators, poor soils, and insect vectors carrying disease (e.g., tsetse fly). Now these problems are being overcome as new grasses, agriculture, and vector controls have been introduced.

SOUTH AMERICAN GRASSLAND

Tropical grasslands of S. America are not as well known to the rest of the world as the savannas of Africa, even though the S. American savannas are wetter and have more species. The Cerrado of Brazil is the nation's second largest biome type and the most biodiverse savanna in the world (including 837 bird species). Few government protections exist for The Cerrado, which has changed dramatically in the 21st century to support soybean and beef exportation. Mass deforestation has occurred. The other major area is Los Llanos in the northern countries of Colombia and Venezuela.

GRASSLAND AND SAVANNA OF AFRICA

The African savanna includes parts of 27 countries and half the continent. In particular, the Serengeti Plain is a grass savanna with nutrient-rich volcanic sand and the largest diversity of hoofed animals in the world. The Serengeti is the only area that still hosts mass migrations, but problems include poaching, bush-meat hunting, climate change, increased agriculture, and development.

AUSTRALIAN GRASSLAND

Grassland is extremely important in Australia. Human colonization and grazing make up the greatest land use, covering 70% of the continent. The northern coast has a variety of tropical tall-grass ecosystems, seasonally dependent on rain. This includes areas with a grassy understory of tropical woodland dominated by eucalyptus. This area blends into a complex of temperate and tropical ecosystems to the east, south, and west.

THINKING QUESTIONS

1. Australia has a similar desert-to-forest-diagonal as the U.S. How is Australia's different?
2. How did the word "prairie" come to be used in describing eastern tallgrass areas in N. America? What is the derivation of "Great Plains?" How did "Palouse" get its name and how is it pronounced?
3. Much of the world's population is dependent on grassland for beef production. What are other ways that humans are dependent on grassland for calories, domestic animals, blood, and fiber, either historically or currently?
4. Some of the largest tropical savanna areas in S. America have been recently transformed. Investigate the current uses for the land and the beneficiaries.
5. Investigate the current demographic and political realities of African countries whose land is mainly tropical savanna.
6. Soils of the Great Plains, especially in the eastern tallgrass region, are extremely fertile, thus highly useful for agriculture. What accounts for the fertility of these soils and the fertility of other temperate grassland soils? What led to the "dust bowl" of the 1930s?

7. How did the invention of barbed wire change demographic factors on prairies?
8. What are the specifics of the Conservation Reserve Program in achieving conservation success for U.S. grassland?
9. What was the history of human inhabitation by the first people in the prairies/savannas of a given continent, for example, the N. American Great Plains?
10. What is the history of Cahokia, a Native American city that was on the Mississippi River?
11. Fire-tolerant species of trees have benefited from frequent fire set intentionally by indigenous people in prairie areas. What fire-tolerant species have benefited for your local area?
12. What are "balds" of the southern Appalachians? How are they different from above timberline ecosystems? What formed the balds?
13. What are ways to conserve natural grasslands now in highly populated areas such as the eastern U.S.? Could airports and powerlines help?
14. Teach yourself to identify grass-like species by beginning with the mnemonic, "sedges have edges, and rushes are round, grasses have nodes from their tips to the ground." How are sedges and rushes related to grasses and what does the mnemonic mean?
15. Grow native tallgrass prairie species like Big Bluestem in a very long glass or clear plastic tube to determine the length and fibrousness of the root.
16. Needham (1916) invites you to visit an old pasture and investigate it like a cow. Cows pull on grasses as they eat, breaking the grasses at their joints or pulling grasses out of their sheaths. In a grassland or pasture, pull on the plants yourself to mimic the grazer. If any plant comes up by its roots it will not survive grazing. Find a plant with stolons, find a rosette, find a taproot plant, and find a fibrous-rooted plant.
17. Investigate grasses on lawns that have low-growing seed heads. Individuals that have been mowed display low-growing seed heads. In the extreme, putting greens may have extremely low-growing seed heads. Unmowed areas nearby may display full-height seed heads. Are low-growing seed heads a result of evolution or do individuals acclimate? Devise an experiment to test your hypothesis.
18. Why mix clover with grass in a lawn? Clover purportedly grows best in parts of the lawn most heavily trafficked. Does this hold in your lawn?
19. As shown in Figure 5.1b, cool season turf grasses have two growth spurts per year. The growth in late summer is when lawn and pasture grasses most frequently make flowers and then seeds. For those seeking a perfect lawn it will look weedy during this phase. Investigate and draw the flowers and seeds of lawn or pasture grasses.
20. Inventory the vegetative species in a pasture. A few non-grasses will survive because they are prickly (teasel and thistle), or poor tasting (mullein, buttercup, daisy, and yarrow), or numerous (dandelions and plantains), or tiny (chickweeds). The grasses persist because they are apparently delicious and digest easily, adding mass to grazers. Grass seed gets deposited and spread by the mammals that carry it in their fur, feces, or hoofs, and that knock around the seed heads as the wind blows.
21. Needham (1916) maintains that when mammalian herbivores consume tall vegetation and trample small succulents on land that is not forest, the outcome is grass if sufficient moisture is present. Without tall plants, grasses are relieved of competition and flourish. If you do not believe it, Needham challenges you to mow almost any weed patch and see if grasses do not take over.
22. Grass is common around human habitations but is kept short. What role does this play in mud control? Would buildings be more prone to fire if we kept our lawns in tall grasses? Did/do aboriginal people living in grasslands control the length of the grass around their habitations?

REFERENCES

Apfelbaum, S.I., and A. Haney. 2010. *Restoring Ecological Health to Your Land*. Island Press, Washington, DC.

Askins, R.A. 2002. *Restoring North America's Birds*. Yale University Press, New Haven, CT.

Lomolino, M.V., B.R. Riddle, and R.J. Whittaker. 2016. *Biogeography*. 5e. Sinauer, Sunderland, MA.

Needham, J.G. 1916. *The Natural History of the Farm*. Comstock Publishing, Ithaca, NY.

Noss, R.F. 2012. *Forgotten Grasslands of the South*. 2e. Island Press, Washington, DC.

Wilsey, C.B., J. Grand, J. Wu, N. Michel, J. Grogan-Brown, and B. Trusty. 2019. *North American Grasslands*. National Audubon Society, New York.

6

Biomes: Shrubland, thickets, and desert

ACTIVITY: Writing a Naturalist Guide

Goals: Awaken curiosity and understand the work of "interpretive naturalists." Even if this is not your career choice, you may find yourself in charge of a scout group or family hike or want to write interpretive trail guides for a nature area. People assume that national park rangers and environmental educators just know everything as they lead a hike, but most naturalists carefully research and prepare a script for themselves, at least early in their career.

Objective: Write entries for a trail guide at a local park and submit them to the local park for their website, pamphlet, or volunteer training.

Step 1: At a park or nature area listen to a naturalist or your instructor lead a hike demonstrating nature interpretation. Read a sample pamphlet.

Step 2: Make a list of ten questions that occur to you during the hike and later as you wander the park on your own.

Step 3: Back home write five short paragraphs appropriate for a nature pamphlet at the park. Address your original questions or the prompts below. Research the answers using the web or library. Include citations. Do not plagiarize. Be factual and interesting.

PROMPTS FOR TOPICS IN AN INTERPRETIVE NATURE GUIDE

- What is the source of the lakes in your region?
- What are the littoral, limnetic, profundal, and benthic areas of a pond?
- What does it mean for a lake to be eutrophic?
- What is the definition of a wetland?
- What kind of rock is most common in your location?
- What is metamorphic rock?
- In what type of rock does one find fossils?
- Which venomous snake species occur? What should you do if you get bitten by a snake?
- Is it true that bats fly into people's hair? What bat species occur in your region? What is White Nose Syndrome?
- Kneel on the ground and smell the Earth. What scents do you detect?
- Which spider species in your location are venomous?
- Can you find any squirrel leaf nests in the trees or signs of other mammals? Are the nests concentrated in one area or randomly placed or systematically placed?
- What does it mean for a stream to be "first order" or "second order?"
- Why does the water temperature in a stream stay relatively consistent over time?
- Why do curves in a stream tend to get exaggerated over time?

DOI: 10.1201/9781003271833-6

WHAT IS MEANT BY SHRUBLAND (SCRUBLAND)?

Shrubland is characterized by the dominant presence of shrubs in areas too wet to be desert but too dry to support a closed canopy of trees (Figure 6.1). In contrast to grassland, rain is more erratic, sometimes greater than in grassland, but occurring in winter. Some shrubland is known as **xeric** (very dry) and found poleward of deserts and on the west side of continents.

WHAT IS THE PLANT LIFE FORM FOR SHRUBLAND?

Shrubs have multiple woody stems branching near the ground and no central trunk (Figure 6.2). Shrubs are usually less than 3 m tall and have a high **root to shoot ratio** (high underground component compared to above ground). This architecture makes shrubs the **tanks of vegetative life**. Their multistem form is the most efficient known for gathering light, moisture, and resisting heat and evaporation. Shrubs regularly withstand conditions that kill other plants.

In xeric conditions, leaves are often small, needlelike, and some have waxy leaves that reflect sun. Some are succulent. Despite their toughness, native shrubs are often considered second class, little appreciated by humans because of their limited agricultural value. Yet, species diversity in shrubland can be very high, with extreme specializations and high uniqueness value among the sort of

odd creatures that evolve there. The sheer extent of land covered by shrubland fosters speciation, and thus appreciation (Figure 6.1).

Shrub species can include oaks, hollies, pines, junipers, olive, cactus, eucalyptus, cycads, and others. When the shrubs are comprised of tree species, the community is sometimes called "woodland" rather than shrubland. In the Rocky Mountain states of the U.S., common shrub species include Rabbitbrush, Sagebrush, Saltbrush, Shadescale, and Greasewood, a combination sometimes called chico brush.

WHAT ARE OTHER CHARACTERISTICS OF SHRUBLAND?

Aromatic herbs (e.g., sage, lavender), woody shrubs, grasses, and short trees (e.g., acacia) can be present. The aromatic herbs emit oils when leaves are touched. The oils conserve water and deter herbivores. Some species (Sagebrush) exhibit **allelopathy**, the secretion of toxic substances by a plant's roots to harm the growth of nearby plants. Thorns are characteristic, which deter herbivores. Many shrub species can sprout stems directly from roots after a disturbance and thus regrow themselves. Some shrubs are flammable and ignite easily. Without fire shrubs do not recycle nutrients to the soil. Little green growth remains in old shrubs, which do not appeal to herbivores. Shrubland has been highly degraded worldwide by development, cattle grazing, irrigation, and too much or too little fire. Fire suppression can eventually result in hot, out-of-control fires that produce erosion, landslides, and death.

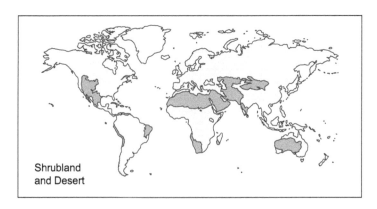

Shrubland and Desert

Figure 6.1 Shrubland is often found poleward of deserts and on the west side of continents (*G. Giacomin*).

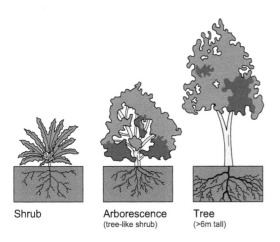

Shrub   Arborescence   Tree
         (tree-like shrub) (>6m tall)

Figure 6.2 Differentiation occurs among life forms of woody-stemmed plants (*G. Giacomin*).

WHAT ANIMALS OCCUR IN SHRUBLAND?

More animal species than expected inhabit native shrubland, given the difficult palatability of shrubs compared to grasses. However, the herbivores native to shrubland are less likely to live in herds. Instead, individual animals hide behind shrubs or within leaf litter and evolve specializations. Such non-herding species as Emus, Tinamous, Roadrunners, Black Mamba snakes, and Aardwolves are shrubland species. Many of Australia's marsupial herbivores, seed-eaters, reptiles, and insects survive in hiding places provided by shrubs.

North of the equator, insects mainly pollinate species of shrubland (Smith 2019). South of the equator, birds mainly pollinate. Thus, flowers on shrubs of the southern hemisphere are heavier. Ants do much of the work of planting seeds in shrubland. Typically, a tasty substance on the outside of a seed entices ants to take seeds underground, leaving the poor-tasting inner part to germinate.

In the U.S., animals of shrubland include migratory birds, Pronghorn, and Mule Deer. Although Pronghorn and Mule Deer also live in grassland, both species are specially adapted to host microorganisms that digest oils. In winter Pronghorn can feed on Sagebrush almost exclusively. Pronghorn are often misidentified as "antelopes" but they are not closely related to them. They are in their own mammalian

group, unique to N. America, and are the second fastest running land animal after Cheetah.

LOCATIONS OF SHRUBLAND AND NOTABLE TYPES

Most natural shrubland is climatically derived (e.g., Great Basin, southwestern Australia). It typically occurs in harsh environments between 20° and 40° latitude. Winters are mild and moist and induce shrubland to flower. Summers are hot, dry, and sometimes have fire. Other shrubland areas are anthropogenic via burning, grazing, removing trees, or introducing crops, which can alter the soil and species permanently (e.g., Japan's dwarf bamboo region). Some of the notable types of shrubland are described next.

MEDITERRANEAN

Mediterranean, the most renowned shrubland type, has cool, moist winters and hot dry, fire-prone summers. Examples are the wine-producing areas: California, Chili, South Africa, and western Australia. In Mediterranean areas of Europe, the shrubland has wetter summers than most shrubland and centuries of human influence, alterations, and domestication (Smith 2019).

THORN SCRUB

Tall (>5 m) open-canopy natural scrubland is found in wide areas on either side of the equator between savanna and desert, particularly in Africa and Australia.

GREAT BASIN OF THE U.S.

The Great Basin occurs west of the Rocky Mountains from Oregon through Utah, thus is temperate, arid, and often considered desert or grassland. Part of it is dominated by sagebrushes and the other part by Saltbrush. Salt builds up in the soil over time in shrublands, which can create an impenetrable layer for plant roots. Saltbrush can take up the salt and crystalize it on its leaves, giving a gray-green hue. Chenopod scrub is another name for Saltbrush scrubland.

Much of southwestern Australia is chenopod scrub.

SOUTHWESTERN AUSTRALIAN

Southwest Australia hosts a highly diverse xeric shrubland. Mallee (Gum Tree) covers a large part of it. The Gum Tree is highly flammable and catches the grass understory on fire. It survives after fire by having a lignotuber underground that stores water like a rain barrel.

FYNBOS – CAPE REGION OF SOUTH AFRICA

This xeric shrubland hosts 8,500 species of plants, equivalent to all of Europe. Nutrient-poor soils cover ancient rocks.

COASTAL SHRUBLAND OF EUROPE

Next to the ocean in the northern countries of Europe and the U.K. shrubbiness sometimes occurs because conditions are too windy and salty for trees.

CHAPARRAL AND CHACO OF THE AMERICAS

In the southwest U.S. and Mexico, chaparral is sometimes called woodland. Chaparral is Spanish for thicket of shrubby evergreen oak, forming a closed canopy. A similar type occurs in southern Chile and is called "chaco."

GRAND CHOCO OF SOUTH AMERICA

A xeric shrubland of S. America, Grand Chaco is larger than France and has multiple species of venomous snakes.

HEATHLAND OF EUROPE

Heathland occurs in cool or cold regions of northwest Europe and is not a xeric shrubland. **Heathland** is defined by a dwarf shrub canopy 1 m or less tall, mostly supporting laurel and blueberry. Heathland also occurs in South Africa, western and southeastern Australia, and in other places, including the U.S., often in Blueberry. It is debatable whether European heathland was human-created or natural. The "moors" of Scotland were of human origin and used for grazing.

THICKETS IN U.S. FORESTS

In eastern N. America openings in the forest are created by windstorms, ice storms, fire, insect infestation, and beavers. These can become invaded by brambly shrubs and trees as part of succession, often happening right after a disturbance, without a grass stage. These thickets are short-lived but birds that inhabit them are adapted to shifting (e.g., Chat, Rufus-sided Towhee, and Orchard Oriole). Because of human development, the species that depend on this shifting patchwork are becoming rare. Some effective conservation strategies include clear-cutting small patches of forest. For 3–12 years after the clear-cut, birds of shifting thicket are present. In power line management, some electric companies create permanent thickets under power lines to prevent invasion of trees. This requires cutting of young tree and spraying stumps individually with herbicide or girdling all trees. Conservation measures in other areas include retention of beaver, wildfire, insect outbreaks, and flooding along wooded streamsides.

CONSERVATION OF SHRUBLAND

Shrubland can be a challenging place for humans to live, yet life was once sustainable for nomadic hunter-gatherers who often used fire to enhance their existence. Clearing for agriculture brought an end to these lifestyles and caused dramatic declines in species diversity. Even shrubland that escaped all disturbance except overgrazing can become monocultures with little conservation or agricultural value. Thus, conservation must be region-specific and informed by local natural history and indigenous knowledge.

WHAT IS "DESERT?"

Desert is better defined by climate than plant life form because some deserts have no vegetation. Thus, **desert** occurs where evaporation exceeds precipitation. More precisely, a **desert** is where the potential for **evapotranspiration** (evaporation, and water evaporating from stomata) is more than precipitation (Quinn 2009). This is opposed to wet

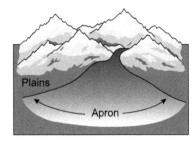

Figure 6.3 Mountain sediments carried by runoff can sometimes produce dramatic alluvial fans (*G. Giacomin*).

climates, where precipitation exceeds evapotranspiration, resulting in excess that runs off as rivers. In simpler terms, desert usually gets less than 10 in. (25 cm) per year of precipitation.

Ironically, when water does fall in deserts, it can cause heavy erosion, even if desert streams remain dry most of the time. Many desert areas have **interior drainage**, meaning that instead of water running to the ocean, the water flows to the lowest point between mountain ranges. The surrounding mountains may bare the marks of heavy erosion and have steep sides, abruptly bordered by gently sloping flats below. Sediment carried by the runoff settles out in a gently sloping area at the base of the cliffs known as an **alluvial fan** (Figure 6.3). Springs may appear at the base of the fan as water percolates out of the sediment toward the low point. A lake may appear at the low point in which salt accumulates as water evaporates, salt that would ordinarily be swept to the ocean. Layers of sodium chloride and calcium carbonate (lime) may accumulate on dry lakebeds in the valley. Plants around the lakebed may be salt-adapted, and no plants may be growing in the heavy salt areas.

When rivers do run through desert, they carry water that has accumulated in wetter climates (Quinn 2009). For large rivers like the Nile or the Colorado, plants near the water may not be xeric-adapted. A similar phenomenon explains why oases have palm trees, which are not xeric-adapted.

Deserts occur on all the continents except Antarctica. Because of interior drainage, desert organisms tend not to be widely distributed to other deserts, but instead evolve from species in nearby wetter climates (Quinn 2009). Amazingly, evolution in separate places tends to produce similar adaptations and look-alike organisms, even though the species are not closely related. This phenomenon is called

"convergent evolution" (see Chapter 10). Examples include Kangaroo Rat and Jerboa, Jackrabbit and Patagonian Hare, Kit Fox and Fennec Fox, and unrelated lizards.

WHAT ARE OTHER CHARACTERISTICS OF DESERT?

Desert covers 20%–35% of the Earth's land surface and more desert occurs in the northern hemisphere, but it is because more land occurs there (Figure 6.1). Some deserts are fairly wet from fog. Some even have thunderstorms. In contrast, the Atacama Desert in Chile had no rainfall between 1919 and 1964 and little since. A similar situation exists for parts of Namibia in Africa, although fog occurs there. The temperatures of deserts throughout the world vary. Low elevations and low latitudes have hot deserts. High elevations and high latitudes have cool deserts. Some deserts have snowfall. Deserts can be classified as warm, cold, or west coast fog, with warm deserts in the subtropics and temperatures above freezing, and cold deserts in the mid-latitudes with winter temperatures below freezing. West coast fog deserts occur in tropical or subtropical regions on the west side of continents.

DESERT FORMATION

Most deserts occur in the mid-latitudes and toward the west and middle of continents. Cold oceanic currents near a coast can cause desert. Additionally, rain shadows create dry habitat on the lee side of mountains leading to desert. The largest factor creating desert is mid-latitudinal desertification (see Chapter 8). Deserts do not occur at the equator where tropical forest is prevalent. Instead, hot, moist air rises from the equator, rains itself out there, then moves north and south creating warm, dry winds. Eventually, these dry air masses reach latitudes at which they cool and drop dry air back on Earth's surface at middle latitudes (20–40° of latitude).

PLANT LIFE FORM IN DESERT

If rainfall is less than 2 cm per year, no vegetation occurs, as in the Saharan sand dunes. Conversely, some deserts have enough moisture to support small trees (e.g., Joshua Tree and Saguaro Cactus of the Sonoran Desert). Most plant species have acquired special adaptations to store water and prevent loss of water or have deep taproots. Note that members of the cactus family are a product of the New World only.

Desert can form **soil crust** on the desert floor. This crust is complex and extremely fragile and consists of a whole community of organisms. Fungi, lichens, bryophytes, algae, and filamentous cyanobacteria (blue-greens) intertwine and derive their energy from photosynthesis. The filaments of the cyanobacteria bind with soil particles on the top millimeters, and some fix nitrogen. The crust alters reflectance of sunlight, stabilizes soil, and holds nutrients for plants. Once broken, the soil crust takes years to recover. Fire, off-road recreation, and even stepping off the trail in a desert can create long-term damage.

ANIMAL LIFE IN DESERT

In general, deserts support few above-ground herbivores. Birds are present but may be migratory. Those animals that do not migrate tend to be diminutive and act as keystone species. In N. America, these include hummingbirds that pollinate cactus, and small mammals that distribute the seed they eat. Some seed eaters are able to get all their water from seeds. They may aestivate during dry periods. Few large mammals can endure desert, but exceptions are the Addax in the Sahara, the camel in Asia, and the Red Kangaroo in Australia. Coyote and Mountain Lion are large predators in N. America that serve as keystone species in deserts.

For deserts that have substantial rivers, most desert life takes place in floodplains and canyons where woodland can occur (Askins 2002). It is flooding, not fire, that creates the disturbance necessary for sustained life in these communities. Floods remove trees and fertilize with silt. Abrupt boundaries occur between upland desert and either steep-walled canyons or steplike terraces. In either case, life is abundant compared to the uplands because of the source of water.

In desert uplands many seeds lie dormant for periods of several years, but germinate, flower, then set seed in quick succession after rain (Sinclair et al. 2006). Insects breed and nomadic birds move in to take advantage of the high seedset.

THE MAJOR ECOLOGICAL PROCESSES OF DESERT

James Brown's pioneering research over the years provides three ecological generalizations

about the desert food chain in general. First, decomposition by microbes is slow and unremarkable because of the lack of water, but invertebrates break down detritus to smaller particles. Second, carnivores are opportunistic and omnivorous, in other words are generalists, not specialists, creating higher complexity in the food web than most other communities. Third, herbivores usually eat only 3%–4% of the above-ground vegetation. Those that eat seeds are often keystone species. They either eat most of the seed, preventing germination, or select the size they like best, which influences plant composition. For instance, in the Chihuahuan Desert, Kangaroo Rat preferentially feeds on large seeds. In the presence of the rats, shrubs with large seeds decline and grasses with small seeds proliferate.

DESERTS OF N. AMERICA

The four major deserts are Great Basin, Mohave, Chihuahuan, and Sonoran. All but the Great Basin are hot deserts. Besides these major areas, smaller deserts occur (Figure 6.4).

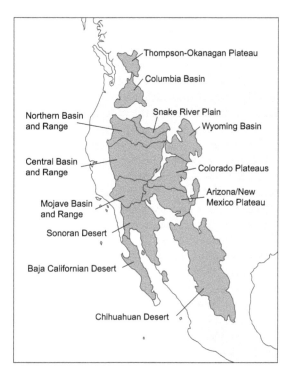

Figure 6.4 Deserts of N. America (*G. Giacomin*).

CONSERVATION PROBLEMS OF DESERT

Large cities deplete the water table leading to desertification. In extreme situations in deserts throughout the world, wars ensue in fights for water, oil, and agriculture. Other persistent problems include lack of natural disturbance caused by damning and irrigation, desertification, channeling of rivers, development, invasive species, overgrazing, poaching, agricultural expansion, erosion, reckless recreation, and mining for resources such as sand, copper, uranium, and salt.

Fields of solar panels provide an alternative to fossil fuels, but they shade out plants underneath. Even so, their impact is less than worldwide oil extraction and its ensuing wars. Ironically, climate change has caused increased precipitation in some deserts including those in N. America, changing the plant composition. Overall, most deserts are extremely fragile and take centuries to recover from human disturbance.

THINKING QUESTIONS

1. The presence of the shrub life form occurs from the arctic to the tropics, from arid regions to coastal salt-spray zones, and on five continents. Does this mean that all shrubs have the same close ancestor? Provide an evolutionary answer to the question of origin for the shrub life form.

2. What explains why so many animals in shrublands and deserts are venomous?

3. Leaves of plants adapted to shrubland and desert come in a variety of forms. Cactus needles are modified leaves with little evaporation. Aloe leaves employ thick leaves that store water. Creosote leaves have a waxy coating. Ocotillo drops its leaves during drought. Some plants have their stomata only on the underside of leaves. Acacia produce tiny compound leaves to prevent wilting. What is it about the presence of tiny compound leaves that makes them less susceptible to wilting?

4. What are C_4 and CAM plants and how are they specially adapted for life in deserts and shrublands?

5. Some shrubland and desert animals drink no water. What are some adaptations that allow animals to acquire water in other ways? Can animals gain water metabolically?

6. Despite the stereotype, sandy dunes occur in only 20% of deserts. What is the origin of sand dunes in deserts?

7. Sand can only be suspended at most 1.5 m (5 ft.) from the ground during a storm, busting the myth that high cliffs can be sand blasted. However, wind erosion can lead to sand-blasted rocks near the ground producing "yardangs." It can also lead to "desert pavement." What are these phenomena and what are their appearances?

8. Of the three desert types (cold, hot, west coast fog), which do you think would have the least variable and most moderate temperatures (neither too hot nor too cold)? Why? Find evidence to support your answer.

9. Rocky soils in deserts tend to have a greater variety of life than sandy or salty regions. What is the major reason?

10. Sand dune areas can actually carry abundant moisture for plants with long roots. What explains the moisture-holding capacity deep in sand dunes?

11. Cactus and euphorbia taxa are not closely related and evolved on different continents, yet they both produce succulents and have common features. Find examples of convergent evolution among these taxa.

12. Poinsettia is a euphorbium with toxic qualities. What symptoms are seen from ingestion of Poinsettia? What about ingestion of holly berries?

13. For cacti, if needles are modified leaves that do not photosynthesize, how does a cactus fix carbon?

14. Large cacti can withstand temperatures of 60°C (140°F). The offspring of these species cannot. How do the offspring survive the brutal desert conditions their parents?

15. What is the meaning of desertification and where are its effects most prominent worldwide? What are the causes?

16. Investigate whether a shrub species in a natural habitat (even a forest) has random, systematic (uniform), or aggregated (bunched) spacing. Either make a map that shows exact placement of each shrub and assess from viewing the map or use distance-to-nearest-neighbor statistical techniques.

    - How might uniform placement be a sign of allelopathy or competition for water or nutrients?
    - If the shrubs you mapped do not have uniform placement what do you think caused the pattern they display?

17. Research information about the structure and composition of desert crust. Collect small samples of soil crust in an undisturbed natural area where you live (even if not a desert). Examine the crust while it is still intact with bare eyes and dissecting scope. Make wet mounts of the surface material. Examine at all magnifications including oil immersion. Draw what you see and label what you can for eukaryotic cells. Do you see filamentous cyanobacteria? Is it *Oscillatoria*?

18. "Taking a bearing" with a compass refers to measuring the degrees from north, as taken in a clockwise direction. It is one thing to walk in the direction of the bearing, but what about when you want to return home by the same route. It requires calculation of a back bearing. For instance, if the original bearing were 140, the back bearing is $140 + 180 = 320°$. if the original bearing were 210, the bearing would be $210 + 180 = 390$, which of course is larger than 360. You would correct anything over 360 by subtracting $390 - 360 = 30$. Another way, rather than doing this math, is simply to turn the compass backward so the direction of travel arrow is now facing you. Practice taking bearings and back bearings for when you might find yourself lost in a desert.

19. Within peer-review literature find two articles on the subject of shrubland, thicket, desert.

REFERENCES

Askins, R.A. 2002. *Restoring North America's Birds*. Yale University Press, New Haven, CT.

Quinn, J.A. 2009. *Desert Biomes (Greenwood Guide to Biomes of the World)*. Greenwood Press, Westport, CT.

Sinclair, A.R.E., J.M. Fryxell, and G. Caughley. 2006. *Wildlife Ecology, Conservation, and Management*. 2e. Blackwell Publishing, Cambridge, MA.

Smith, J.M.B. 2019. Scrubland: Ecology. Brittanica. com.

7

Biomes: Savanna and forest

ACTIVITY: Vegetation Sampling in Forests

Goals: Get practice in algebra, graphing, statistical analysis, sampling, and developing an analysis from statistical results. "Veg sampling" includes some of the most common techniques in ecology for assessing quantities of plants (and sessile animals).

Objective: Practice

- setting up 20 m×20 m plots.
- using quadrat and point-quarter methods of vegetation sampling to determine tree density.
- testing two hypotheses via paired and unpaired t-tests and statistical software.

Step 1 (Questions): Write the difference between truly "random sampling" and "haphazard sampling." Write legitimate ways to generate truly random numbers for choosing coordinates in a grid. Calculate how many m^2 are in a 20 m×20 m plot.

Step 2 (Warmup algebra): The size of 1 ac. is often difficult to perceive other than it is approximately the size of an American football field. Calculate exactly what percent of an American football field (without end zones) is 1 ac. In your calculation cancel units and zeros through factor analysis and unit conversions. Use the correct number of significant figures and rounding. Show your work.

Step 3 (Know): The words quadrat and plot are sometimes used interchangeably, but there are traditional differences. **Plots** can be rectangular, square, or round. **Quadrats** are traditionally square.

Field Exercise 1: Each group of three students should place three 20 m×20 m quadrats in the forest randomly through the use of random coordinates, then count the total number of trees within and record. In addition, the density of large (dbh>5 cm) versus small (dbh<5 cm but taller than 1 m) trees in the 20 m×20 m quadrats should be counted. (dbh is diameter at breast height.) Back in the lab an unpaired t-test should be used to test the null hypothesis that the mean number of large trees is equal to the mean number of small trees in the forest using the plots as replicates. Write a conclusion in which you offer a biological explanation for why large or small trees were more prevalent.

Field Exercise 2: In the very same place where each quadrat was placed estimate the density of trees per 400 m^2 using the point-quarter method. Use a paired t-test to test the null hypothesis that mean density of trees counted in the quadrat method is equal to the mean density of trees estimated by the point-quarter method. Research the point-quarter method and its biases. Write a conclusion about why the point-quarter method may have given the same or different results as the quadrat method.

DOI: 10.1201/9781003271833-7

Figure 7.1 Trees of a savanna (*Zebras, Serengeti savanna plains, Tanzania.jpg Wiki Commons*).

INTRODUCTION

Trees are woody plants that have a single main trunk and branches well above ground (Kuennecke 2008). Trees develop anywhere adequate moisture exists to support them. They need more water than grass or shrubs because of their large size and greater proportion of above-ground mass, which is subject to drying.

WHAT IS SAVANNA?

In areas too wet for grassland or shrubland, but not wet enough to support a closed canopy of trees, savanna occurs. **Savanna** is characterized by scattered trees not forming a closed canopy, with grass and forbs between (Figure 7.1). Either small clusters of trees grow here and there within the greater matrix of grass, or single trees are interspersed throughout the grass matrix. "Closed canopy" refers to trees that are so close to each other that their leaves form a ceiling with little direct light penetration to the forest floor.

WHAT ARE CHARACTERISTICS OF SAVANNA?

Savanna generally has more precipitation, humidity, and higher temperature than grassland, but is subject to extreme drought. It is disturbance-adapted to the point where many savannas are transitioning to closed-canopy forest because fire or grazing has been suppressed. Poor soils generally characterize savanna except right under the trees where large animals rest in the shade. Most of the moisture, animal life, and nutrient cycling all happen in these shady places where manure provides nutrients and a seed base.

Existing savanna is often human-altered and semi-natural. Humans seem to favor savanna above all else, preferring it in yards and lawns, perhaps a throwback to the savanna origins of *Homo sapiens*. In India, savanna is all that is left of original forest, a result of human degradation. The most extensive savannas in the world are in Africa, but the most biodiverse are in S. America (the Cerrado). Large savannas can also be found in India, northern Australia, Malaysia, N. America, and parts of S. America other than the Cerrado.

ANIMALS OF SAVANNA?

In S. America, Marsh Deer and Capybara (large rodents that occur in herds) populate some savannas. In the African savanna, Wildebeest, Zebra, and Impala are common herbivores. Giraffe, gazelle, and Kudo are woody browsers. Carnivores of the African savanna include Lion, Leopard, Cheetah,

Figure 7.2 Ponderosa Pine near Payson, Arizona.

hyena, and wild dog. Scavengers include vulture and jackal. Mound-building termites proliferate.

In savanna throughout the world, large herbivores interact with fire proneness (Bowman and Murphy 2010). For example, in southern Africa heavy grazing produces short grass that is resistant to burning. Woody plants develop that would ordinarily be killed by fire. White Rhinoceros, Wildebeest, Impala, Warthog, and Zebra graze these short lawns. With light grazing, tall bunch grasses are prone to fire and support a less diverse herbivore assemblage.

CONSERVATION OF SAVANNA

In the Cerredo of S. America, the savanna and associated forests make up one of the worldwide hotspots of biodiversity. Bolivia and Paraguay have established national parks, but in Brazil, where most of the land is located, soybean and rice plantations have been established.

The African and Australian savanna have been severely degraded from overgrazing, development, and human population growth. In the sub-Saharan region, settlements have largely decimated the trees, causing desert encroachment. Poaching for

horns and hides remains a problem. In parts of Africa, savanna is being converted entirely to pine and eucalyptus for paper pulp and wood, and to corn, pineapple, or grazing for domestic animals. This occurs even though the native savanna supports five times the animal biomass of domestic animals. The domestic animals have more heat-related and disease problems.

DOES N. AMERICA HAVE SAVANNA? YES

Oak-dominated savanna was once common between forest and prairie of the Great Plains, perhaps kept open by Bison (Apfelbaum and Haney 2010). Oak "barrens" also occurred within the oak-hickory region of the eastern U.S. on south-facing slopes with dry soils, and in California.

Ponderosa Pine forests along the Rocky Mountains and Pinyon Pine/juniper forests in the southwest U.S. could also be considered savanna (Figure 7.2). Ponderosa Pine is found in high altitudes and trees are straight with massive trunks and fire-resistant bark. Grassy habitat between trees creates a meadow. Fire keeps the understory open.

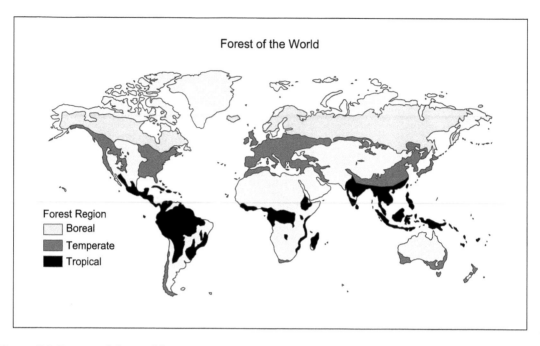

Figure 7.3 Forests of the world (*G. Giacomin*).

WHAT IS MEANT BY FOREST?

Forest refers to a closed canopy of trees. Forests can be temperate or tropical (Figure 7.3). Tree species can be **coniferous** (cone bearing) or **broadleafed** (flower bearing). They can be **evergreen** or **deciduous** (lose their leaves before every winter or dry season). Generally, coniferous trees are considered **softwoods**, and broadleafed trees are considered **hardwoods**. Forests can span the gradient from dry to rainforest. A **mixed** forest is one with both coniferous and broadleafed trees.

WHAT IS THE PLANT LIFE FORMS OF FOREST?

While trees are the dominant life form, additional forms found in forests are shrubs, vines, epiphytes, grasses, forbs, ferns, mosses, and fungi.

CHARACTERISTICS OF TEMPERATE DECIDUOUS BROADLEAF FOREST

Broadleaf forest once dominated the eastern U.S., parts of western Europe, eastern Asia, and parts of eastern Australia (Figure 7.4). In Europe most of the large tracts have been cut except for pockets in the mountains. The Asiatic broadleaf forests are as diverse as in N. America.

Temperate deciduous broadleaf trees sustain a temporary loss of nutrients each fall from leaf drop. For this reason deciduous species can survive only if they occur in rich soils. In poor soils conifers outcompete deciduous broadleaf trees because conifers can retain nutrients in their leaves for (usually) 3 years before dropping them.

Leaflessness for much of the year characterizes temperate deciduous trees, in some places for more months than with leaves (Yahner et al. 2000). For the winter, trees send nutrients from the leaves and branches to their roots for storage until spring, then drop their leaves as an adaptation to winter freezing (Sinclair et al. 2006). Re-developing these leaves requires a long growing season of 4–6 months.

Leaflessness and gaps in the forest canopy allow a rich community of plant species on the forest floor. When a gap in the canopy opens because of disturbance (e.g., downburst, tornado, or ice storm), shade-intolerant plants compete for light and grow rapidly. This process is called **release**. Grasses and forbs on the soil surface are stimulated by the heat from the light and begin to grow.

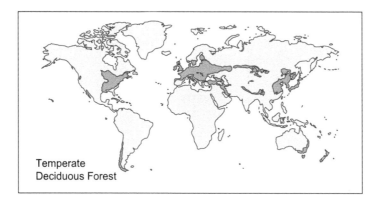

Figure 7.4 Temperate deciduous forests of the world (*G. Giacomin*).

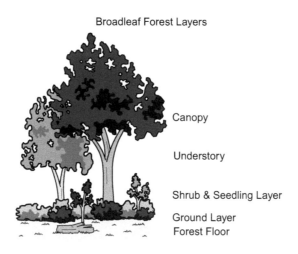

Figure 7.5 Strata (layers) of the eastern forests of the U.S. (*G. Giacomin*).

STRATIFICATION IN TEMPERATE DECIDUOUS BROADLEAF FORESTS

Stratification (layering) characterizes naturally occurring temperate deciduous broadleafed forests (Figure 7.5). Note that forests are volumes, not areas (McCombs 2015). The layers are not always well defined, but more or less include those described below. Forest age, location, and season may influence the presence of a stratum:

- an **overstory canopy** of tall trees,
- an **understory canopy** of shorter tree species,
- **shrubs** that are woody with no central trunk, that also include nonwoody herbaceous plants, and ferns just above the forest floor,

- a **log, moss, and litter layer**, also known as **ground layer** or **litter layer** of decomposing material,
- occasionally, an additional layer, an **emergent layer** occurs with supersized trees that emerge above the canopy even in temperate areas.

LIGHT VARIES SEASONALLY WITHIN THE TEMPERATE DECIDUOUS BROADLEAF FOREST

Sunlight on the forest floor is greatest during spring before full leaf-out, coinciding with spring wildflower bloom (Figure 7.6). Spring wildflowers conserve important nutrients in the soil by accumulating potassium and nitrogen as they grow,

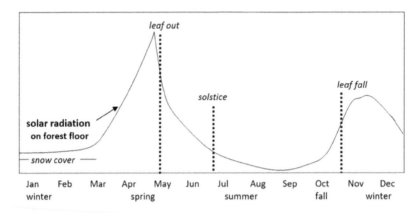

Figure 7.6 Seasonal solar radiation can be seen on the floor of the deciduous broadleafed forest of the U.S.

then returning these nutrients to the soil and to their mass underground as they **senesce** (die back) over the summer and fall (Yahner et al. 2000). The second greatest light period is just after leaf fall. During winter a high percentage of the forest floor is illuminated, but the sunbeams are at an angle and often on snow. The darkest period is midsummer. Overall, light in a deciduous forest is gappy and the presence of plants is heavily influenced by the presence of sunflecks.

TYPES OF DECIDUOUS BROADLEAF FOREST IN EASTERN N. AMERICA

Bottomland hardwood includes floodplains of waterways adapted to regular flooding with wet soils. Historically, these were the most productive forests in N. America, with massive individual growth and richness in animal species (Apfelbaum and Haney 2010). Remnants in their uncut growth are still present in the 21st century.

Northern hardwood forests are dominated by Sugar Maple and Red Maple, American Beech, birch, Basswood, and White Ash. These occupy southeast Canada and the northern U.S. from the Great Plains to New England and show the greatest brilliance of orange, red, and yellow color during leaf fall.

Oak-hickory is the most common forest type in the eastern U.S. and southeast Canada. At one time much of this area was dominated by American Chestnut but is now dominated by oak species. Hickory species include Pignut and Mockernut in the south and Shagbark in the north. Pioneer tree species include Sassafras, Tulip Poplar, sumac, Eastern Red Cedar, and some pines.

TEMPERATE EVERGREEN FORESTS

Cone-bearing gymnosperms outside the taiga in N. America include the coastal pine forests of the U.S. Southeast, the rain forests of the Pacific Northwest, and the coniferous forests of the Rocky, Sierra Nevada, and Cascade mountains (Figure 7.7).

Longleaf Pine forest once covered 90 million ac. of the South Atlantic and Gulf states in the U.S., but is now only in small fragments. One of the problems is that Longleaf Pine must be maintained by fire to burn out invading hardwoods. With fire, pines maintain a competitive advantage, surviving better on nutrient-poor, dry, sandy soils. On these soils, wiregrass grows in the understory, burning easily and carrying fire. Longleaf Pine forest supports 30 threatened or endangered species, and can have as many as 124 plant species in one $10\,m^2$ plot (Apfelbaum and Haney 2010). Besides lack of regular burning, other conservation threats are agriculture, development, logging, and conversion to Loblolly Pine for timber.

OTHER TEMPERATE CONIFEROUS FORESTS IN N. AMERICA

Coniferous rainforest of the Pacific Northwest stretches from southern Alaska and British Columbia to northern California into the Sierra Nevada Mountains. Annual precipitation is heavy (between 60 and 140 in., 150 and 360 cm) (Apfelbaum and Haney 2010). The rain comes from winds that move in from the northern Pacific and dump on the western slopes. The Pacific Northwest forest differs from the northern boreal forest. On

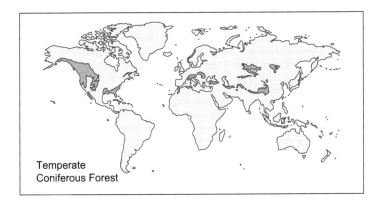

Figure 7.7 Temperate coniferous forests of the world (*G. Giacomin*).

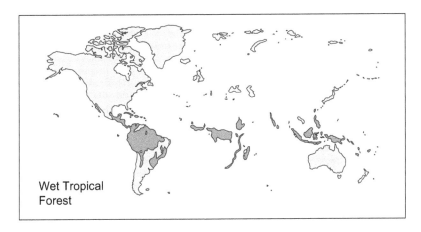

Figure 7.8 Wet tropical forests of the world (*G. Giacomin*).

the mountain slopes the air is cool with heavy fog that settles on the trees then drips off. This wetness supports an understory of mosses, epiphytes, and ferns. Immense, but slow-growing hemlock, Silver Fir, Sitka Spruce, or Redwood trees dominate. This community makes up the tallest trees in N. America. Coniferous trees of the Pacific Northwest rarely burn. High species diversity is supported within a complex combination of vegetative layers and a dense canopy with little light penetration (Askins 2002). Large cavities in the trees develop amidst broken treetops, intertwined debris, and decomposing moss-covered fallen logs.

The forests of the Rocky Mountains and U.S. southwest at low slopes have savanna-like stands of Ponderosa Pine and pinyon/juniper (Askins 2002). In higher areas, Lodgepole Pine dominates beginning at mid-elevation in dense stands unlike the more open Ponderosa Pine stands below. Even

higher in the Rocky Mountains Douglas Fir forest is found. Also present are White Fir, Blue Spruce, Limber Pine, and Quaking Aspen. Higher still is Engelmann Spruce, and higher still is subalpine fir forest. Lodgepole Pine at mid-elevations is a highly flammable, early successional species (Askins 2002). It was the chief tree species that burned in Yellowstone National Park in 1988 when nearly three-quarters of a million ac. (290,000 ha) burned.

TROPICAL FORESTS

Between the Tropic of Cancer (23°N) and Tropic of Capricorn (23°S) lie "**the tropics**." Within the tropics is the equatorial zone where most of the tropical rainforest occurs (Figure 7.8). In this equatorial zone, day length, temperature, and rainfall remain consistent throughout the year. To the north and south of the equatorial zone but within the tropics, tropical seasonal forests occur with wet and dry seasons. The greater the distance from the equator, the more seasonal the forests become.

Tropical forest covers 7% of the Earth (5 million ac., 2.1 ha). Of this, tropical rainforest makes up 86% and seasonal forest 14%. Most of the tropical forest is in the Americas (45%), followed by African regions and Madagascar (30%), and Asia Pacific, New Guinea, and Australia (25%). Tropical forest has the highest biodiversity of all biomes and supports more than half of Earth's species, less than half of which have been named. Some of the largest trees, smallest land animals, and greatest diversity of mammals, birds, herptiles, and insects are native to tropical forests, as are all the Great Ape species.

SOILS OF TROPICAL FOREST

Typically, soils are ancient and highly weathered in the tropics owing to years of high heat and heavy rain. The soil has high acidity, low nutrients, and low organic matter. This explains why forests cannot always be restored after deforestation in the tropics. Some volcanic and other areas have younger soils, which can be fertile.

HUMANS IN THE TROPICAL FOREST AND CONSERVATION

Beyond a reservoir of biodiversity, tropical forests provide homes to indigenous people, and supply natural products that have traditionally included spices, cork, rubber, medicines, and food. More than 20% of the world's oxygen and 20% of the world's freshwater are produced in the Amazon rainforest. Tropical forests store great quantities of carbon if left intact. Thus, the fate of the tropical forests affects the entire world. Deforestation has an amplified effect because of the concentration of species lost, the oxygen production lost, the carbon storage lost, and changes to global climate. Forest burning in the tropics either for deforestation or through climate change releases huge quantities of carbon, can permanently affect the soil, and deforestation in general increases reflectance, which alters rainfall. Fragmentation of the forest leads to biodiversity loss, and introduction of non-native species. Clearing for palm oil production, gold mining, mineral extraction, livestock ranching,

road building, oil drilling, economic development, overpopulation of humans, and political unrest are all pressing and devastating problems for tropical forests.

TROPICAL RAINFOREST LIFE FORMS

The dominant life form in tropical rainforest is broadleaved evergreen trees. Emergent trees are scattered and rise above the upper tree canopy. They receive the brunt of sun exposure and wind and thus have thickly buttressed bases to hold them in place. The upper canopy and several more vegetative layers below compete for sunlight. Plants have ever larger leaves the lower they are from the top. Near the ground short trees occur.

Throughout the canopy are vines, epiphytes (plants that grow on trees), and carnivorous plants. If the canopy is dense, plant growth on the ground will be sparse because of lack of light. Instead, the floor is often thickly laced with roots that can become massive. Buttresses come off trees to function as props for shallow-rooted trees. The forest floor is wet because leaves have drain tips, pointed ends that direct water off the leaves. This prevents rotting on the leaves themselves.

ANIMALS OF TROPICAL RAINFOREST

The large animals of tropical forests stay mainly in the canopy rather than on the forest floor. The greatest diversity of primates occurs in tropical forests. Birds reach their greatest diversity in the Amazonian forests. Many birds and bat species are adapted to the flowering periodicities of their preferred tree species.

TROPICAL SEASONAL FOREST

The dominant life form is broadleaved deciduous trees or evergreen trees (e.g., eucalyptus in Australia) adapted to long dry seasons (Figures 7.9 and 7.10). The canopy is less complex than in rainforest, and a ground layer develops if the trees lose their leaves for a season and sunlight reaches the forest floor. Fires are frequent, leading to fire-resistant adaptations for understory plants. The dry season lasts 4–7 months, but rain can be monsoonal during the wet season.

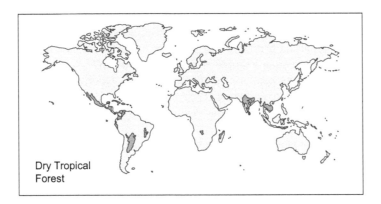

Figure 7.9 Dry tropical forests of the world (*G. Giacomin*).

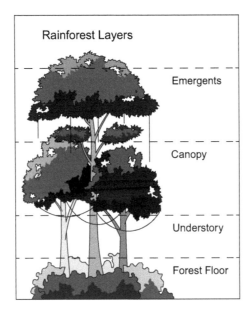

Figure 7.10 Tropical forests sometimes have an emergent layer above the rest of the canopy (*G. Giacomin*).

THINKING QUESTIONS

1. Some ecologists have asserted that humans instinctually love savanna owing to their early evolution within African savanna. Does the hypothesis have credible evidence? Investigate biophilia.
2. What is it about being adjacent to mountains, oceans, deserts, or tropical forests that causes dramatic wet and dry seasons in tropical savanna?
3. For years, tropical savanna was sparsely populated by humans, but many of the problems are being overcome. How? Look into specific regions across the globe.
4. Chapter 7 did not characterize the major animal life of temperate broadleaved deciduous forest. How would you describe it?
5. Where is temperature the highest in a temperate broadleaved deciduous forest in summer? In winter?

6. What determines whether a nut-producing tree has a bountiful year? For instance, how does date of last frost affect production of acorns in oak species of N. America?

7. What is the difference between a simple and a compound leaf?

8. What is the difference in the appearance of hemlock, pine, Douglas-firs, yews, junipers, spruce, and the difficult group known as "cedars?"

9. What is the taxonomic difference between gymnosperm and angiosperm trees, and how does each reproduce?

10. Not all coniferous trees have a conical appearance. Review the shapes of coniferous trees and their locations on the continent.

11. The aroma of conifers, the smell of pine wood, for example, is different from broadleafed trees. Conifers emit a turpentine-like smell, with terpenoids dissolved in their resin. What service do these terpenoids provide to the trees? Angiosperm trees appear in the fossil record 160 million years after gymnosperms appeared. How do angiosperm trees handle the same problems that are handled by terpenoids?

12. From a cell and tissue perspective in trees and shrubs, what makes something woody?

13. How are trees able to lift water to their highest reaches?

14. What are the functions of xylem and phloem, and what makes maple sap run during certain times of the year?

15. What is responsible physiologically for growth rings visible on a cleanly cut stump? Do all trees worldwide have growth rings?

16. Areas of New England in the U.S. have actually gained substantially more trees over the last 100 years. What accounts for this?

17. Investigate the work of environmental non-governmental organizations that are working to conserve tropical forests. What accomplishments have they made?

18. Collect twigs from different species of hardwood trees with alternate branching, opposite branching, and whorled branching. Most tree species of N. America have alternate branching, but a few have opposite. Catalpa has whorled. The opposite branching group can be remembered by the MAD Horse acronym (maples, ash, dogwood, horse chestnut). (Horse chestnut is buckeye group.) Examine local shrub species, determine which have alternate or opposite, and think of a new acronym.

19. In summer and early fall investigate the prevalence of common insect pests in trees and record what size and species of tree are affected. Look for cicada damage on trees when cicadas make a slit in a small branch in which the eggs are laid. This produces a clump of dead, brown leaves at the end of branches high in oak trees. Look too for the webby nests of webworms and tent caterpillars.

20. Tree age in conifers can be determined by counting the rungs of branches from bottom to top. Try it. Does the same technique work for hardwoods?

21. During the late summer and early fall, place acorn "traps" under oaks. The traps look like square hammocks of fine netting that are suspended. Use the traps to compare acorn production under different individuals, species, or growing conditions.

22. Spruce needle tea: Collect spruce needles. Once at home or in the lab boil these needles in a little water then smell this spruce tea. Is it good? What is chemically responsible for the aroma?

23. Apical meristem damage in pine trees: Pines and other coniferous tree species sometimes get their apical (central) meristem bud damaged by insects. A nearby branch may then assume the role of the central meristem or two branches compete for that role (see photo). However, the lumber quality decreases for this tree. How common is this? Find trees with this damage in pine plantations or coniferous stands or use a *Coleus* house plant to watch the branching take place when the tip of the plant is repeatedly snipped.

24. Worroll and Houghton (2018) suggest several whittling and carving exercises to become familiar with hard and softwood species. They provide specifics of carving tent pegs, elves, and walking sticks.
25. Densitometers are instruments that can measure the percent open canopy in a forest. They can be purchased or one can measure the same thing by taking a picture of the canopy, printing out the photo and placing a grid on a clear plastic over the photo.

REFERENCES

Apfelbaum, S.I., and A. Haney. 2010. *Restoring Ecological Health to Your Land.* Island Press, Washington, DC.

Askins, R.A. 2002. *Restoring North America's Birds.* Yale University Press, New Haven, CT.

Bowman, D.M.J.S., and B.P. Murphy. 2010. Fire and biodiversity. In: N.S. Sodhi and P.R. Ehrlich (eds.), *Conservation Biology for All.* Oxford University Press, Oxford, UK.

Kuennecke, B.H. 2008. *Temperate Forest Biomes (Greenwood Guide to Biomes of the World).* Greenwood Press, Westport, CT.

McComb, B. 2015. *Wildlife Habitat Management: Concepts and Applications in Forestry.* CRC Press, Boca Raton, FL.

Sinclair, A.R.E., J.M. Fryxell, and G. Caughley. 2006. *Wildlife Ecology, Conservation, and Management.* 2e. Blackwell Publishing, Hoboken, NJ.

Worroll, J., and P. Houghton. 2018. *A Year of Forest School: Outdoor Play and Skill-Building Fun for Every Season.* Watkins Publishing, London, UK.

Yahner, R.H., C.G. Mahan, and A.D. Rodewald. 2012. Managing forests for wildlife. In: N.J. Silvy (ed.), *The Wildlife Techniques Manual: Management*, 7e, Volume 2. The Johns Hopkins University Press, Baltimore, MD.

8

Why are biomes where they are?

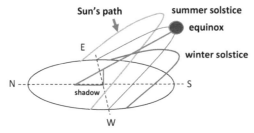
INTRODUCTION

It is one thing to categorize biomes and ecoregions by their plant life form. It is another to determine why these biomes occur where they do. The subject of this chapter is the climatic and geologic processes that maintain *life forms* at the global level. In the next chapter, processes that determine *number of species* across the globe will be reviewed, a subtly different but related topic.

WHAT DETERMINES LIFE FORM AT THE GLOBAL LEVEL?

At the biome scale, animal life form is largely dependent on the type of plant life form that occurs. Plant life form is largely dependent on climate, in particular, the plant's need for water (Figure 8.1). More precisely, plants have a problem with water loss when their stomata are open for gas exchange. Let's sort this out step by step starting with photosynthesis.

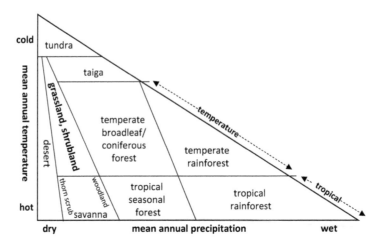

Figure 8.1 Modified "Whittaker diagram." Tundra here includes arctic and alpine.

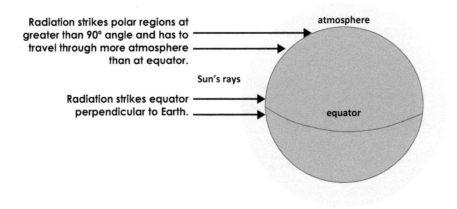

Figure 8.2 Why polar areas are cold and the equator is hot.

Just to remind yourself of the components, write a simple chemical equation for photosynthesis in this space.

Think about how plants get the necessary CO_2 to the inside of their tissues. Their stomata must be open for gas exchange, creating the problem of water diffusing out. As long as the plant has plenty of water, the stomata stay open, but on hot dry days, the stomata close, which conserves water. With closed stomata, the leaves cannot exchange gases, like the nose of the plant being plugged. Because of the stomata problem, different plant life forms evolved to meet the evapotranspiration problem. The succulent life form is a way to meet the driest conditions. Tall trees match the wettest conditions.

WHAT DETERMINES CLIMATE?

If plant life form and ultimately biome distribution is a function of climate, what determines climate? Unlike weather, which refers to current atmospheric conditions, climate incorporates the average conditions over a designated time for a particular place. The driving factor for climate is the sun. Think about what causes different temperatures on the Earth's surface. The amount of sun energy depends on latitude, the distance north or south of the equator. Because Earth is round, the sun strikes at different angles (Figure 8.2). Near the equator, it strikes perpendicularly with more solar energy received per unit area than in higher latitudes. The area with the most direct solar energy produces the most photosynthesis and has the most plant biomass. Toward

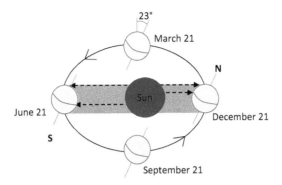

Figure 8.3 Seasons occur because the Earth tilts on its axis. The northern hemisphere has summer beginning in June because that is when the Earth is tipped to receive more light in the northern half.

the poles the light must pass through more atmosphere and has more reflection. Distance to the sun is not the primary cause of seasonality.

DETERMINANTS OF LIFE FORM AT THE GLOBAL LEVEL – SEASONS

Seasons occur because the Earth is tilted on its axis, currently by 23.5° (Figure 8.3). When the northern hemisphere is most tilted toward the sun from May to September, summer is in the north. When the southern hemisphere is most tilted toward the sun from November to February, summer is in the south.

Distance to the sun does not explain seasonality. The Earth is 3.1 million miles (3.3%) closer to the sun in January compared to July. This may be partially why the southern hemisphere summer is a little warmer than the northern hemisphere summer, but there are other factors at work. The southern hemisphere has more extensive ocean, which moderates temperatures by providing a heat sink. Thus, summers are cooler and winters are warmer in the southern hemisphere.

DETERMINANTS OF LIFE FORM AT THE GLOBAL LEVEL – AIR MASSES

A major factor influencing climate is **midlatitudinal desertification**. Many of the world's deserts are located at the midlatitudes – approximately 20°–40° north or south of the equator (Figure 8.4). Deserts are not directly on the equator where moist tropical forest occurs instead.

To understand this desert phenomenon, realize that solar radiation strikes the Earth most directly on the equator. This creates an uplift of warm, wet air rising above the tropical forests and oceans (Figure 8.5). As this saturated air expands and cools at high altitudes, the air rains itself out like a towel being wrung. Precipitation falls back to Earth directly on the tropics below.

Once the air at the equator has risen and expanded as high as gravity will allow, the continued air flow coming from ground level drives the air masses away from the equator toward the poles. As it moves away from the equator, in what is known as a Hadley cell, the air cools. The air sinks back to the Earth's surface at approximately 30° north and south of the equator. This sinking air is still dry from having rained itself out and warms as it contracts in the low atmosphere. The result as it comes back to Earth's surface is a dry wind that creates and maintains desert.

DETERMINANTS OF LIFE FORM AT THE GLOBAL LEVEL – PREVAILING WINDS

To begin to understand prevailing winds, first write which way the Earth is rotating, north, south, east, or west?

The air moving away from the equator sets up air circulation patterns that influence the rest of the world. However, the air moving toward the poles does not travel directly north or south. It begins to circulate clockwise in the northern hemisphere, and counterclockwise in the southern hemisphere (Figure 8.6). To understand why the air circulates, realize that air at the equator does not move as fast as the Earth is turning, producing the effect of a wind blowing toward the west.

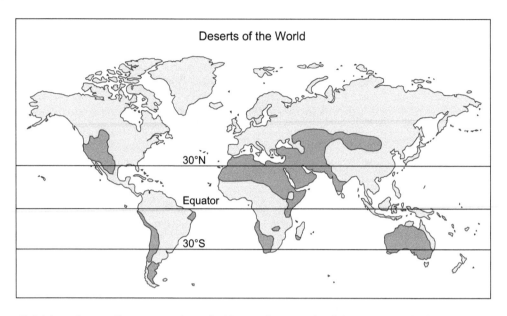

Figure 8.4 Most deserts lie at approximately 30° north or south of the equator (*G. Giacomin*).

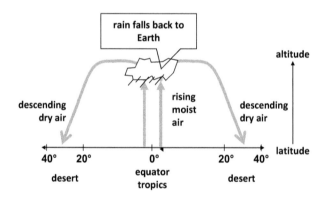

Figure 8.5 Air masses creating midlatitudinal desertification.

Realize too that as the air from the tropics moves north and south from the tropics it encounters a round Earth, which causes circulation.

DETERMINANTS OF LIFE FORM AT THE GLOBAL LEVEL – TRADE WINDS, ITCZ, DOLDRUMS, MONSOONS

The circulating air in the northern and southern hemispheres creates **trade winds**, most prominently resulting in east to west flow at low latitudes. Trade winds were named because they enabled ship travel from Europe and Africa to the Americas (Figure 8.7). During certain times of the year, flows bring dust and hurricanes from Africa to the Americas in the subtropics.

Where the trade winds from the northern and southern hemispheres come together is a cloudy area known as the **Intertropical Convergence Zone (ITCZ)**. Generally, this area is near the equator with cloud cover visible in a line from space. Sometimes the ITCZ results in no wind at all above the oceans of the tropics, known as the **doldrums** by mariners. It is in these low-latitude areas that seasonal shifts of air from land to ocean take place near the time of the solstices. These bring strong winds and heavy rains called **monsoons** to some of the tropical seasonal forests (20–30 in., 50–75 cm per day).

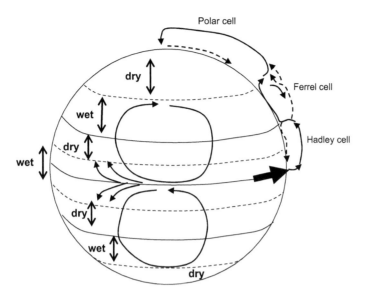

Figure 8.6 Prevailing winds occur because of the clockwise circulation of air and water in the north and counterclockwise in the south. The turning of the Earth toward the east means the air at the equator appears to move toward the west.

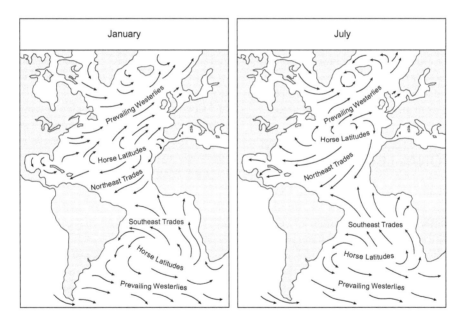

Figure 8.7 Trade winds and currents in the Atlantic Ocean (*G. Giacomin*).

DETERMINANTS OF LIFE FORM AT THE GLOBAL LEVEL – EL NINO

A phenomenon that affects both ocean currents and winds is the **El Nino-Southern Oscillations**. It affects the tropics and the world, but especially affects the southern hemisphere. During El Nino, warm waters of the central equatorial Pacific are pushed eastward toward the normally cold currents of the western coast of S. America. This produces rain in areas normally dry in S. America, and drought and tropical forest fires in Asia and Australia that usually receives rain from the warm water. Climate change appears to be increasing the frequency of El Nino events.

DETERMINANTS OF LIFE FORM AT THE CONTINENTAL LEVEL – PREVAILING WESTERLIES IN TEMPERATE LATITUDES

In temperate latitudes of both the northern and southern hemisphere the circulating air that began in the tropics produces **prevailing westerlies**, winds primarily from the west. The prevailing westerlies lead to rain shadows where dry biomes occur, described later in this chapter.

The circulating air originating in the tropics explains why the temperate regions have the most variable weather in the world and the most storms. Some of the air from the equator travels all the way to the poles (Figure 8.6) but sinks as it cools above the coldest areas creating polar cells and jet streams. The most complex interactions occur in the temperate regions where "Ferrel cells" develop as a product of the more influential polar and Hadley cells. As a result, the most variable and adverse weather occurs in temperate regions. Polar cell winds interface with temperate zone air over water and land. Various pressure and temperature changes release energy creating tornados, thunderstorms, and blizzards in temperate zones, most dramatically demonstrated in the U.S. which gets the most storms of any large continental area.

DETERMINANTS OF LIFE FORM AT THE REGIONAL LEVEL – ALTITUDE

Why is it usually colder overall on the top of a mountain even though it is closer to the sun than a valley? **Adiabatic cooling** occurs when air cools by expansion (which uses energy). The same phenomenon is at work when air is let out of a tire and feels cool. A release of energy associated with expanding gases produces the feeling of coolness. With thin air on the tops of mountains, lower overall temperatures occur at high altitudes. Additionally, thin air allows more solar radiation to reach the ground at midday (Bailey 2014). At night, rapid heat radiation from the ground occurs, losing more energy to the night sky when it is clear than in valleys.

WHY DO VEGETATION PATTERNS CHANGE WITH ALTITUDE?

Plant species at high elevations are often the same as those hundreds of miles closer to the poles.

Examine a map of N. America. From his work in Arizona mountains, C. Hart Merriam estimated that each mile of altitude was equivalent to 800 miles of latitude.

Mount Mitchell, just north of Asheville, North Carolina, is the highest peak in eastern N. America at 6,300 ft. (1,920 m). It is covered by firs and spruces, the same vegetation one begins to see in the town of Groundhog River in Ontario, Canada. Was Merriam's formula correct? Provide your answer in miles and kilometers then check the distance to see if he was right.

DETERMINANTS OF LIFE FORM AT THE REGIONAL LEVEL – ASPECT

When studying zones of vegetation in the Smoky Mountains, Robert Whittaker (1920–1980) noticed that spruce and fir did not start at the same elevation on every mountain. The **aspect** (direction the slope faced) made a difference. Equator-facing slopes had vegetation characteristic of more southern latitudes and warmer than nearby flatlands (Bailey 2014). Pole-facing slopes had vegetation characteristic of more northern latitudes. Pole-facing slopes are colder than nearby flatlands and snow tends to persist. The snow reflects light (creating "albedo"), leading to even lower temperatures.

In the northern hemisphere, south slopes tend to have the warmest temperatures. Which direction would have the warmest slope in the southern hemisphere? Explain why.

DETERMINANTS OF LIFE FORM AT THE REGIONAL LEVEL – RAIN SHADOW

Rain shadows influence climate in mountain areas (Figure 8.8). If mountain ranges are located perpendicular to prevailing winds, air masses of moist warm air are deflected upward as they meet the mountains. On its way over the mountains, the air cools and expands, causing moisture to drop out as precipitation. The opposite is true on the lee side. It is dry because it receives air coming down from the mountaintop after it has rained itself out. Thus, desert and grassland are often found downwind of the mountain chain.

In the U.S. prevailing winds are from the west for most of the nation. The largest of the mountain regions are parallel to the coasts, setting up areas for

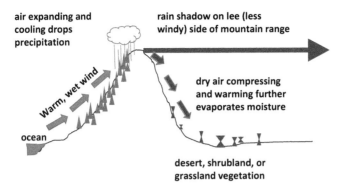

Figure 8.8 Rain shadow occurs when prevailing winds on one side of the mountain range dump water, enhancing dryness on the other.

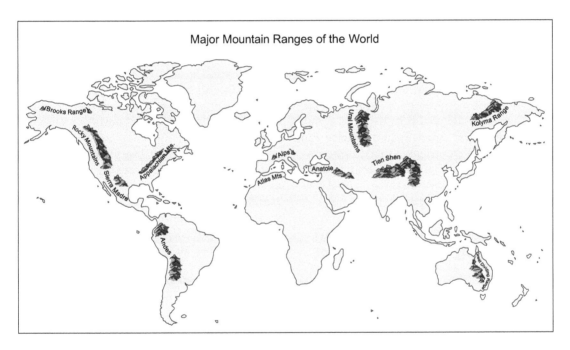

Figure 8.9 Mountain ranges of the world, showing a preponderance of ranges that run primarily north-south, including most of those in N. America (*G. Giacomin*).

large rain shadows (Figure 8.9). Thus, moisture drops out on the western slopes of the Rockies. The effect is more dramatic in the Rocky Mountains than the Appalachians largely because of the much higher elevation in the Rockies. The rain shadow created by the Rockies extends 1,500 km from Colorado to Illinois.

The absence of prevailing westerlies explains why less of a rain shadow occurs in the Appalachians than the Rockies. In the southern Appalachians, winds often come from the south off the Gulf of Mexico, bringing moist ocean air especially in the summer, canceling any rain shadow effect from the west.

DETERMINANTS OF LIFE FORM AT THE REGIONAL LEVEL – VALLEY INFLUENCES

When comparing air temperature in a valley to air temperature on a mountaintop, the pattern is sometimes the opposite of what one would expect. Temperatures at night may be lower in valleys than on peaks. Likewise, snowfall may be greater in the valley.

For example, Miles City, Montana, recorded one of the lowest temperatures on record for the

U.S. (−54°C), lying in a deep valley (Bailey 2014). In contrast, Pikes Peak 3,400 m higher has never recorded a temperature below −40°C. In fact, some mountain valleys are too cold to support tree growth and have only grasses, while the surrounding slopes have trees (Bailey 2014).

The cold valley, warm peak phenomenon is explained by thermal inversion. When air warms and rises during the day in a valley, it sinks as it cools during the night. A **thermal inversion** can occur if the cold air gets trapped below warmer air around the peaks the next day. Particularly if the next day is cloudy, the pocket of cold air may never reach high enough temperatures to rise through air stuck near the peaks. The result can be foggy, smoky, or polluted air in the valley. However, no two valleys are the same. A valley more open on one side will have more circulation than one that is closed.

Thermal belts, also known as isothermal belts or verdant zones, may form midway up the slope. Where they occur, thermal belts have the overall highest temperatures especially in the spring and fall when thermal inversions occur. Fruit growers including apple and grape growers intentionally grow their products within these thermal belts rather than on peaks or in valleys.

DETERMINANTS OF LIFE FORM AT THE REGIONAL LEVEL – MARITIME INFLUENCES

Maritime (ocean) climates are moderate compared to **continental** (interior) areas. Large bodies of water alter the air temperature on a coast because the water acts as a heat reservoir. In other words, the water gives off some of its stored heat to the land during the winter and at night. Likewise, the water acts as an air conditioner during the summer, absorbing some of the heat from the nearby land.

Without these large water bodies, continental interior areas are subject to large temperature fluctuations. In North Dakota temperatures historically reached 105°F (40°C) in summer, and −30°F in winter. In contrast, Juneau, on the Alaskan coast, historically had average temperatures of 70°F (21°C) in summer and 17°F (−8.3°C) in winter.

The ocean creates its own wind, which can vary by season and time of day (Chapin et al. 2002). During winter, the air over land is colder than over the adjacent sea. Warm moist air over the water rises, drawing in cold dense air flowing from land to ocean. In summer, the air over land is warmer than over sea. The land air rises, drawing in cool moist surface air from the ocean. Within this rising moist air, condensation of water vapor produces abundant precipitation known as monsoons in East Asia.

On a daily basis, sea breezes occur because strong heating over land causes air to rise, drawing in cool, moist air from the ocean setting up a circulation cell (Chapin et al. 2002). At night the breeze blows toward the ocean because the water is warmer than the land. Air rises over the ocean, drawing in cooler air from land.

DETERMINANTS OF LIFE FORM AT THE LOCAL LEVEL

Several types of physical factors may cause patchiness in life forms (Bailey 2014). Soil may be very shallow or non-existent on the slopes of exposed

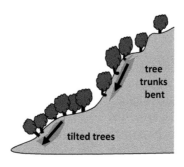

Figure 8.10 Soil slumping and landslides are evident by trees that tilt downslope or tree trunks that are all bent in the same direction with only fast-growing species in the slide areas (*E. Robbins*).

bedrock in mountainous areas. Only lichens, mosses, and flowering plants tolerant of shallow soils may thrive near the tops of mountains.

Lower on mountains, gravity may create unstable sites on steep slopes (Bailey 2014). Slide areas from rockfalls, avalanches, mudflows, and soil slumping occur. Soil slumping can be detected by the characteristic leaning of trees toward the downslope or trunks that are bent (Figure 8.10). In these unstable areas, slow-growing species will not prevail. Instead, fast-growing species like aspen occur, which reproduce through root suckers. Soil movement disturbs the roots, which stimulates sprouting.

THINKING QUESTIONS

1. Do other planets have seasons?
2. When the Moon is very bright, is any heat detectable within moonbeams?
3. After a candle is lit, what is the direct source of energy that fuels the flame?
4. What explains why the Earth's mountain ranges mainly have an orientation that runs north-south?
5. Chickens are said to oil their feathers just before rain, dipping their beaks into their preen glands and rubbing it on their feathers. Can you confirm this? Do other bird species oil just before rain?
6. Chemically, what is the smell of rain? Some say it is ozone, but ozone does not have an odor. What is it?
8. What would be the effect on Earth if there were no Moon?
9. Sound, light, and radio waves gets bounced between the top of the cool lower layer and the ground, according to Gooley (2014). Thus, sounds travel farther and louder during thermal inversions. Mirages have a different appearance. Usually, mirages make objects look short and fat, then hover. An inversion makes objects look tall and thin. It also improves the chances of seeing the "green flash," a phenomenon in which a green flash appears at sunset. Do these explanations hold up? Investigate.

REFERENCES

Bailey, R.G. 2014. *Ecoregions: The Ecosystem Geography of the Oceans and Continents.* 2e. Springer, New York.

Chapin III, F.S., P.A. Matson, and H.A. Mooney. 2002. *Principles of Terrestrial Ecosystem Ecology.* Springer, New York.

Gooley, T. 2014. *The Lost Art of Using Nature's Signs.* The Experiment, LLC, New York.

9

Why are individual species where they are?

ACTIVITY: Drawn to Nature

Goal: Practice in drawing, slowing down, and observing.

Exercise 1: Sketch the basic shapes of trees found outside. Try to find at least five different shapes and write the location of each tree and the species if possible.

Exercise 2: Watch birds and squirrels at a feeder. Draw the postures of at least five individuals. Try to determine the common name of the species. Add notes about location, time of day, and other details you notice. It is okay to leave traces of your simple shapes in the final drawings.

Exercise 3: Make more detailed drawings of trees and annotate them.

Exercise 4: Draw a map of part of campus.

Exercise 5: Sketch a landscape view from somewhere on campus. In other words, draw as much of the scene and its features as you can see looking in one direction. Add interesting annotations and side drawings noting interesting science features if you wish. In other words, take a major observation or feature (e.g., unusually smooth ice, what the lake must look like in cross section, downed logs within the shallows) and draw it. You might draw the molecular or weather-related explanation that explains to a naïve reader what produces extra smooth ice. Add these notations and drawings next to your landscape.

WHAT CHARACTERIZES SPECIES DIVERSITY PATTERNS THROUGHOUT THE EARTH?

Earlier chapters considered patterns of life form diversity. This chapter examines biogeographic patterns of species diversity, including aspects of **species richness** (number of species) and **endemism** (the number of native species restricted to a particular region). One of the most dramatic patterns in ecology is the **species richness gradient with latitude** (Figure 9.1). This is the increasing number of species from the poles toward the equator. Figure 9.1 displays the pattern in general that fits for species richness in ants (Economo et al. 2019), three vertebrate species (Saupe et al. 2019), and many other plant, terrestrial, and aquatic species.

Maps showing "global hotspots of biodiversity" depict areas that have some of the greatest concentrations of native species (Figure 9.2, Table 9.1). Hotspots are the Earth's most biologically rich yet threatened terrestrial areas. They were first identified by the nonprofit, Conservation International, and are now overseen by the Critical Ecosystem Partnership Fund that recognizes 35 hotspots.

DOI: 10.1201/9781003271833-9

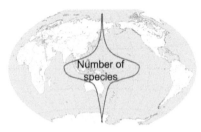

Figure 9.1 Species richness increases in a gradient from the poles to the equator.

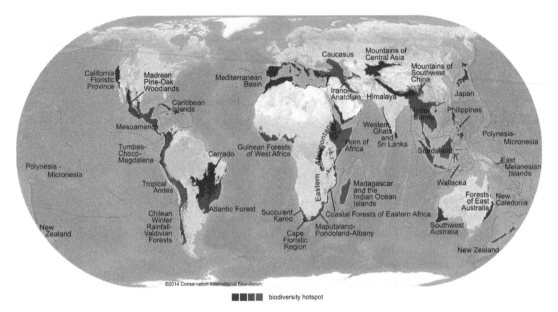

biodiversity hotspot

Conservation International (conservation.org) defines 35 biodiversity hotspots — extraordinary places that harbor vast numbers of plant and animal species found nowhere else. All are heavily threatened by habitat loss and degradation, making their conservation crucial to protecting nature for the benefit of all life on Earth.

Figure 9.2 Hotspots of biodiversity. To be included, a region must have at least 1,500 vascular plant species and lost at least 70% of its primary native vegetation (cepf.net) (*Conservation International, Creative Commons*).

A hotspot requirement is that the area has lost 70% of its native vegetation. Thus, the Amazon Basin, the most biodiverse place on Earth, is not included. Still, the hotspot map applies names to each relevant area, bringing them to life, calling out highly diverse oceanic islands and showing patterns not evident in Figure 9.1.

THE SPECIES DISTRIBUTION CONSTRAINT HIERARCHY

The biogeography of species diversity is more complex than for life form. The areas of greatest species occurrence follow a set of rules that can be organized in a chart known as a **constraint hierarchy** (organized from information in Eliot 2018, Lomolino et al. 2016). Figure 9.3 is an organizing scheme that shows the limits of membership within nested groups, in this case spatial scales. The constraint hierarchy can be used to understand why species tend to accumulate in certain areas. The way to read the chart is to start with the main menu at the global scale. It explains that a plant species could occur anywhere it tolerates existing levels of common physical and chemical limiting factors, such as temperature, frost line, pH, and salinity.

Table 9.1 Hotspots of biodiversity by name and number. In all, 50% of the world's plant species and 42% of the terrestrial vertebrate species reside in these hotspots, even though they make up only 2.3% of the world's land surface

Number	Location	Plant species	Number	Location	Plant species
1	Tropical Andes	30,000	19	California Floristic Province	3,488
2	Sundaland (Borneo, Sumatra)	25,000	20	Cape Floral Region South Africa	9,000
3	Mediterranean Basin	22,500	21	New Zealand	2,300
4	Madagascar and Indian Ocean	13,000	22	Chilean Temperate Forest	3,892
5	Indo-Burma	13,500	23	Caucasus of Middle East	6,400
6	Caribbean Islands	13,000	24	Wallacea (Indonesia, Timor-Leste)	10,000
7	Atlantic Forests of Brazil	20,000	25	Mexican Pine-oak Woodlands	5,300
8	Philippines	9, 253	26	South African Karoo	6,356
9	Eastern South Africa	8,100	27	East African Coastal Forests	4,000
10	Mesoamerica of Central America	17,000	28	East Afromontane	7,598
11	Brazilian Cerrado	10,000	29	Horn of Africa	5,000
12	Southwest Australia	5,571	30	Iran-Anatolia	6,000
13	Mountains of Central Asia	5,500	31	Himalaya	10,000
14	East Melanesian Islands	8,000	32	Mountains of Southwest China	12,000
15	New Caledonia of South Pacific	3,270	33	Japan	5,600
16	Western Colombia and Ecuador	3,270	34	Micronesia/ Polynesia/Hawaii	5,330
17	Guinean forests of West Africa	9,000	35	Forests of East Australia	2,100
18	Western Ghats and Sri Lanka	5,916	36	N. Am. Coastal Plain of SW U.S.	1,500

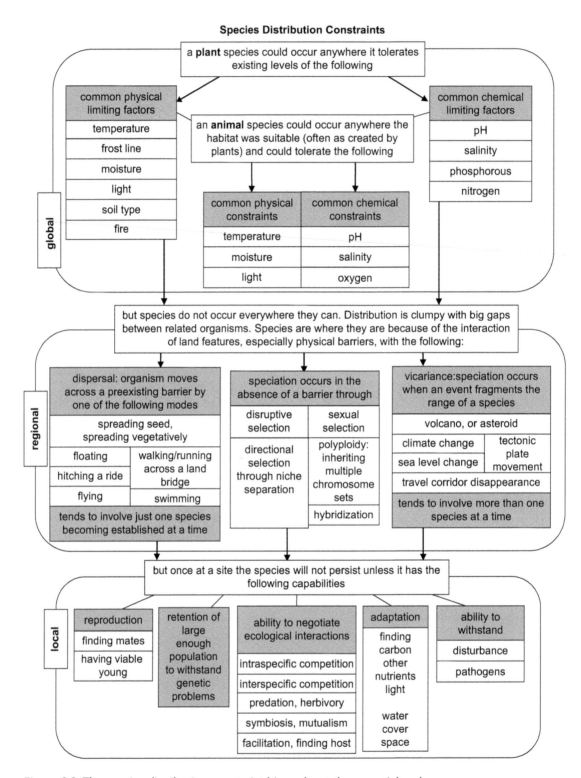

Figure 9.3 The species distribution constraint hierarchy at three spatial scales.

At the regional scale the menu further explains that species do not occur everywhere they can. Distribution is clumpy and gappy. Species are where they are because of the interaction of land features with dispersal and speciation mechanisms. It is only at the local level that factors like reproduction, competition, and predation play a part.

As an example of how the hierarchy works, consider the case of Kudzu (*Pueraria lobata*), a vine introduced to the southeast U.S., now so aggressive that entire houses have been covered by it. Kudzu was introduced in the U.S. from Japan mainly in the 1940s to control erosion along railways and roadcuts. In the hierarchy at the global level, Kudzu could occur anywhere in the world where moderate to high precipitation is present, if there are mild winters and moderately fertile soils. At the regional scale Kudzu occurs in the U.S. only where it has been introduced by humans. At the local level Kudzu tolerates sunny to slightly shady areas but does poorly in deep shade. When in the sun, it outcompetes other plant species by having high growth rates. Cows and Japanese Kudzu Bug graze it (Finch 2015), although other herbivores cannot tolerate the latex-like compounds in the vine. To control Kudzu in the southeast U.S., managers could allow trees to form a canopy, shading out the vine, or introduce certain grazers.

WORKING THROUGH THE CONSTRAINT HIERARCHY AT THE GLOBAL LEVEL

At global levels, species are constrained by **the law of the minimum** formulated by German agricultural chemist, Justus Liebig (1803–1873). According to Liebig's "law," the rate of any process is limited by the least, or slowest, factor affecting it (Figure 9.4). In other words, an organism is constrained by the weakest link in its requirements.

Take terrestrial plants. The limiting factor, the factor in least supply, is generally nitrogen in a form that can be taken up by plant roots. For plants in lakes and streams, phosphorous is generally the nutrient in least supply. Without more of this, plants cannot grow, even if all their other needs are met. Since Liebig's time, the reality is more complicated. The growth of a population at a particular level of one resource often depends on the level of one or more other resources. For instance, a plant may be able to use increasing levels of light only if it has increasing levels of nitrogen and phosphorous to build more photosynthetic cellular structures. Besides phosphate, aquatic plants are often limited by nitrate, silicate, and/or iron (Dybzinski and Tilman 2009). Besides nitrogen, terrestrial plants are often limited by potassium, calcium, or other trace metals.

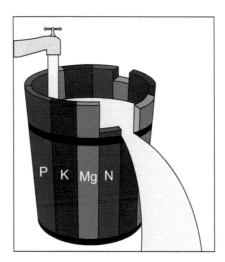

Figure 9.4 Liebig's law of the minimum. The capacity of the barrel is limited by the length of the shortest stave. Capacity can only be increased by lengthening that stave. When the stave is lengthened, another nutrient becomes the limiting factor (*G. Giacomin*).

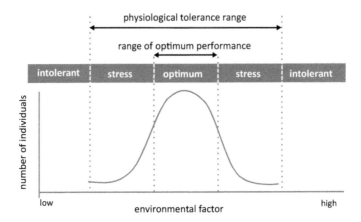

Figure 9.5 Shelford's law of tolerance. A species is most abundant within an optimum range of values for an environmental condition.

TOLERANCE LIMITS

University of Illinois physiological ecologist Victor Shelford (1877–1968), co-creator of the biome idea, elaborated on limiting factors by introducing the idea of **tolerance limits** (Figure 9.5). Not only did he recognize a minimum for each essential factor, a maximum also occurred, according to Shelford. Written another way, **the dose makes the poison**, meaning that anything can be toxic if the quantity is too great. Minimum and maximum tolerances for an organism are set by evolutionary factors. Within these limits, an optimum exists in which an organism grows and reproduces best. A narrow tolerance range produces a **specialist**. A broad tolerance range produces a **generalist**. For the constraint hierarchy, organisms occur where they do because conditions are neither too great nor too little in that location.

SPECIES GRADIENTS

Species richness increases toward the equator for plants because precipitation, high temperature, and absence of frost lead to the best potential in the world for sustainably maintaining high evapotranspiration rates. Abundant plant diversity occurs at the equator, and animal diversity follows (Lomolino et al. 2016). Thus, tropical forests host more than 50% of the world's species, while occupying only 7% of the land surface.

Other species gradients exist besides the latitudinal gradient (Lomolino et al. 2016). Mountain regions generally support more species than flatlands because mountains have more environmental heterogeneity. Gradients also exist for endemism (Lomolino et al. 2016). Peninsulas have higher endemism than continental areas, although fewer species overall. Isolation encourages speciation. Mediterranean and island ecosystems have high endemism. For islands, the larger the area the higher the endemism. For example, 80% of the 8,000 plants in Madagascar are found nowhere else; 90% of 9,000 flowering plants of New Guinea are endemic; and 76% of New Caledonia's 3,250 vascular plants are endemic.

On the ocean floor, hydrothermal vents, trenches, and seeps become the equivalent of isolated islands and have high endemism. Within ocean water in general, marine species tend to be more widely distributed, and therefore have less endemism than terrestrial. Oceans have fewer physical barriers and therefore less isolation that would lead to speciation.

THE CONSTRAINT HIERARCHY AT THE REGIONAL SPATIAL SCALE

At the regional scale the hierarchy demonstrates that species could colonize many more areas than they do, but three factors affect distribution: dispersal, speciation, and vicariance mechanisms, and these interact with land features.

DISPERSAL

Not all species make good colonizers. Species with a better chance of colonization are said to be **vagile**, i.e., more capable of successful, long-distance movements, or able to withstand unsuitable conditions better than others (Morrison et al. 2006). Certain mollusk species, for example, disperse particularly well via boat hulls. Some spider species disperse well by having the ability to balloon across the ocean on air webs. Some gnat-like insect species fly into high winds aloft and travel several hundred miles downwind. According to a study in Hawaii the most successful dispersal methods for plant species were through the actions of birds (75%), oceanic drift (23%), and air flotation (1%) (Carlquist 1974).

Beyond modes of dispersal, other factors play a role in colonization (Morrison et al. 2006). Colonization rates increase if patches are connected by corridors or habitat remnants are scattered throughout the intervening matrix lands. When sites can act as a stepping stone for dispersal, colonization is enhanced, even if the stepping stone is suboptimal and occupancy is intermittent. Colonization rates increase with number of founders. A low number of founders makes the population vulnerable to genetic bottlenecks and perhaps inbreeding depression.

For oceanic islands, colonization rates increase with island size and proximity to dispersal sites if all things are equal. Based on proximity, we would expect the most species diversity in order of New Zealand > Galapagos > Chatham Islands > Hawaii. Yet, this is not the case. Hawaii has the most species by a wide margin. Why?

All things are not equal (Morrison et al. 2006). The Galapagos have closer proximity to land and a regular flow of strong ocean currents, so they should have more species than Hawaii. Their colonization rate is one species per 3,000 years compared to one species per 70,000 years in Hawaii. By this measure, the Galapagos should have 23 times the species diversity of Hawaii, but it does not because of confounding variables.

Hawaii is 63 million years older than the Galapagos, which formed only 3 million years ago. Hawaii has twice the landmass of the Galapagos, presenting a larger target for arriving colonists. Hawaii has a greater variety of environments and a more diverse climate. The Galapagos have a harsh environment for all living beings on account of their location near cold upwelling zones, which discourages precipitation. Thus, Hawaii has the most biodiversity.

SPECIATION

At least three major mechanisms foster the formation of new species (Lomolino et al. 2016). A few individuals may undergo a **founder event** (establishment of a new population by a few individuals). A population then becomes geographically isolated. Because gene flow is retarded in the isolated area, speciation occurs. The process can happen quickly if population numbers are small and genetic drift is exaggerated.

In founder events, speciation is called **allopatric** (literally "different land") **speciation**. Two subtypes are recognized. In mode I, an environmental event creates an isolating barrier, splitting a group from the rest of its species, a process called **vicariance**. The barrier can be sea level rise, volcanic eruption, or tectonic plate movement (Figure 9.6).

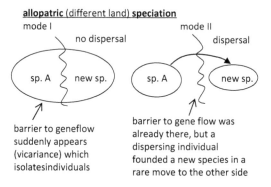

allopatric (different land) **speciation**

Figure 9.6 Allopatric speciation occurring by different mechanisms. Further speciation can occur beyond one new species. (*Adapted from Lomolino et al. 2016.*)

One clue that vicariance had taken place is if multiple speciations occurred at the same time. For example, when Gondwanaland broke into several areas, including India and Madagascar moving away from the African continent, it fostered the emergence of hundreds of new species of the same age. Allopatric speciation mode II occurred when individuals dispersed across an existing barrier to colonize a previously uninhabited region (Figure 9.6). Unlike the result in vicariance, mode II involved just one species colonizing at a time. For example, turtles and finches dispersed through separate events to the Galapagos Islands.

Speciation in the absence of a geographic barrier can result from disruptive selection such as when a parasite colonizes a different host species. It can also occur through chromosomal changes most often found in plants. In **polyploidy**, an entire additional set of chromosomes is passed on, which can triple the haploid number of chromosomes. In **aneuploidy**, a single chromosome breaks or fuses with another to change the number of chromosomes by plus or minus one. Polyploidy tends to occur in plants because of self-pollination. It may account for the evolution of 70%–80% of angiosperm species (Briggs and Walters 1984).

THE CONSTRAINT HIERARCHY AT THE LOCAL SPATIAL SCALE

Classic ecological interactions come to the forefront at the local scale. Competition, predation, herbivory, symbiosis, mutualism, facilitation, and parasitism must be negotiated by a species once it finds itself at a site. A species will not persist unless it can find mates, produce viable young, find food, build an immune system against local threats, and retain a large enough population to prevent genetic problems. These subjects are largely what is covered in conventional ecology. Population, community, and even ecosystem ecology have mainly focused on these subjects throughout ecology's history, yet they dominate only at local spatial scales. For conservation purposes, the most fundamental realities apply at global and regional scales interplaying with evolution. It makes the point that most research in ecology has been at spatial (and temporal) scales too small to elucidate what may be the most overarching theories in nature. The next section of this book addresses evolution and begins to address these issues.

NEXT STEPS

Once a more thorough understanding of ecology has been presented, Chapter 31 will circle back to new perspectives in our understanding of hotspot dynamics. The next section on evolution not only provides the basics of micro and macro evolution, it also presents new ideas from post-modern evolution. Post-modern evolution? More has been learned about genetics and evolution in the last 25 years than in the previous 150. There is much to cover regarding this new understanding.

THINKING QUESTIONS

1. Using the section in this chapter titled "Species gradients," describe the environment you would expect to be the hottest of all hotspots in species diversity. After doing so, try to think of a real geographic area that might match your description.
2. European culture endured ten or more centuries of teaching in which evolution was thought to be non-existent. Do you think this has affected how we view ecology now at the macroecology scale?

REFERENCES

Briggs, D., and S.M. Walters. 1984. *Plant Variation and Evolution*. 2e. Cambridge University Press, Cambridge, UK.

Carlquist, S. 1974. *Island Biology*. Columbia University Press, New York.

Dybzinski, R. and D. Tilman. 2009. Competition and coexistence in plant communities. In: S.A. Levin (ed.), *Princeton Guide to Ecology*. Princeton University Press, Princeton, NJ.

Economo, E.P., J. Huang, G. Fischer, E.M. Sarnat, N. Narula, M. Janda, B. Guénard, J.T. Longino, and L.L. Knowles. 2019. Evolution of the latitudinal diversity gradient in the hyperdiverse ant genus *Pheidole*. Global Ecology and Biogeography. https://doi.org/10.1111/geb.12867

Eliot, C.H. 2018. Ecological interdependence via constraints. *Philosophy of Science* 85:1115–1126.

Finch, B. 2015. The true story of Kudzu, the vine that never truly ate the South. *Smithsonian Magazine*, September 2015.

Lomolino, M.V., B.R. Riddle, and R.J. Whittaker. 2016. *Biogeography*. 5e. Sinauer, Sunderland, MA.

Morrison, M.L., B.G. Marcot, and R.W. Mannan. 2006. *Wildlife-Habitat Relationships Concepts and Applications*. 3e. Island Press, Washington, DC.

Saupe, E.E., C.E. Myers, A. Townsend Peterson, J. Soberon, J. Singararyer, P. Valdes, and H. Qiao. 2019. Spatio-temporal climate change contributes to latitudinal diversity gradients. *Nature, Ecology, and Evolution* 3:1419–1429. doi.org/10.1038/s41559-019-0962-7

10

Introduction to evolution: The modern synthesis

ACTIVITY: Learning Wilderness Survival Skills (priorities/shelter)

Introduction: Imagine you are on vacation with your family in Las Vegas for Christmas. You decide to take a side trip to the Grand Canyon one afternoon. The weather forecast missed a surprise snowfall. The GPS in the vehicle you rented leads you down a dead-end. When you try to turn around, the car gets hopelessly mired in a muddy ditch. Visibility is zero and so is cell service. Night falls. One of the members of the family is a marathon runner and decides to go for help by running to the main entrance of the park. Good idea? Think this could never happen to you? Use the internet to read the outcome of the Karen Klein story.

Even in our own homes, natural disasters can lead to the unexpected need for basic wilderness skills. Fire building, shelter building, and wilderness first aid provide life-saving advice. They build confidence for time spent outdoors.

Priorities: If you had to spend the night unexpectedly in a wilderness situation, what would be your first priority? Food, cell service, shelter, water, fire, first aid? The first rule is to S.T.O.P. (Sit; Think; Observe; Plan). What do you have? What do you need? How can you maximize your chances of getting what you need? Your most essential equipment is between your ears. Take a few minutes to assess your situation, with whomever is with you. What do you need most?... to be found?... to be extricated from wherever you are?... to evade a threat?... to survive the storm?... to treat an injury?... to rest until you can go farther?

If there are no injuries the priority should be shelter. Keep in mind the Rule of Threes. In general, humans can live 3 minutes without air, 3 hours without shelter (heat/shade), 3 days without water, 3 weeks without food, 3 months without companionship. Signs of hyper-/hypothermia include the -umbles: fumble, stumble, mumble, and grumble. Inappropriate laughing or crying can be a warning. Frostbite will be evident when fingers, toes, and nose begin to turn firm, numb, and white. Gradual (30 minutes) rewarming with water slightly above body temperature and prevention of re-freezing is in order. Attempts to signal must take place knowing that a cell phone battery is best used to send text, not a call. Climbing to the tallest nearby hill to get a signal might help. Shelter for the night should still be the first priority. Stay with the vehicle, stay together, and get under a makeshift roof to reduce heat loss to the night sky.

Activity: Make a lean-to shelter by piling vegetation against a downed log or other brace. Build a low roof to reduce space within the shelter so that body heat is not wasted. Use a ground cover of vegetation or something you brought. Once built, crawl in and experience the thermal protection afforded by lying next to the ground but protected from moisture, wind, and sky. Measure the temperature inside versus out.

DOI: 10.1201/9781003271833-10

THREE PERIODS FOR UNDERSTANDING BIOLOGICAL EVOLUTION

To understand the major ideas in evolution, let us frame the 160-year history since *Origin of Species* into three divisions:

- the Darwinian synthesis,
- the modern synthesis,
- the postmodern era.

WHAT WERE THE MAIN FEATURES OF THE DARWINIAN SYNTHESIS?

In 1859 *Origin of Species* sold out within a few days of publication. The book went through innumerable reprintings and Darwin wrote several editions making improvements to his arguments each time. He wrote 12 other books after *Origin*, including *The Descent of Man and Selection in Relation to Sex*, published in 1871.

Darwin's ideas were considered some of the most important in history because they provided a scientific alternative to the Aristotelian paradigm held for ten centuries. The direct hand of God was no longer necessary for explaining the vast species diversity of Earth. The ape ancestor argument by Darwin suggested that humans were mere animals.

The weakness in Darwin's argument was an inability to describe the details of heredity. Interestingly, Gregor Mendel was alive and working out the idea of recessive and dominant genes at the time Darwin was preparing *Origin*. From 1856 to 1863 in the garden of his monetary, Mendel grew 28,000 pea plants and hand-pollinating them. He carefully wrapped them while they were still living to prevent insect pollination. Mendel demonstrated that parental traits did not always blend as commonly believed at the time. His discovery was published in a journal not widely read in 1865, but the article resurfaced 50 years later. One of the great unsolved mysteries of biology history is why Mendel never contacted Darwin. Read more about this fascinating detective story in Galton (2009) and Fairbanks (2020).

WHAT WERE THE MAIN FEATURES OF THE MODERN SYNTHESIS?

In the late 1920s and early 1930s, J.B.S. Haldane, Sewall Wright, and Ronald Fisher formed mathematical descriptions for the genetic makeup of populations. In doing so they reinvigorated the subject of evolution by incorporating natural selection with Mendel's insights on how traits passed from parent to offspring (Whitfield 2008). Specifically, Mendel's Law of Segregation and the Law of Independent Assortment became known as "Mendel's Laws of Inheritance." "**Modern synthesis**" was coined by Julian Huxley in 1942 to represent the blending of Mendel and Darwin's ideas. Most of the promotion of the modern synthesis came in the 1950s especially during the centennial celebration of *Origin*.

THREE FURTHER CATEGORIES FOR UNDERSTANDING THE MODERN SYNTHESIS

A helpful framework for understanding the modern synthesis is to recognize molecular evolution, microevolution, and macroevolution as categories. Molecular evolution and microevolution are covered in this chapter as understood through the modern synthesis. A postmodern explanation of molecular and microevolution is covered in Chapter 11. Macroevolution is more fully described in Chapter 12.

WHAT IS MOLECULAR EVOLUTION?

Molecular evolution investigates the coding of information in organisms and the changes in these codes over time. The code is present through nucleic acids in DNA and RNA specifically as A, G, T, C, and U (adenine, guanine, thymine, cytosine, and uracil). These nucleic acids code for proteins that determine function and structure throughout the organism's body.

Since 1995 when it became possible to investigate the entire gene structure for a species, our understanding of molecular evolution has been in a period of dizzying growth, with scientists struggling to keep up. This has included advances in **systematics**, the study of classification for organisms, as opposed to **taxonomy**, the naming of species.

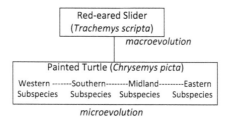

Figure 10.1 Microevolution is changes within a species. Macroevolution is transition to a new species.

WHAT IS MEANT BY MICROEVOLUTION?

Microevolution is evolution occurring at the population level within a species (Figure 10.1). **Macroevolution** refers to evolutionary patterns and processes above the species level. It encompasses evolution in the fossil record under the principles of dispersal, extinction, and speciation.

In the classic **microevolution definition**, evolution is a change in allele frequency. Microevolution usually occurs at small scales, perhaps at the level of industrial melanism, such as the change in moth color because of pollution. In the classic modern synthesis understanding of microevolution, the mechanisms responsible for a change in allele frequency are natural selection, mutation, genetic drift, and gene flow.

WHAT ARE THE BASICS OF MICROEVOLUTION AND POPULATION GENETICS?

Recall that **phenotype** is the physical expression of a gene, the observable appearance of an organism. **Genotype** refers to the genetic constitution of an organism as expressed by the order of nucleotides such as AGA CTA TAC. You can think of it this way: the genotype resides in nucleic acids that serve an informational function. The phenotype is largely manifested when proteins are made in an organism's body.

At the genotypic level, **alleles** are different forms of a gene at one locus on homologous chromosomes. If B and b are two heritable forms of the gene for eye color and if the dominant allele produces brown eyes, this is represented by BB (homozygous dominant) or Bb (heterozygous).

Two recessive alleles produce blue eyes represented by bb (homozygous recessive).

The microevolution definition of evolution as a change in allele frequency provides a way to determine whether evolution has occurred. Theoretically we would measure allele frequency within a population at two times. For example, say the proportions of B and b alleles present in a population at one time were B=56% and b=44%. Later, after several births and deaths in the population the allele frequency changed to B=58% and b=42%. Because the allele frequency changed, evolution occurred.

When alleles occur in frequencies different from Hardy-Weinberg expectations, we conclude that processes such as drift and selection as well as nonrandom mating influence gene frequencies. The Hardy-Weinberg principle is akin to a neutral model showing us what to expect in the absence of drift, mutation, or selection. It provides a standard against which deviations can be identified.

WHY DO CHANGES IN ALLELE FREQUENCY OCCUR?

In classic modern synthesis thinking, changes can occur because of:

- **mutations**: substitutions in single nucleotides of a gene or chromosome occur through deletions, insertions, inversions, or frameshift errors.
- **genetic drift**: random shifts in allele frequency over the course of generations. This is recombination due entirely to chance during mating. Genetic drift is a weak force occurring in large populations as readily as small, but genetic drift tends to become exaggerated in small populations because of loss and fixation (Figure 10.2).

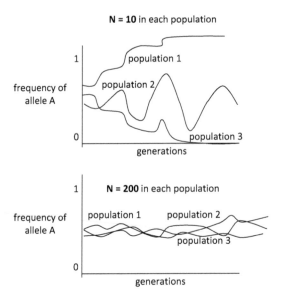

Figure 10.2 Genetic drift effects are less dramatic when the population size (N) is large.

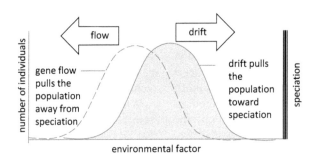

Figure 10.3 Gene flow counters the effects of genetic drift and natural selection. Geographically isolated populations have the tendency to drift toward speciation if there is no gene flow.

- **gene flow**: loss or gain of alleles through emigration or immigration in the population. This occurs through animal migration, wind pollination, or other ways. Gene flow tends to counter the effects of genetic drift and natural selection by retarding the development of geographically isolated traits (Figure 10.3). In other words, even a little bit of gene flow can swamp out genetic drift and natural selection that tends to develop in isolated conditions. This explains why speciation does not occur more often.
- **natural selection**: Darwin's great contribution. Natural selection offered the mechanism to explain why organisms with better fitness proliferate. Note that selection and evolution are not synonymous. Selection is a process that results in evolution, whereas evolution is the historical pattern of change through time.

NATURAL SELECTION IN MORE DETAIL

Natural selection was described by Darwin as survival of the fittest. It is best defined now as the process whereby organisms better adapted to their environment survive and produce more offspring. Another way of phrasing it is that heritable phenotypes leaving the most offspring will predominate.

Summarizing Darwin's ideas, natural selection occurs because of the following phenomena:

1. More individuals are produced than can survive.
2. There is a struggle for existence.
3. Individuals show variation.
4. Those with genetically advantageous features proliferate if these features lead to greater reproduction and survival. Unfavorable traits become less prevalent.

Keep in mind that adaptations must be genetic (heritable) for natural selection to proceed. For example, a population of Groundhogs (Woodchucks, Whistle-pigs, Marmots, *Marmota monax*) might have individuals of various body fat percentages. The fatter ones are more likely to survive the winter but each summer the same percentage are fat and lean. This could occur because some individuals have more food volume by chance, not because of genetics or evolution. For evolution to occur, a genetic trait would need to be present that conferred an advantage such as better digestive enzymes that were passed on to offspring.

WHAT ARE ADAPTATIONS?

Favorable traits have a special name called **adaptations**, heritable features of organisms that increase their fitness. In this context, **fitness** is the genetic contribution of an individual to the next generation of a population. The fittest individual is the one that passes the most adaptations to the next generation.

It helps to think of adaptations as evolutionary solutions to ecological challenges. Stephen J. Gould called them "contraptions," some evolutionarily derived coping method that deals with a challenge. Think of adaptations in three categories:

- **morphological adaptation**: "morphology" refers to form or looks. Think of the short, powerful legs with sharp claws that make digging easy in Groundhogs.
- **physiological adaptation**: this includes growth and development, temperature and pH control, or other aspects of homeostasis. In Groundhogs, think of their ability to slow their metabolism in mid-summer to begin gaining weight for the winter. The pound of vegetation they eat per day turns largely into fat stores, without the need to drink water other than dew.
- **behavioral adaptation**: this includes foraging, mating, and showing aggression. For Groundhogs, think of their ability to withstand forest fires and predators by burrowing deep in the ground during danger (Figure 10.4).

Figure 10.4 This Groundhog survived unscathed during a prescribed burn, burrowing deep in a tunnel.

NATURAL SELECTION DOES NOT CREATE ADAPTATIONS

A subtle, but key idea of evolution is that natural selection does not produce adaptations. The adaptations are produced by new traits entering a population through mutation, genetic drift, sexual recombination, or gene flow (and see other ideas in Chapter 11). In other words, natural selection is an eliminator, not a producer. Despite the name, "*The Origin of Species*," Darwin's focus was on natural selection, the eliminator, not the producer.

SPECIFIC TYPES OF EVOLUTION AND NATURAL SELECTION

Convergent evolution occurs when the phenotype at one extreme among species has a selective advantage (Figure 10.5). Convergent evolution, for example, occurs among taxonomically different species of desert plants throughout the world. Despite being only distantly related, species have evolved leaves in the form of spines, adaptive because it reduces water loss. These include cacti in one family, stoneworts in another, sunflowers in another, and species in the poinsettia family such as the Baseball Plant, spurges, and the Sandbox Tree with exploding seed capsules.

Sexual selection refers to the selection of traits based purely on the maximization of mating success. Members of one sex (usually the female) become fixated on a trait in the other sex because it represents superior mating quality. The trait becomes exaggerated over time as in the case of the Peafowl (peacock) tail becoming larger with subsequent generations, or male Bullfrogs developing deeper, louder voices over time. Run-away selection occurs for traits that may seem to confer a negative advantage. Peacocks with large tails would seemingly run slower in the face of predation, and Bullfrogs would seemingly be found more often by predators if they had a louder voice.

The adaptations are usually found in males because females are usually the choosier sex. Females have larger gametes and often more responsibility for raising the offspring. Thus, the females have more to lose if they choose the wrong partner. Sexual selection can be extremely strong. Darwin expressed great interest and conjecture about sexual selection, and it has been a fascination of ecologists since.

Gametic selection is natural selection acting at the unit of the gamete within a species. It occurs because some sperm swim faster and are more likely to fertilize an egg. Likewise, some pollen grains grow faster pollen tubes and are first to fertilize an egg. In other words, this is survival of the fittest gamete, not survival of the fittest individual.

A type of natural selection not acting at the individual or gamete unit is **species selection,**

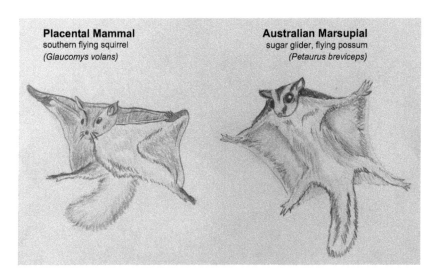

Placental Mammal
southern flying squirrel
(Glaucomys volans)

Australian Marsupial
sugar glider, flying possum
(Petaurus breviceps)

Figure 10.5 Convergent evolution takes place when a phenotype at one extreme has a selective advantage among species that are not closely related (*E. Tauber-Steffen*).

survival of the fittest species not the fittest individual or gamete. This idea can be applied only when examining fossil records. For instance, marine animal species with direct development are more likely to go extinct than those with a planktonic stage (Ridley 1996). Not only do the direct development individuals stay in one place for most of their lives; their descendants stay there too. If conditions in the ocean affect their local region, the species goes extinct. In contrast, planktonic larval species have wider geographic dispersal and are less vulnerable to local circumstances. They are less likely to go extinct, but less likely to speciate too.

An even grander example of species selection involves the mammals and birds after the asteroid struck the Earth at the end of the Mesozoic. Mammals and birds had a selective advantage over non-homeo-thermic reptiles as the climate became dramatically colder and wetter. With the ability to maintain a constant body temperature, mammals and birds thrived while many reptile species succumbed.

KIN SELECTION

Kin selection is sacrificing one's self for the good of one's close relatives. It appears to be **altruistic**, sacrificing self for the good of the group, but nearly every act of apparent altruism studied in detail has increased the altruist's fitness. The individual either receives a reciprocal favor because of the apparent altruism, or the benefit goes to a genetically related member of its group.

The way to understand kin selection is through an example. When a Belding's Ground Squirrel (also called Pot Gut, Sage Rat, or Picket Pin, *Urocitellus beldingi* by the way) spots a predator, it gives a squeak (alarm call) just before it scurries down its hole. By squeaking it puts itself at risk to the predator but saves the lives of other nearby squirrels. It appears to be altruistic, but the squirrel squeaks only in the presence of its close relatives.

The modern synthesis explanation is via the **selfish gene concept** in which the unit of selection is the gene (Dawkins 1976). J.B.S. Haldane in 1955 explains: parents and offspring are related by 50%, siblings by 50%, cousins by 12.5%. Nieces and nephews are related to their aunts and uncles by 25%. Therefore, saving two offspring is equivalent to saving one's self. This led Haldane to say he would die for two of his brothers or eight of his cousins.

Inclusive fitness is a way to quantify the sum of an individual's total contribution of genes to the next generation. It is gained by personal reproduction and the non-direct fitness gained by helping non-descendant kin (Pfennig and Kingsolver 2009). In the squirrel example, the squeaking only makes sense if the nearby squirrels are closely related. If it were done purely as an altruistic act, that individual would probably not survive and reproduce. The gene for altruism would not be passed on. Selfish individuals would proliferate instead. An altruistic gene might benefit the species, but it would harm the individual and therefore not be passed on.

CAN KIN SELECTION BE APPLIED TO HUMANS?

Sociobiology focuses on human social behavior as derived through evolution. Applying kin selection to humans is within its venue, which was famously championed in a book by E.O. Wilson (1975) titled *Sociobiology: The New Synthesis*. Controversial questions are sometimes addressed within sociobiology, like whether more homicides occur between stepchildren and their parents than between biological children and their parents.

The book introduced such a controversial premise that the author was doused with a pitcher of ice water and told, "Wilson, you're all wet," when he stood up to address the American Association for the Advancement of Science meeting in 1978. Sociobiology ideas question whether altruism and unconditional love can really occur, or whether we are just slaves, only acting when it confers some advantage to the potential reproductive success of our genes.

THE EVOLUTION OF AN IDEA

While the concepts of the modern synthesis in this chapter are widely taught, they only tell part of the story in biological evolution. Most of the ideas described so far had their origins in the first half of the 20th century. Since 1950, an explosion of new discoveries has provided insight at even basic levels. The next chapter tells that story.

THINKING QUESTIONS

1. Explain the difference among "adapt," "evolve," and "adaptation."
2. Read more about the basic concepts of sociobiology and then answer the question, do you think humans can ever act in a way that is truly altruistic? Give an example if yes.
3. Do you think selection can ever work at the level of the group in "group selection?"
4. Before moving into 21st-century ideas about evolution, review the myths tied to Darwinian-derived concepts.

   - Evolution runs counter to the second law of thermodynamics.
   - Natural selection creates new types of organisms.
   - Humans are the pinnacle of evolution.
   - Evolution occurs for the good of the species.
   - Every species has a purpose.
   - Because there are so few intermediate forms (missing links), evolution is not valid.
   - There is no evidence for evolution.

REFERENCES

Dawkins, R. 1976. *The Selfish* Gene. Oxford University Press, New York and London.

Fairbanks, D.J. 2020. Mendel and Darwin: untangling a persistent enigma. *Heredity* 124:263–273.

Galton, D. 2009. Did Darwin read Mendel? *QJM* 102:587–589. doi:10.1093/qjmed/hcp024

Pfennig, D.W., and J.G. Kingsolver. 2009. Phenotypic selection. In: S.A. Levin (ed.), *The Princeton Guide to Ecology*. Princeton University Press, Princeton and Oxford.

Ridley, M. 1996. *Evolution*. Blackwell Science, Cambridge, MA.

Whitfield, J. 2008. Postmodern evolution? *Nature* 455:281–284.

Wilson, E.O. 1975. *Sociobiology: The New Synthesis*. Belknap Press: An Imprint of Harvard University Press, Boston, MA.

11

Advances in microevolution, molecular evolution, and evo-devo

---

ACTIVITY: Learning Wilderness Survival Skills (fire)

Introduction to fire: Building a fire in a wilderness situation can be lifesaving for multiple reasons: warmth, prevention of shock, cooking, water purification, signaling, camaraderie, comfort, and stress relief. This is an essential skill, but if someone fell through the ice, was rescued, and needed a fire immediately, could you get it started? What if it were a rainy day, or in the snow, or even in a camp situation where a fireplace was nearby and matches were available? Could you find enough firewood in a wet forest?

The essentials for fire include fuel (tinder, kindling, and larger sticks and logs), oxygen, and ignition. It helps to have an accelerant that burns easily, which you unexpectedly may have with you in the form of a waxy lip balm, an alcohol-based hand sanitizer, or even oil-rich Doritos and Cheetos.

Activity: Each student must make their own small, controlled fire in an open-field situation, using natural sources for fuels that are found outdoors. Students can use anything they have with them in their book bag or purse. Matches should be provided and perhaps even small candles if in a wet environment. Each fire should be graded on an A to F basis. The fire must sustain itself for 10 minutes and then be extinguished. Students have an hour total to collect the fuels and sustain their fire. If residing in a fire-prone location, obviously this exercise should not be done. Even in a wet environment, fire extinguishers and buckets of water should be immediately available.

The most common issues that present problems:

- **Failure to block the wind around the fire:** Dig a depression, mound up snow, or find a natural fire break before placing the fire.
- **Failure to collect enough tinder and kindling:** Tinder is dry fine fibers or shavings that can easily catch fire such as the cottony fibers from the inside of a milkweed pod, or drier lint that is kept in a baggy in one's backpack for this purpose. Paper is not good especially on a wet day because paper absorbs humidity. Kindling is tiny sticks just larger than tinder that can be placed one at a time if necessary on the fire once the tinder is ignited. Have on hand far more tinder and kindling than what seems necessary. Separate these into piles within arm's reach of the fire.
- **Failure to collect enough larger sticks that are dry:** Once the kindling is burning, add progressively larger sticks. Collect sticks not from the ground, but from dead limbs that are

DOI: 10.1201/9781003271833-11

stuck in trees or shrubs. The stick will be burnable if it passes the snap test – break the twig and listen for a clear snap. If the twig bends but does not break it is too green.

- **Trying to start a fire in a puddle:** If the ground is wet, a little platform of sticks will need to be built to keep the tinder and kindling out of the moisture.
- **Failure to provide air flow:** Novices tend to throw together a jumbled pile of burnable materials on the ground and attempt to light it. Such a pile will not have enough flow. A better idea is to start with a base either in a triangle and build into a teepee or to build a log cabin sort of structure with abundant space between sticks for air. Placing kindling on the tinder stick by stick will be necessary on wet days and patience will be necessary.

THE REVOLUTION IN EVOLUTION

Since 1950, the science of evolutionary biology has been the evolution of an idea. Advances in molecular biology, developmental biology, cell biology, and genomics have brought startling discoveries that substantially increased our understanding of evolution. **Developmental biology** is the study of embryonic development in animals. **Genomics** is the sequencing of whole genomes for species and the mapping of their genes. The pace of learning in genomics has increased exponentially since 1995 when the first prokaryote was sequenced.

While tenets of the modern synthesis have been taught universally with a mainly unified front for over a century, a growing movement is advocating for greater recognition of discoveries since the 1950s even if this shifts the focus considerably (Corning 2020, 2022, Heng 2019, Koonin 2009, 2011, 2016). The danger in changing the narrative is the appearance of disagreement among scientists when presented to evolution-deniers. The danger in not changing the narrative is continuing to teach a paradigm that is known to be incomplete. New discoveries show evolution to be even more astounding than thought, with an understanding emerging that is more directed, cooperative, and systematic than something based on chance alone. This requires an explanation.

The modern synthesis taught the following gene-centered view: individual genes in DNA are the underlying cause of biological variation. All evolution is the accumulation of small genetic changes guided mainly by natural selection. Under these assumptions, changes in gene frequencies over time provide the evidence for evolution. Because genes are selfish and self-preserving, the unit of evolution is the gene.

Other major features of the modern synthesis can be summarized with the following tenets (Koonin 2011):

Random mutations are the only source of evolutionarily relevant variability: This was purported, even though Darwin left room for some Lamarckian-type processes. Additionally, ideas on **genetic drift** (random changes in allele frequency for a population over time) had been presented by Wright in the 1930s, but not emphasized.

Beneficial changes that are fixed by natural selection are infinitesimally small, and **thus evolution occurs via gradualism** (the gradual accumulation of tiny modifications): This was an idea insisted upon by Darwin.

Evolutionary processes have remained at essentially the same rate throughout the history of life, which is **the idea of uniformitarianism**,

Macroevolution is governed by the same mechanisms as microevolution: Dobzhansky and his definition of **evolution** (a change in allele frequency) was the chief proponent of this principle.

The evolution of life can be represented as a single tree of life: It was believed that all extant diversity of life forms evolved from a single common ancestor, now dubbed the **last universal common ancestor (LUCA)**.

Nearly all parts of the genome (all nucleotides) have a specific function: Now we know there are long sequences of non-coding portions of DNA, especially in higher organisms.

In summary, the modern synthesis indoctrinated gradualism, uniformitarianism, and the monopoly of natural selection as the main route to evolution.

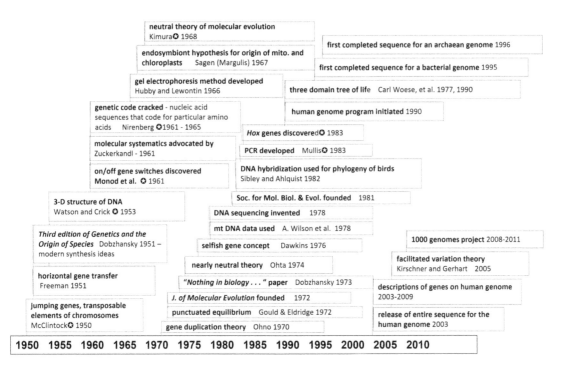

Figure 11.1 The evolution of an idea in molecular evolution over the last 60+ years is summarized here. ✪ indicates Nobel Prize.

WHAT ARE THE FEATURES OF POSTMODERN THINKING?

Contemporary discoveries from molecular evolution, cell biology, and developmental biology have brought about **postmodern thinking** (Koonin 2011) or an **inclusive biological synthesis** (Corning 2022). Key achievements since 1950 are outlined in Figure 11.1 and summarized in the rest of this chapter.

1. **Transposable elements** (jumping genes) were discovered by Barbara McClintock (1902–1992) in the mid-1940s and reported in 1951, but not appreciated for 30 years until she won the Nobel in 1981. **Transposable elements** are sections of DNA that switch positions on the chromosome. This dispels the idea of genes fixed in their position on a chromosome.
2. Similarly, **horizontal gene transfer** was first described by Victor Freeman while doing diphtheria research in 1951. Its significance was not widely appreciated until the 1980s. **Horizontal gene transfer** is the transfer of genes from one organism to another other than through

reproduction (which would be vertical transfer, adult to offspring). For instance, one species of bacteria can transfer to another a gene for antibiotic resistance. Genes can be transferred by viruses, or artificially via genetic engineering. A virus can insert its genes into the genome of an organism it infects, and the infected cell can reproduce these genes. Horizontal gene transfer appears to be exceedingly common in bacteria and unicellular eukaryotic organisms. It occurs in eukaryotes as they pass genes from mitochondria and chloroplast genomes to nuclear DNA.

3. **The 3-D structure of DNA was proposed by Watson and Crick in 1953**: This focused studies at the molecular level from the 1950s into the 1960s and later. Notable were new techniques developed by the 1980s in protein electrophoresis, ribosomal RNA techniques, DNA sequencing, DNA hybridization, and PCR (polymerase chain reaction).
4. The **endosymbiont hypothesis** describes the origin of mitochondria and chloroplasts as organelles in eukaryotic cells, conceived by Lynn Margulis (1938–2011) in 1967. She

proposed the idea that mitochondria and chloroplasts were once whole organisms that became engulfed within another organism and then lived symbiotically within it. The cell itself then reproduced the mitochondria as organelles. It explains why mitochondria and chloroplasts have their own DNA sequences different from nuclear DNA.

5. The neutral theory of molecular evolution was proposed by Motoo Kimura (1924–1994) in 1968. In opposition to the idea of natural selection as the dominant mechanism in evolution, the **neutral theory** posits that most changes in allele frequency occur because of mutations that are neither beneficial nor harmful, with random drift acting on neutral alleles.

In other words, genes evolve that do not have a biological function. These neutral molecular changes do not influence the fitness of the individual organism and are not subject to natural selection. These changes happen, for instance, when single-nucleotide point mutations encode the same amino acid (e.g., GCC and GCA both encode alanine). Even whole new alleles may be created through drift or mutation, but these alleles are neutral in their effect.

Because there is no selection acting either negatively or positively on a trait encoded by these changes, they remain in the genome and accumulate. Assuming they are produced at the same rate throughout history, researchers can use the quantity of changes in neutral genes to estimate the time since a taxon diverged from an ancestor. This is known as a **molecular clock** and has provided the opportunity to more accurately estimate evolutionary age for a species group.

6. Subsequently, Tomoko Ohta, a student of Kimura, proposed the "nearly neutral theory" in 1974. One aspect of her theory is that the size of the breeding population (**effective population size**) determines whether it is drift or natural selection that dominates changes in allele frequency. When the breeding population is large, evolution is governed primarily by selection. When the breeding population is small, drift becomes a prominent factor. In actual evolution, almost all lines of descent pass through multiple population bottlenecks, which are the phases when evolution is dominated by random drift. Hence, it appears that drift is the major contribution of evolution for all organisms.

As a result of these neutral theories, debate in the 1980s ensued over whether it was selection or drift that dominated evolution. Some in the creationist debate claimed that evolutionists were questioning whether natural selection occurred at all. Evolutionists were not debating that. Advocates of the neutral theory acknowledged that natural selection can and does take place with positive effects on the phenotype (Koonin 2011). Even so, it appears that neutral mutations and random drift predominate, making directional and stabilizing selection within natural selection comparatively rare.

7. The theory of evolution through **gene duplication** was developed by Susumu Ohno (1928–2000) in 1970. **Gene duplication** may create an extra copy of a gene, a chromosome (aneuploidy), or the entire genome for an organism (polyploidy). Plants are the most frequent group to undergo gene duplication of their entire genome. In the case of wheat, six copies of the genome are in the nucleus of each cell. This can produce a different phenotype, even to the point of producing a new species.

8. Through a series of publications in 1977 and 1990 Carl Woese (1928–2012) (pronounced "woes") proposed the **three-domain tree of life** for all cellular life forms (Figure 11.2). This challenged the five-kingdom classification system (Monera, Protista, Fungi, Animalia, Plantae).

In the **three-domain tree of life**, prokaryotes are split into two domains, Archaea and Bacteria, and a third domain Eukarya (eukaryotes). In this scheme archaea are thought to be more closely related to eukaryotes than bacteria. It was thought at the time that archaea had distinctly different habitats than known bacteria. Archaea were mainly found in extreme environments such as hot vents on the ocean floor, volcanic areas, or in extremely acidic conditions. Now we know that archaea can be found in many of the same non-extreme habitats as bacteria.

Not all scientists agree with the Woese model, because RNA is just one of many ways molecular biologists could form categories among organisms. An alternative would use mtDNA and DNA synthesis, with cell

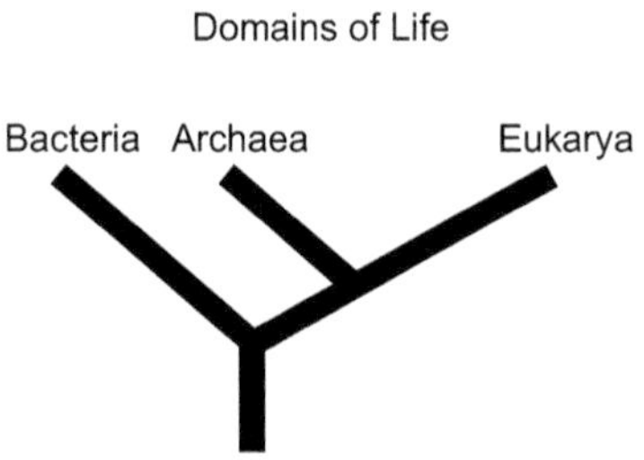

Domains of Life

Bacteria Archaea   Eukarya

Figure 11.2 The three-domain system of life developed by Woese and colleagues (*G. Giacomin*).

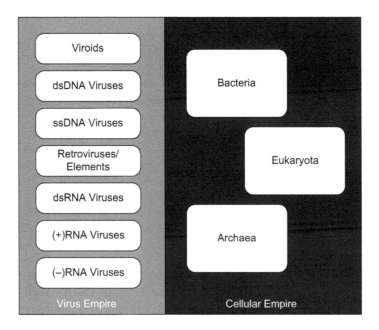

Figure 11.3 Koonin (2011) would divide life into two "empires" recognizing viruses as alive. It seems clear that viruses largely shape the evolution of cellular genomes and vice versa (*G. Giacomin*).

morphology and geographic distribution as corollaries. The scheme is based solely on the sequences for small-subunit ribosomal RNA, which is why Margulis and Chapman (2010) and others think a broader range of factors should be taken into account before creating a new classification system.

9. **Molecular evolution discoveries regarding viruses**

The modern synthesis focused almost exclusively on the evolution of animals and plants, which are multicellular eukaryotes that mostly reproduce sexually. Protists, prokaryotes, and viruses with non-sexual reproduction were generally considered unimportant. In postmodernism, the microbial world is central.

For years, biologists regarded viruses as not living because they cannot reproduce without being in the cell of an organism. This does not make them unimportant in evolution. Viruses and cells exchange genetic material in a dramatic example of a Darwinian-type struggle for existence (Herron and Freeman 2014). In fact, Koonin (2011, 2016) suggests that the microbial world lies at the base of the tree of life, describing it as a single, vast, interconnected gene pool.

Because of the importance of viruses, Koonin (2011) is a proponent of dividing life into two empires (Figure 11.3) even if viruses cannot reproduce outside cells. Viruses are the most abundant biological entities in the world.

The total number of virus particles exceeds the number of cells by at least an order of magnitude. Because the gene composition of the overall DNA is dramatically different than known bacteria, understanding more about viruses could lead to substantial new understandings of evolution. The most difficult question yet to be answered in evolution is the origin of life. Origin from a single common ancestral virus appears to be unrealistic.

10. **The human genome sequence was released in 2003**: One of the astonishing conclusions from the human genome project is that humans have only 23,000 genes, less than the total for many plant and nematode (round worm) species. Furthermore, most human genes can be found in other animal species, including 98.6% similarity with the Chimpanzee. If this is the case, why do such dramatic differences occur between other species and humans, mentally and physically? See the section on evo-devo later in this chapter.

11. **Because of new techniques in molecular biology, we now know that the greater the phylogenetic distance between organisms, the less the genetic sequences on DNA match one another**: This is the definitive evidence for evolution sought by the modern synthesis.

POSTMODERN DISCOVERIES IN DEVELOPMENTAL BIOLOGY

The combination of evolution and developmental biology is known as **evo-devo**. It tries to discover how processes evolved in embryonic development. It examines how novel features develop such as the evolution of limbs and feathers.

12. The **diversity of body plans among closely related species is not always reflected in the diversity of genes**: For example, within the millipede-centipede group, the species that have developed more pairs of legs do not have more genetic variation. Changes in the regulatory system could account for this rather than differences in genes.

Evidence for this regulatory theory came in the 1980s when genes in a cluster known as Hox were discovered in vertebrates. Hox genes determine where limbs and other body

segments grow in a developing embryo or larva. Genes for the development of legs can be turned off, as in snakes. The implication is that large evolutionary changes in body morphology are associated with changes in gene regulation turned on and off by switches, rather than the evolution of new genes. Humans having only 23,000 genes yet possess great differences between themselves and other animals can be understood by the idea of on/off switches for genes.

PHENOTYPIC PLASTICITY WITHIN EVO-DEVO

13. **Phenotypic plasticity provides an ever-increasing understanding of how adaptations actually come about**: In traditional ideas about natural selection, it is not natural selection that creates new adaptations. It is mutation or drift or draft (neutral or slightly deleterious genes spread or get fixed in genomes by hitchhiking onto beneficial genes). It is natural selection that acts upon the phenotypic manifestations of these genes, sort of like a sieve. However, the process can also happen differently. Rather than natural selection waiting for genetic mutations and recombination followed by the elimination of the less fit, the environment itself can produce a more fit individual through phenotypic plasticity. Once produced, selection proceeds and these new changes can get fixed in the genome later. By this process, genes follow adaptation rather than lead it.

Let's sort this out one step at a time. **Phenotypic plasticity** is the capacity of a single genotype to produce different phenotypes in response to environmental variation. These phenotypic changes occur over the lifespan of an individual, and can be anticipatory or responsive, reversible or not. For example, phenotypic plasticity includes:

- tree leaves that are smaller and less thick at the top of the canopy than leaves on the same tree near the ground (Figure 11.4). Small treetop leaves have enough surface area for photosynthesis but leaves at the bottom or in young trees must be very large for photosynthesis in the shade.

Figure 11.4 Phenotypic plasticity is apparent in common tree species. These large hickory leaves are characteristic of branches near the ground under a canopy, not at the top of the canopy.

- Skin that becomes darker in response to sunshine.
- Dandelion leaves that have different shapes depending on whether they grow in the sun (deeply lobed) or shade (barely lobed).
- A caterpillar developing into a mimic of an oak catkin while feeding on one tree species, and a second from the same egg mass developing into a mimic of a twig while eating from another source (Whitman and Agrawal 2009).
- Individual Red Knot, long-distance migrating shorebirds, making several changes in their gut according to what prey are available at different sites along the migration route (Piersma and Gils 2011). Where only soft worms occur, the size of the bird gizzard shrinks. Where only clams with a hardened shell occur, the gizzard and other digestive organs change size dramatically.
- Human beings who run regularly. With weight-bearing and frequent use, muscle cells may align differently on bone, and blood vessels may grow toward areas low in oxygen. In young adult runners, the bones of the leg actually change shape, becoming straighter and stronger through the process

of remodeling. Calcium is taken up from the convex side of the curve and given to the concave side.
- Immune responses, antipredator defenses, acclimatization, dietary shifts, and learning all qualify as phenotypic plasticity. Once we know what it is, phenotypic plasticity is everywhere.

HOW DOES PHENOTYPIC PLASTICITY AFFECT EVOLUTION?

In the modern synthesis, mutation was thought to come first, and natural selection acted on that mutation. In the postmodern view adaptation may come first as a response to the environment (Figure 11.5). If natural selection now acts on the expression of this new phenotype, and the new adaptation allows greater fitness, the genes in the population over several generations may become fixed within this new phenotype.

For example, suppose that during a flood, individuals of a fish species dispersed to a stream with faster flow than what they previously encountered. Within a few weeks they built up their muscle strength to live normally in the faster flow and

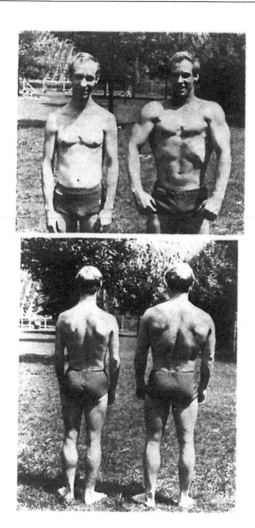

Figure 11.5 Phenotypic plasticity is evident in identical twins Otto and Ewald from Germany. Otto was a distance runner and Ewald was a field athlete (Creative Commons.)

find new kinds of prey. The fish reproduced and had offspring hatched into the faster flow. The new generation of fish even produced an additional fin that helped them survive in the new environment, yet genetic tests showed minimal if any changes in the genome. This scenario is known as **facilitated variation**. Gene sequences already coded in the DNA generate phenotypic variation upon which selection acts to produce evolution. It was possible for a new fin to grow because the genetic information to produce appendages is modular, as in a suite of genes. Evolvability has been built into the genome, and this molecular machinery facilitates the incorporation of new changes (Kirschner and Gerhart 2005).

In other words, the ability to produce new appendages and many other abilities lie dormant in the genome (Williams 2008). It is to an organism's advantage to retain this suite of genes in case it needs them. These dormant modules make up what was once thought to be junk DNA. These genes can rearrange and shuffle but move as a unit and have switches that turn them on or off (Williams 2008). The phenotypic change does not have to be invented through mutation as proposed by the modern synthesis. Intermediate forms do not need to come about. The suite of genes is already there and can be activated or repressed. In fact, the cells go to great lengths to ensure there are minimal mistakes in copying or translation,

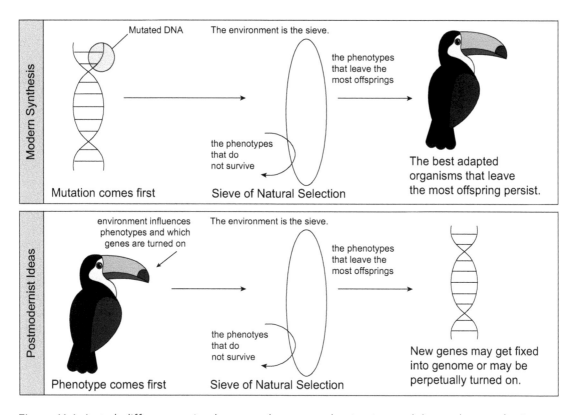

Figure 11.6 A stark difference exists between the postmodernist view and the modern synthesis regarding the primary source of variability on which natural selection acts. In the postmodern view the environment both creates the phenotype and acts on it through natural selection (G. Giacomin).

retaining five to seven copies of the genome on the chromosomes in humans.

Piersma and van Gils (2011) argue that facilitated variation occurs the most in organisms with the most phenotypic plasticity in their physiology, behavior, and learning. Thus, variation happens the most in organisms at the highest levels of biological organization. By this argument, behavioral change and learning can be the motor of evolution for an organism (Figure 11.6).

Facilitated variation explains how evolution can occur very quickly (over two to three generations), not gradually as often argued. Even the descendants of clones could show wildly different evolutionary lines if they had plasticity built into their genome. With such fast-acting evolution, it is easier to understand how adaptive radiation can dramatically occur in a group like the Hawaiian Honeycreeper family. Kirshner and Gerhart (2005) use Darwin's Galapagos Finch as an additional

example. The modern synthesis taught that selection waited for tiny variation in the size of the finch beak, with natural selection sieving out the smallest beaks. Over many generations the size of the beak became substantially larger. In facilitated variation, just two regulatory changes in the genes need to take place in embryogenesis for the beak to become larger. The two together allow for the evolution of larger beaks in only one reproductive event.

The idea of phenotypic plasticity represents a major change in how we view evolution. The environment can trigger more phenotypic variation than what is produced through mutation or new gene combinations. Through facilitated variation, evolution can take place without changes in the allele frequency. Dobzhansky's definition of evolution – a change in gene (allele) frequency – would no longer describe what we know about evolution. This is such a dramatically different way to think

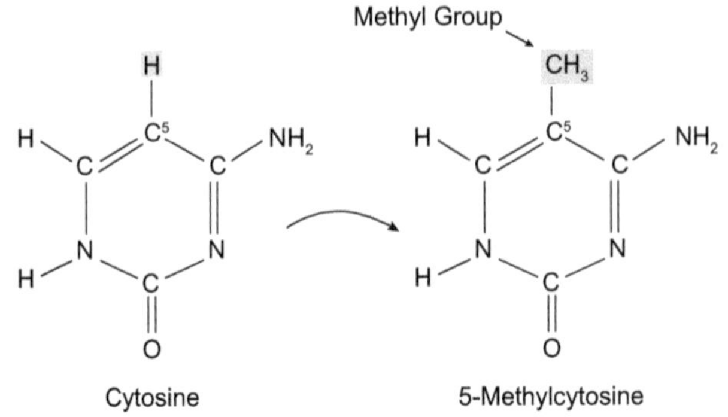

Figure 11.7 Adding a methyl group inactivates some genes. When this nucleic acid is methylated, it stops gene expression (*G. Giacomin*).

about evolution, the question now may be how species stay the same from one generation to the next.

14. **The environment can directly change genes**: It now appears that parental experience really can be passed on to offspring as suggested by Lamarck. We already know that environmental effects such as radioactivity can cause mutations in sperm and eggs that are passed to offspring. Traits can be passed without directly changing the genome of an organism, known as **epigenetic inheritance** (when evolution proceeds without a mutation or new genetic combination). Epigenetic inheritance occurs in facilitated variation, known by the adage, **old genes, new tricks**.

15. **Epigenetic inheritance occurs through tags that turn on and off genes, called epigenetic marks** (Williams 2008). Commonly, the epigenetic mark is a methyl group, which fastens to DNA (Figure 11.7). Once attached, it turns the gene off because it blocks the attachment of proteins that normally turn the gene on. For instance, famine can produce health effects in the children and grandchildren of individuals who had restricted diets, resulting in obesity among the offspring. Normally each new generation has genes that are reset. However, at times the methyl groups escape the reprogramming process, passing on the turned-off gene to the offspring.

16. **Epigenetic inheritance can be induced by environmental stressors such as temperature, diet, or environmental chemicals**: For instance, exposure to cocaine in mice can affect their offspring, making them resistant to the beneficial effects. It means we have a responsibility to curb and reverse these changes if human-induced issues are causing the problems.

THE EMERGING PARADIGM

In the postmodern understanding of evolutionary biology, natural selection can shape evolution, but neutral processes such as genetic drift and draft are more prominent. Viruses are central to the evolution of life. Horizontal gene transfer is widespread in prokaryotes, and exchange of genetic material between hosts and parasites is important. In developmental biology, the abilities of cells to self-organize into complex structures can lead to major evolutionary innovations. This may include the origin of vertebrate limbs with perhaps little or no genetic change.

Heng (2019), Corning (2022), and others rejected the idea of the gene as the object of selection, as was taught in the modern synthesis and as proposed in the popular 1976 book by Richard Dawkins, *The Selfish Gene*. Heng (2019) quotes both Ernst Mayr and Lynn Margulis, champions of evolutionary thought in the 20th century, by emphasizing that a gene is not visible to natural selection. It is the whole self and an entire genotype that is visible to natural selection. The modern synthesis emphasized the gene in the 1930s and 1940s because it was quantifiable, but Heng (2019) points out that even Mendel had inconsistencies in his very controlled pea plant experiment and could not demonstrate his work in species other than peas.

Heng (2019) suggests that geneticists step outside the single-gene theory and look at chromosomal-level arrangements. His argument is that genomics has largely failed to deliver on promises to cure cancer since the human genome was sequenced in 2003 because scientists have concentrated on single-gene mutations. Even though it is true that dozens of gene mutations have been linked to types of cancers that occur in families, these discoveries have not led to dramatic new cures because more is happening to produce cancer than single-gene mutations. Furthermore, heart disease, obesity, diabetes, and other common diseases do not show tight relationships with single-gene mutations. Meanwhile, scientists might be ignoring obvious alternatives because they were misled when taught the modern synthesis. Heng's (2019) research focused on watching cancer cells grow. He observed wholesale changes in chromosomal rearrangements as cancer cells struggled for survival. The chromosomal-level changes were so dramatic that he likened it to a new species growing inside tumors. He argues that the genome is the level of selection, and evolution in the cancer cells occurs in dramatic short spurts at the chromosomal level. This should be the focus for cancer research.

Corning (2022) takes this further, proposing the synergism hypothesis, the idea that evolution is largely a symbiotic-cooperative relationship among organisms. That organisms drive each other to evolution and to large groups of beneficial species. In human evolution, it is possible to see many examples of this as we drive up large populations of evolving species to provide for our shelter, clothing, medicine, and antibiotic capabilities. We have even made our own species more lactose tolerant over time, owing to our symbiotic relationship with milk-producing cows. Other species do similar things. They enhance large populations of the species with which they are symbiotic, all the while sharing genes widely with every microbial infection passed among them. In this respect, evolution is self-directed. This viewpoint does not deny that natural selection can and does happen. Synergistic processes should be recognized for their overwhelming importance in providing a decisive survival advantage. By fundamentally changing our view of evolution, it may signal a change in the entire conceptual framework of biology and, with it, ecology. Threads of this new philosophy can be found in all subdisciplines of ecology as described in this book.

THINKING QUESTIONS

1. Explain what the following means:
   "… we now know that the greater the phylogenetic distance between organisms, the less the genetic sequences on DNA match one other."
2. In two sentences summarize the hardened form of the modern synthesis.
3. Choose three of the advances in molecular biology during the postmodern era and explain their relevance.
4. What was really being argued in the neutral theory?
5. Which two molecular biology postmodern ideas surprised you the most when compared with the modern synthesis? Provide a little overview.
6. Give an example not listed in the book, but one you find to be dramatic, for phenotypic plasticity.
7. For facilitated variation and epigenetic inheritance, explain each as if speaking with family members who are not biologists, perhaps using the phrase, "old genes, new tricks."
8. How do you think a new understanding of evolution might change ecology, regarding phenotypic plasticity, facilitated variation, and epigenetic inheritance?

REFERENCES

Corning, P. 2020. Beyond the modern synthesis: a framework for a more inclusive biological synthesis. *Progress in Biophysics and Molecular Biology* 153:5–12. https://doi.org/10.1016/j.pbiomolbio.2020.02.002

Corning, P. 2022. A systems theory of biological evolution. *Biosystems* 214:1–5. https://doi.org/10.1016/j.biosystems.2022.104630

Heng, H.H. 2019. *Genome Chaos: Rethinking Genetics, Evolution, and Molecular Medicine*. Academic Press, London, UK.

Herron, J.C., and S. Freeman. 2014. *Evolutionary Analysis*. 5e. Pearson, Boston, MA.

Kirshner, M.W., and J.C. Gerhart. 2005. *The Plausibility of Life: Resolving Darwin's Dilemma*. Yale University Press, New Haven, CT.

Koonin, E.V. 2009. Towards a postmodern synthesis of evolutionary biology. *Cell Cycle* 8:799–800.

Koonin, E.V. 2011. *The Logic of Chance: The Nature and Origin of Biological Evolution*. FT Press, Science, Upper Saddle River, NJ.

Koonin, E.V. 2016. Viruses and mobile elements as drivers of evolutionary transitions. *Philosophical Transactions of the Royal Society London, Series A B* 371:20150442. https://doi.org/10.1098/ rstb.2015.0442

Margulis, L., and M.J. Chapman. 2010. *Kingdoms and Domains: An Illustrated Guide to the Phyla of Life on Earth*. Academic Press, London, UK.

Piersma, T., and J.A. van Gils. 2011. *The Flexible Phenotype*. Oxford University Press, Oxford, UK.

Whitman, D.W., and A.A. Agrawal. 2009. What is phenotypic plasticity and why is it important? In: D.W. Whitman and T.N. Ananthakrishna (eds.), *Phenotypic Plasticity of Insects: Mechanisms and Consequences*. Science Publishers, Inc., Enfield, NH.

Williams, A. 2008. Facilitated variation: a new paradigm emerges in biology. *Journal of Creation* 22:85–92.

Woese, C.R., and G.E. Fox. 1977. Phylogenetic structure of the prokaryotic domain: the primary kingdoms. *Proceedings of the National Academy of Sciences* 74:5088–5090. https://doi.org/10.1073/pnas.74.11.5088

12

An autobiography of the Earth

ACTIVITY: Learning Wilderness First Aid (by Helen Weber-McReynolds and Louise Weber)

The basics: Organizations around the world offer certification courses in Wilderness First Aid, Wilderness First Responder, and Wilderness Emergency Medical Technician.

Activity: If a certified instructor is available to lead an introduction to wilderness first aid, a few basics could be reviewed in a lecture format. Students could then break into small groups and one person per group could be assigned a mock injury. Students could then practice basic techniques, including an evacuation. One person in the group acts as scribe, documenting each step taken. This serves as the basis for assessment. The following would be reasonable to cover in a 1-hour lecture and 1.5-hour outdoor activity.

Scene size up: This includes initial assessment of life threats that may be present before someone attempts first aid. If the responder or bystanders face severe risk by helping the patient, first aid may not be possible. Questions that should be asked during scene size up: what is the mechanism of injury or illness?… can materials on hand be used as a barrier (for blood, microbes, rain)?… how many patients are there?… what resources are available?… what is the initial impression of the scene by the responder?

Physical exam of the patient: This includes taking vital signs and history (if possible). Can a bystander provide identification of the patient? Can the patient provide consent? Is the patient responsive? Spine control should be assessed. An ABCDE assessment should be done, which includes: **Airway** – open, clear. **Breathing** – assist breathing if necessary. **Circulation** – pulse, CPR, bleeding (quick full-body visualization). **Disability** – nervous system: spine stabilization and brain function assessment. **Environment** – first priorities to prevent further injury, shock, and discomfort. A head-to-toe physical exam should take place, recalling that shock and anxiety can mask pain and symptoms initially.

Vital signs exam include: Level of responsiveness, heart rate, respiratory rate, skin color, temperature, moisture, blood pressure as estimated by skin perfusion and peripheral pulses, pupil response.

Patient history includes: Age, gender, chief complaint, mechanism of injury, history of present illness, medications, events preceding accident or illness.

Possible evacuation includes: Can the patient walk out with the responder? Can EMS be called? Should a friend be called? Should a runner be sent? Is a helicopter necessary? Are injuries simple enough to say goodbye and good luck?

Interventions, treatments, and monitoring: A plan for each injury needs to be made. Splints can be made of a cut-up mattress pad, tied around a leg with strips of a tee shirt. Cuts and

DOI: 10.1201/9781003271833-12

abrasions can be irrigated with clean water poured into a Ziplock bag, one corner cut out and a stream of water created under pressure. Self-adhesive athletic wraps can secure a makeshift bandage. A role of athletic wrap, butterfly bandages, superglue, pocket knife, and a small length of parachute cord should be a part of every responder's first aid kit, along with the usual Band-Aids, Tylenol, antihistamines, and antibiotic cream.

Makeshift stretcher: A rope and sleeping bag make an excellent stretcher. The rope can be zigzagged under the sleeping bag so that six or eight people can grab the rope and lift the patient. Try it.

GETTING A HANDLE ON A 4.5-BILLION-YEAR-OLD AUTOBIOGRAPHY

Why should anyone even attempt to learn a story that is 4,500,000,000 years old? For one, it provides background in understanding our current climate change issue. For two, it provides fascinating stories – of dragon-like creatures that once roamed the Earth, of how human beings evolved from ancestors before us, of how the astounding biodiversity of our planet evolved.

If you break Earth's history into a few major ideas, the task of learning this history will seem easier. The first step is reviewing the theory behind radioactive dating and the definition of a fossil. The second is reviewing the major domains and kingdoms of life. The third is recalling the basic chemical formulas for photosynthesis and respiration in preparation for learning how Earth's atmosphere formed. The fourth is breaking down the 600-million-year-old geologic history of life intro three major periods: the Age of Plants and Marine Invertebrates (Paleozoic), the Age of Reptiles (Mesozoic), and the Age of Mammals (Cenozoic). The final step is memorizing the names of the periods within eras and the major organisms to evolve during each.

THE MOST IMPORTANT EVENTS WITHIN MACROEVOLUTION

Before 1850, most people believed the Earth was only a few thousand years old. Even after 1850, physicists underestimated the Earth's age until radioactive dating was developed in 1902. Today we understand the following important events:

- big bang, 14 bya,
- Earth's formation, first occurring as nebulous gas 4.5 bya,
- prokaryotic life on Earth began 3.5 bya,
- oxygen atmosphere began 2.5 bya,
- eukaryotic life began 2 bya,
- Age of Plants and Marine Invertebrates (Paleozoic) began 600 mya,
- Age of Reptiles (Mesozoic) began 250 mya,
- Age of Mammals (Cenozoic) began 65 mya,
- genus *Homo* began 2 mya,
- *Homo sapiens* began 200,000 years ago.

THE 14-BILLION-YEAR TIME SCALE

The first temporal scale to envision is the 14-billion-year history of the universe, which encompasses the 4.5-billion-year history of the Earth (Figure 12.1). Radioactive dating has provided the dates.

Radioactive dating is possible because of the phenomenon of radioactive decay. Radioactivity occurs through one of three processes. An atom can lose protons that fly out of the nucleus, thereby forming a new element (**alpha radiation**). An atom can lose electrons (**beta radiation**). An atom can give off wavelengths of energy (**gamma radiation**).

Carbon dating is one form of radioactive dating, used to date fossils less than 50,000 years old (Table 12.1). Ordinarily, carbon atoms are not radioactive but some are as ^{12}C or ^{13}C (with the numeral referring to the weight of the atomic nucleus). The forms ^{12}C or ^{13}C are isotopes (different forms of an element with a different number of neutrons in the nucleus). ^{14}C occurs in the upper atmosphere as a result of cosmic radiation. Note that ^{12}C can also be written as carbon-12.

Small concentrations of ^{14}C are taken up by plants through photosynthesis when the plants are alive. The ratio of ^{14}C to ^{12}C remains stable as long as the plant is alive. Once dead or eaten by an herbivore, the ratio declines because no more ^{14}C

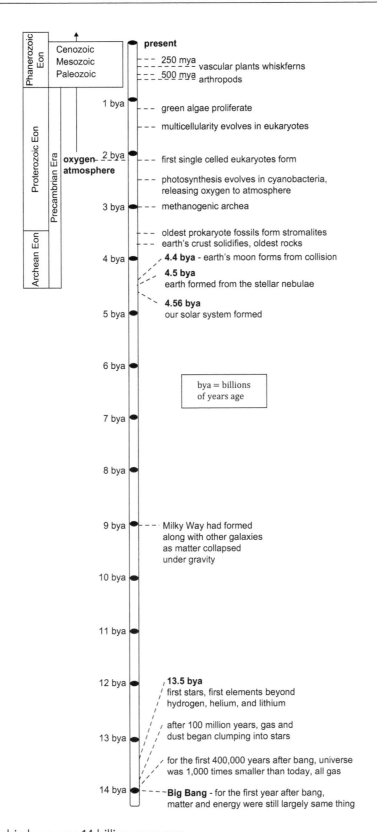

Figure 12.1 The big bang was 14 billion years ago.

Table 12.1 14C and other radioisotopes decay at predictable rates known as half-lives. A half-life is the time it takes for half the mass to decay. By calculating the ratio of undecayed isotope to decayed product we can determine the age of fossils and rocks

Radioisotope	Half-life years	Decay product	Dating range (years)
Carbon-14	5,700	Nitrogen-14	100–50,000
Potassium-40	1.3 billion	Argon-40	10 million–4.5 billion
Uranium-238	4.5 billion	Lead 206	10 million–4.5 billion

Source: Modified from Sadava et al. (2014).

is being taken up through photosynthesis, and the ^{14}C already there begins to decay.

Knowing that carbon-14 (^{14}C) has a half-life of 5,730 years, the percentage remaining of ^{14}C is fit to a curve, allowing age to be determined for fossils of plants or herbivores. The method can be calibrated and verified by counting the rings in fossilized trees, although tree rings can only tell the story for the last 10,000 years. ^{40}P and $^{238}Uranium$ have longer half-lives, and their decay is useful for dating fossils older than 50,000 years.

WHAT IS A FOSSIL?

Fossils are the preserved remains of ancient organisms, including hard body parts like shells, bones, seeds, and teeth. Trace fossils are footprints and burrows formally inhabited by living things. They must be at least 10,000 years old to qualify. Fossils provide some of the strongest evidence for evolution in the following ways:

1. the more distance between the location of fossils, the more radiation and speciation within a taxonomic group;
2. the more recent the layer of rock where a fossil is found, the more likely it is to look like a modern organism.

Note that fossils form only from organisms that have not decayed. Most living things decay or are eaten after death, making it unlikely that fossil formation will proceed in the first place. The material in fossils is usually not the original. Hard parts of an organism act like a mold. The original material dissolves, and minerals from the surrounding rocks slowly impregnate the space. Thus, fossils are not likely to contain DNA, although some do.

THE HISTORY OF LIFE AND SYSTEMATICS

Adjust your temporal scale to begin 5 billion years ago, a time when the conditions were right for the first organisms, prokaryotic bacteria (Figure 12.2). Picture several massive asteroids hitting the Earth during the first 1.5 billion years. Evidence for asteroids is provided by iridium, a rare Earth element, but a component of asteroids (Janke 2004).

Asteroid energy provided the conditions for prokaryotic formation approximately 3.5 bya. **Prokaryotes** differ from **eukaryotes** in their genetic material (RNA instead of DNA), which is generally not confined to a nucleus. **Eukaryotic** (nucleus bearing) organisms enclose their DNA within a nucleus.

Note that **archaea** have characteristics of both prokaryotes and eukaryotes and include a large number of microscopic species. Some are able to generate methane and some live in extremely harsh conditions like hot springs or highly saline areas. These archaea are known as extremophiles. Other archaea live in ordinary circumstances.

The first prokaryotes were **anaerobic** (able to live without oxygen) and **heterotrophic** (not photosynthesizing) deriving their energy from fermentation. By this process they burped CO_2, providing the conditions for the first photosynthesizing blue-greens (cyanobacteria) to evolve, happening in ocean waterways 2.5 bya (Figure 12.3). Photosynthesis is thought to have evolved just once, and possible because of the availability of CO_2.

The blue-greens gave off oxygen, making it possible for eukaryotic organisms to evolve 2 bya (Biello 2009). Early eukaryotes were protistans consisting of protozoa, algae, and slime molds,

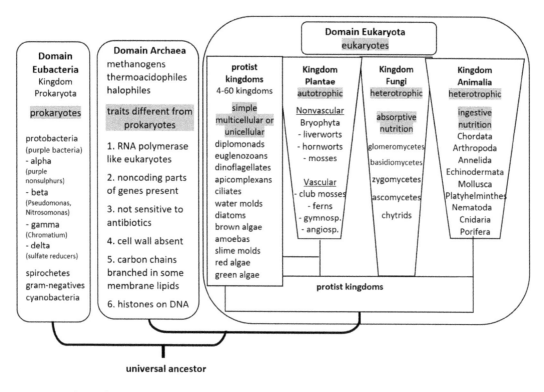

Figure 12.2 Three domains and several kingdoms of life on Earth are recognized based on Woese's three-domain system derived from ribosomal RNA analysis.

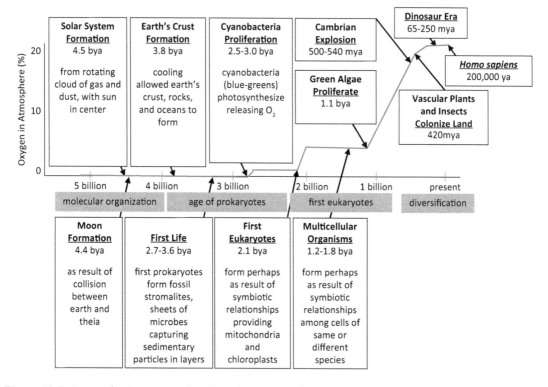

Figure 12.3 Atmospheric oxygen developed over a 5-billion-year period.

becoming the precursors for animals, plants, and fungus respectively. Note that the atmosphere did not immediately reach its now stable concentration of 21% oxygen. It took a billion years (known as the boring billion) between the evolution of cyanobacteria and multicellularity before oxygen levels were assumedly not high enough for multicellularity (Biello 2009).

Think of the atmosphere progressions as:

- prokaryotes (C in CO_2 out) →
- blue-greens (CO_2 in O_2 out) →
- protistans (O_2 in CO_2 out) →
- multicellularity
= animals, fungus (O_2 in CO_2 out)
= and plants (CO_2 in and O_2 out and O_2 in CO_2 out)

By 1.5 bya, the first multicellular animal organisms were present. The most primitive were sponges (Phylum Porifera) evolved from colonial flagellates, a type of protozoan. Other early animal types were sea anemones, jellyfish, and corals.

GEOLOGIC HISTORY SINCE 600 MYA (LOMOLINO ET AL. 2016)

Now think of a temporal scale of approximately 600 mya. This was the **Paleozoic Period**, the age of marine invertebrates and plants (Figure 12.4). All the animal phyla existent today evolved during this time. Specifically, the greatest period of speciation was during a span of less than 10 million years in the middle of the Cambrian, the first era within the Paleozoic. The Paleozoic went a long span without asteroid collisions, giving multicellular organisms the chance to evolve. Thus, most speciation and extinction in the Paleozoic occurred as a product of tectonic plate movement, not asteroids.

TECTONIC PLATES

Before the idea of tectonic plates became widely accepted in the 1970s, the continents were thought to be in the shape and relative positions they are now. Perhaps sea levels changed, mountains rose, land bridges came and went, and continents moved

a little, but the idea of floating plates riding on the surface of the Earth had not been taken seriously. Few ideas have advanced evolution as much as tectonic plate theory.

Alfred Wegener (1880–1930) proposed the idea of continental drift in 1916, but it was not investigated further until World War II and shortly after. At that time observations of the ocean floor produced the astonishing discovery that ocean bottoms were not flat (Lomolino et al. 2016). Rising magna from midoceanic ridges created centers of spreading (Figure 12.5). This produced mountain ranges under the water, sometimes visible above water as volcanic islands and island chains (Lomolino et al. 2016). Underwater areas without mountains had deep trenches (valleys) 10 km deep where parts of the Earth's crust were very thin. In these trenches, matter was being pulled downward and reincorporated into the mantle, setting off occasional earthquakes as crustal slabs were pulled into the molten goo.

GEOLOGIC HISTORY SINCE 600 MYA

At the beginning of the Paleozoic **Gondwanaland** formed as a large continent in the southern hemisphere (Figure 12.5). It was its formation, movement, and splitting apart, not asteroids that accounted for repeated waves of speciation and extinction during the Paleozoic. In thinking about continents forming and moving, envision the 16 tectonic plates that exist now (Figure 12.5). They are 100 km thick and move 2–12 cm per year powered by heat, gravitational compression, and nuclear reactions in the Earth's core.

Some of the plates are continental, thus lighter and older than oceanic plates that are heavier and younger sitting on the ocean floor. When two plates of equal mass collide, they either rub, which may produce earthquakes, or violently uplift to form mountain ranges. Uplifting occurs when dense oceanic plates collide with lighter continental plates. The denser one sinks below the continent, lifting it. This is what formed the modern-day Himalayas when the Indian plate collided with the Eurasian plate.

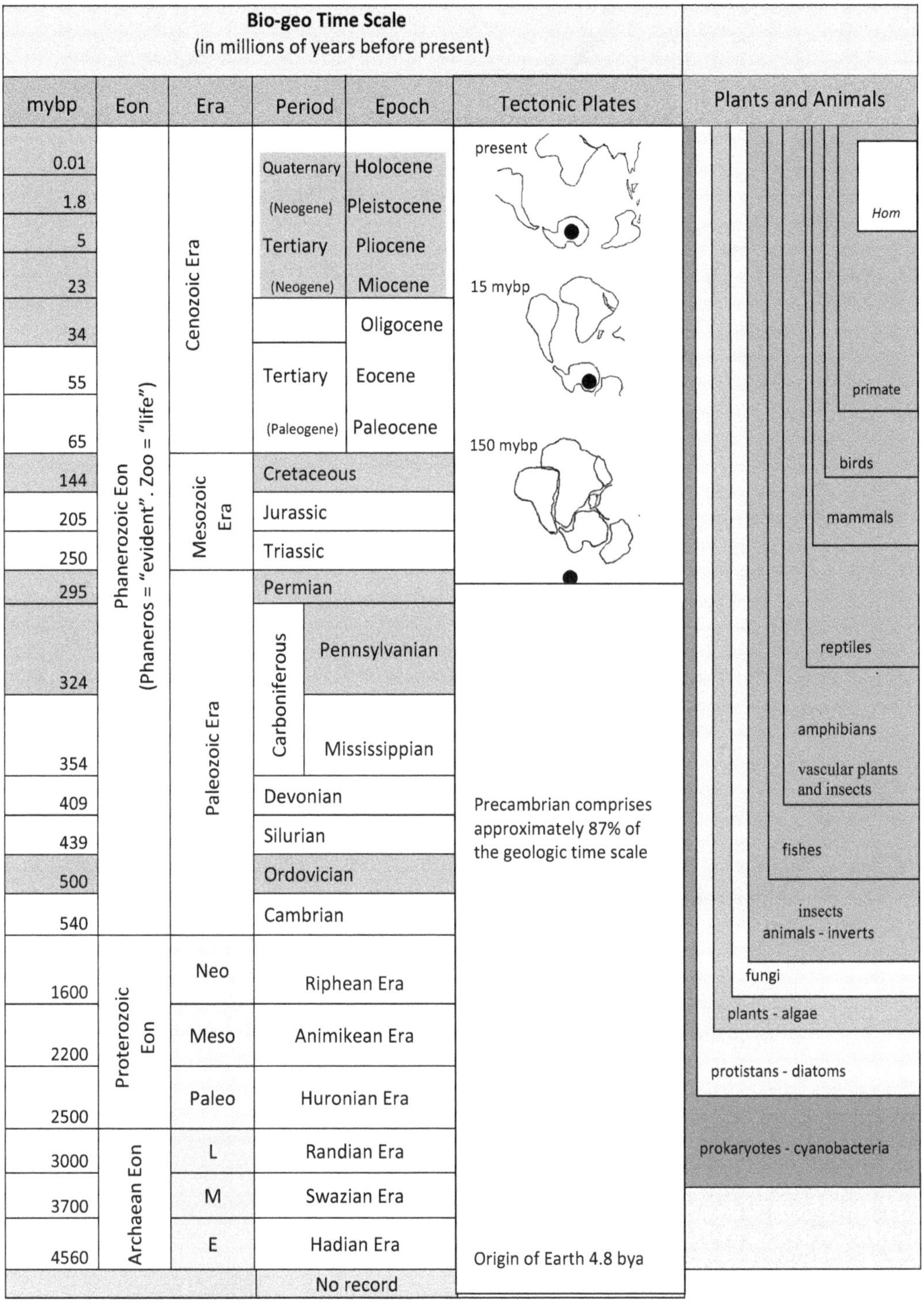

mybp	Eon	Era	Period	Epoch	Tectonic Plates	Plants and Animals
Bio-geo Time Scale (in millions of years before present)							

mybp	Eon	Era	Period	Epoch
0.01			Quaternary	Holocene
1.8			(Neogene)	Pleistocene
5			Tertiary	Pliocene
23		Cenozoic Era	(Neogene)	Miocene
34				Oligocene
55			Tertiary	Eocene
65			(Paleogene)	Paleocene
144		Mesozoic Era	Cretaceous	
205	Phanerozoic Eon (Phaneros = "evident". Zoo = "life")		Jurassic	
250			Triassic	
295			Permian	
324		Paleozoic Era	Carboniferous	Pennsylvanian
354				Mississippian
409			Devonian	
439			Silurian	
500			Ordovician	
540			Cambrian	
1600	Proterozoic Eon	Neo	Riphean Era	
2200		Meso	Animikean Era	
2500		Paleo	Huronian Era	
3000	Archaean Eon	L	Randian Era	
3700		M	Swazian Era	
4560		E	Hadian Era	
			No record	

Tectonic Plates:
present
15 mybp
150 mybp
Precambrian comprises approximately 87% of the geologic time scale
Origin of Earth 4.8 bya

Plants and Animals:
Hom
primate
birds
mammals
reptiles
amphibians
vascular plants and insects
fishes
insects animals - inverts
fungi
plants - algae
protistans - diatoms
prokaryotes - cyanobacteria

Figure 12.4 The bio-geo time scale in millions of years before present. Black dots in insert maps show location of South Pole. (*Adapted by Weber from Janke 2004.*)

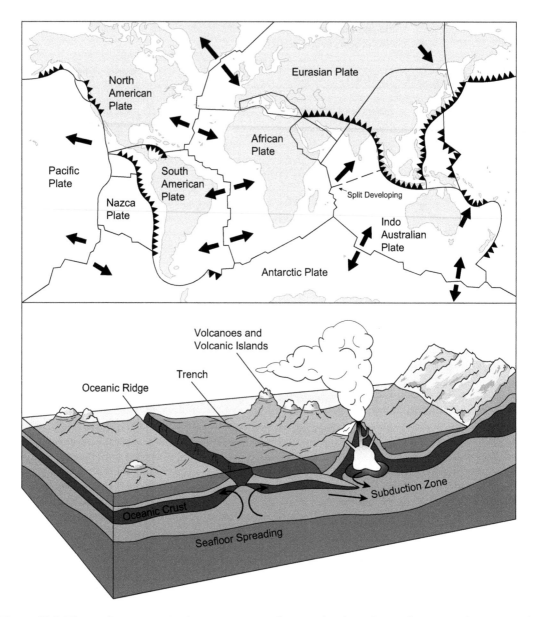

Figure 12.5 The seafloor tectonic plates move away from each other. The seafloor spread is powered by heat, gravitational compression, and nuclear reactions in the Earth's core (*G. Giacomin*).

EVOLUTION DURING THE PALEOZOIC – CAMBRIAN

At the time Gondwanaland was present an explosion of evolution occurred. In the Cambrian animal phyla included trilobites, mollusks, sea stars, and others. The earliest known vertebrates evolved during this time, the conodonts, resembling hagfish. The end of the Cambrian was marked by mass extinction 500 mya as climate changed, causing one of the five great extinctions in the last 600 my (Figure 12.6).

EVOLUTION DURING THE PALEOZOIC – ORDOVICIAN

The trilobites mainly succumbed at the end of the Cambrian, but re-diversified to reach their peak then to succumb again in a second great period of extinction during the Ordovician.

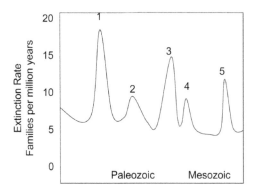

Figure 12.6 At least five previous dramatic extinction periods have occurred.

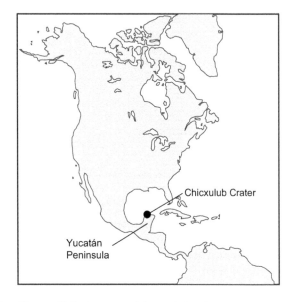

Figure 12.7 An asteroid struck in what is now the Gulf of Mexico at the end of the Mesozoic 65 mya (*G. Giacomin*).

EVOLUTION DURING THE PALEOZOIC – SILURIAN

Vascular plants evolved in seawater during the third era, the Silurian, 430 mya (Figure 12.4). **Vascular plants** have xylem, phloem and include the ferns, horsetails, conifers, and flowering plants. During the next era, the Devonian (Figure 12.4), the non-vascular plants (whisk ferns, mosses, liverworts) developed on land. The Devonian is also known for its radiation of fishes and the first vertebrates on land.

EVOLUTION DURING THE PALEOZOIC – CARBONIFEROUS

During the Carboniferous, 354 mya (Figure 12.4), Gondwanaland was in the south and several smaller continents were in the north. Gondwanaland had a tropical climate, which fostered extensive swamp forests, preserved as coal beds, today's fossil fuels. Winged insects were present at this time and the first reptiles evolved.

The supercontinent Pangea formed at the end of the Paleozoic, which was a time of several massive volcanic eruptions. The volcanoes changed the climate, dropping sea levels and drying shallow seas. This caused 90% of all Earth's species to go extinct (Figure 12.4). Only some of the insects, amphibians, reptiles, and small mammals survived. Pangea formed as a single world supercontinent made up of Laurasia in the north and Gondwanaland in the south (Figure 12.4). Laurasia was the basis for what is now northern Europe and N. America. Gondwanaland was already grouped as a large

continent and made up the basis for present-day S. America, Africa, Madagascar, India, Australia, and Antarctica. As Pangea formed, the oceans became united as one contiguous body.

THE MESOZOIC

The **Mesozoic Period** is the age of reptiles, but the earliest mammals evolved during this time (Figure 12.4). Pangea broke up in the Mesozoic. Great evolutionary radiation took place because of reduced gene flow and increased isolation. Modern corals and bony fish developed.

EVOLUTION IN THE TRIASSIC, JURASSIC, AND CRETACEOUS

Gymnosperms (coniferous plants) first appeared in the mid-Triassic (Figure 12.4). Birds evolved as warm-blooded reptiles in the Jurassic, the middle era. Another mass extinction occurred when N. and S. America split during the Jurassic. Angiosperms (flowering plants) developed during the Mesozoic. At the end of the Mesozoic, the best studied extinction period took place, caused by an impact that hit the Earth in what is now the Gulf of Mexico (Figure 12.7). High levels of iridium found in rocks that had been formed at that time provide

Table 12.2 Cenozoic Era demarcated in millions of years ago

Period	Epoch	mya
Neogene	Holocene	0.01 (10,000 y)
	Pleistocene	1.8
	Pliocene	5
	Miocene	23
Paleogene	Oligocene	34
	Eocene	55
	Paleocene	65

the best evidence for the impact. Half the marine species and most of the dinosaurs were exterminated, especially those in the western hemisphere. This is known as the K-T (Cretaceous-Tertiary boundary) event.

THE CENOZOIC

The **Cenozoic Period** is the age of mammals and birds, the warm-blooded organisms that have done well in a cold, wet climate with low sea levels (Figure 12.4). The Cenozoic was traditionally divided into the Tertiary and Quaternary periods and these names may still appear in some literature. Now, scientists divide the Cenozoic into the Paleogene and Neogene, which more equally divides the span.

EVOLUTION IN THE CENOZOIC – PALEOGENE

The Cenozoic began 65 mya (Figure 12.4 and Table 12.2). In the Paleogene, India collided with Eurasia causing the Himalayas to uplift. Madagascar and the Seychelles separated from India.

EVOLUTION IN THE CENOZOIC – NEOGENE

In the Neogene, the Central American land bridge emerged. There was mass biotic exchange between the Nearctic and Neotropical as a result.

EVOLUTION IN THE PLEISTOCENE ON A 2-MILLION-YEAR SCALE

During the last 2 million years (the Pleistocene of the Neogene), tectonic plate movement was not the cause of extinctions or speciations (Table 12.2).

The extinctions of the Pleistocene (Ice Age) were associated with a cycle of more than 20 glaciations in which the tectonic plates were relatively stable. These glacial cycles were more dramatically variable than in earlier times and had profound effects on the current distribution of species.

Glaciers were 2–3 km thick and so heavy they compressed parts of the underlying Earth by 200–300 m. Up to one-third of Earth's terrestrial surface was covered with ice at times, most of it occurring in the northern hemisphere. Australia and most of Africa and S. America were unglaciated except for high mountains. The glaciations of the Pleistocene were caused by three variations in the Earth's orbit. The changes are the result of Milankovitch cycles. In short, the diameter of Earth's elliptical path around the sun has changed on a 100,000 year cycle. The Earth's orientation of north to the north star has varied on a 22,000 year cycle. The Earth has wobbled on its axis on a 41,000 year cycle. However, this wobble is stabilized by our Moon (Kricher 2009). Without the Moon, the Earth would have wobbled too unpredictably, sending the climate into more frequent and severe oscillations of temperature as experienced by Mercury, Venus, and Mars. Life would probably not have been possible. We can also thank Jupiter for our existence (Kricher 2009). Like a vacuum cleaner in our solar system, its gravitational force is large enough to attract and subsume many dangerous asteroids and comets. This was the fate of the gigantic comet Shoemaker-Levy, fragmented in 1992, protecting the rest of us.

The last glacial period was called the Wisconsinan in N. America and Wurmian in Europe, with the cold reaching its peak 18,000 years ago (Table 12.2). We are currently in a warm period between glaciations. Periods as warm as now occupied less than 10% of the Pleistocene. In just the last 1,000 years, global temperatures have varied,

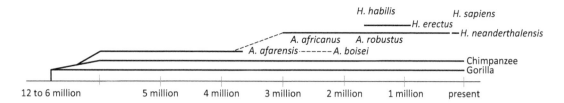

Figure 12.8 Humans evolved from the gorilla-chimpanzee group on a scale of several million years.

producing cold temperatures even during the last 10,000 years, in the Middle Ages (1100–1400 A.D.), and other times of Little Ice Ages including one from 1650 to 1850 A.D. Tectonic plates have been relatively stable during this time, yet the climate and sea level changes have transformed biogeography perhaps more than any time in Earth's history. Overall, glacial cycles created alternating waves of biotic exchange and isolation, pumping the evolutionary process.

The glaciers were the largest force in lake production in the northern hemisphere, with most of the existing lakes forming 12,000–11,000 years ago and peaking in extent 10,000–9,000 years ago. Meltwater created rivers and lakes of great volume compared to current situations but explain the current existence of some wide river valleys. As glaciers retreated (melted), lakes formed as the compressed Earth filled with meltwater. The glaciers sometimes blocked drainage to the sea. Some of these lakes had cataclysmic endings when their ice dams broke, including Lake Missoula, as large as Lake Ontario, emptying in less than 2 weeks.

Pothole lakes of the prairies and Midwest in N. America were created when melting glaciers left depressions deep enough to gouge out groundwater springs. This keeps the basins filled now and in existence because they are deep enough to contact the springs. Pluvial (rainfall) lakes developed in flat areas between mountains in what are now deserts. The largest lake covered parts of Nevada, Utah, and Idaho where the existing Great Salt Lake is a remnant. Death Valley in California was a lake at one time, but now a desert, 93 m below sea level, the lowest elevation and most extreme desert on the N. American continent.

HISTORY OF HUMANS

Order Primates has been in existence since at least the early Cenozoic, 65 mya. During their first 20–30 million years, the primates diversified and included lemurs, lorises, tarsiers, and others (Table 12.2). By 23.8 mya the New World monkeys, the Old World monkeys, and apes had diverged (Kumar et al. 2005). Within the great apes, gorillas and chimpanzees diverged between 6–12 mya and the human and chimpanzee lineages split 5–10 mya (Herron and Freeman 2014).

The first fossils of Family Hominidae are from 3.7 mya and are called *Australopithecus afarensis* (Figure 12.8). The first fossils from Genus *Homo* are from 1.8 mya and are *Homo habilis* who were capable of making tools. By 1.4 mya a hominid, *Homo erectus*, evolved and was larger-brained, more dexterous, had more tools and social skills, was able to hunt better, and might have used fire.

With the advance of fire, *Homo* could modify its environment, protect against predators, hunt game, cook small prey, and eat a wider variety of foods. By 1.5 mya and perhaps 2 mya *H. erectus* may have expanded its range outside Africa and into Asia and Europe. *H. erectus* colonized eastern Asia and western Europe by 500,000 years ago. *H. sapiens* probably evolved not from within Europe and Asia, but from *H. erectus* on the savanna in Africa. *H. sapiens* came into being 100,000–200,000 years ago, first crossing the Sahara, then Eurasia.

By 100,000 years ago *H. sapiens* was beyond Africa and had replaced *H. erectus*. *H. sapiens* had a more advanced brain, more language, more group activities, more tools, and more use of fire. By 50,000–40,000 years ago *H. sapiens* had expanded to Siberia. Once in Siberia, *H. sapiens* first crossed into present-day Alaska probably between 25,000 and 16,600 years ago. At that time, northern areas were glaciated, which lowered sea levels by 100 m permitting the crossing on the 1,500 km wide land bridge.

The humans probably followed game species across the bridge. Humans either moved around

Table 12.3 These now extinct mammal species >40 kg existed in the western U.S. and northern Mexico 12,000 years ago (Martin and Szuter 1999)

Giant Short-faced Bear	Shrub Ox
Steppe Bison	Big-tongued Ground Sloth
Bonnet-headed Musk Ox	Glyptodont
Western Camel	Long-eared Llama
Dire Wolf	American Mastodon
Mexican Horse	Columbian Mammoth
Western Horse	Jefferson's Mammoth
Other horses, asses	Woolly Mammoth

the glaciers on the unglaciated seaward side or more likely after the glaciers began to recede. The human population eventually expanded southward all the way to Tierra del Fuego at the southern tip of S. America within just a few thousand years. Human movement between 12,000 and 10,000 years ago after the glaciers retreated is coincident with megafaunal extinction in N. America.

THE OVERKILL HYPOTHESIS

Evidence indicates that humans caused the wave of megafauna extinctions in N. America from 12,000 to 10,000 years ago. The overkill hypothesis was proposed by Paul S. Martin in 1967, explaining the extinction in N. America of mastodons, mammoths, camels, llamas, horses, ground sloths, cave bears, ungulates and others (Table 12.3).

The predators of these herbivores went extinct at this time too including hyenas, dire wolves and other canids, saber-toothed tigers, lions, large raptors, and scavengers including the Teratorn with a 5 m wingspan. The fossil beds indicate these species were present in the Pleistocene just before human colonization, but then suddenly disappeared. Megafauna such as Caribou, Moose, Dall and Bighorn Sheep that survived the extinctions had dispersed from Eurasia and were already adapted to and wary of humans.

Mass extinctions occurred on other continents at the times when humans colonized. Fossil evidence shows arrow points in carcasses and remains of massive animal kills beneath cliffs. The same pattern occurs with the colonization of the Pacific Islands. What humans brought with them everywhere was fire to manage the habitat, competition for resources, introduction of species, diseases, and hunting.

SUMMARY OF MACROEVOLUTION EVENTS

For us today, the three most overarching influences over time that are responsible for the Earth's current biodiversity patterns include:

- species diversification in the Cambrian 600 mya,
- the meteor event at the end of the Mesozoic 65 mya,
- the last glacial maximum 10,000 years ago.

EVIDENCE FOR EVOLUTION

Evolution at small scales is hard for anyone to deny. Within a human lifespan we can see plant species with resistance to herbicides, insect species with resistance to pesticides, and bacteria species with resistance to antibiotics. Grant and Grant (2006) provide evidence for a 4% change in beak size of finch birds during just a short drought on Galapagos. Clearly, evolution operates at small scales.

Other evidence for evolution includes similarities between species that are not functionally necessary, known as **homologous similarities**. For example, amphibians, reptiles, birds, and mammals have five digits in the embryonic stage. There is no reason why they should have five rather than 3, 7, or 12. In horses and lizards, there are less than five in the adult stage, but this is derived after the embryonic stage. Creationists will say that same architect designed them all. Evolutionists counter that **molecular homologies** occur too in the form of base triplets within DNA that code for amino acids in proteins. For instance, the nucleic acid

UUG always codes for amino acid leucine. All organisms talk the same language in this code.

Other evidence is provided by vestiges of organs no longer functional (**vestigial organs**). For instance, the lateral toe on a pig or dog is high off the ground when the individual is standing. It no longer functions as a working toe. Likewise, the pectoral and pelvic girdles remain even in animals without limbs. In whales these function in sexual behavior.

Other evidence is provided by diatoms, planktonic protists with glass-like pillboxes, which have a substantial **fossil record**, good enough to reveal the origin of new species approximately 3 mya. Creationists say the mere existence of fossils does not give evidence for evolution. Several rounds of creation could have occurred, and most fossil records have large gaps. What is clear is that the order of evolution was fish ---- amphibians ---- reptiles ---- birds. Radioactive carbon dating supports this order.

THINKING QUESTIONS

1. Without just listing the reasons written in this chapter, why is it important for you personally that you understand the basics of macroevolution?
2. Evolutionists argue that some of the most radical changes in the evolution of animals resulted from paedomorphosis. What is that? How does it happen in tunicates?
3. Greenhouse gas emissions have affected the composition of Earth's atmosphere in what quantitative ways? Is the percentage of oxygen still at 21%? If it has remained stable, what accounts for this stability? If it has been altered by modern society, how might decreases or increases affect life on Earth? What about CO_2? Has it remained stable? How much has it changed and in what timeframe?
4. Unlike most mammals that evolved from nocturnal ancestors, primates evolved from shrew-like day-active organisms. Research the details of the primate lineage. Be aware that most mammals cannot see color or three dimensions like primates. Our eyesight is much more like birds, which are mainly day-active. What can most mammals see that primates cannot? What are other benefits of evolving from day-active eyesight?
5. Gooley (2014) writes that almost all warm-blooded animals instantly notice the ape profile and are alarmed by it. Why would they not? It is an oddity among mammals, which are usually four-legged. Besides being bipedal, apes, including humans, have shoulders, constructed of long clavicles. Almost no other type of mammal besides ape has prominent shoulders. Four-leggeds have reduced or no clavicles, which accounts for their ability to slide through tight places. Even a bear standing on two legs has little shoulder profile. (Examine a photo to be convinced.) Additionally, most four-legged predators keep their head below the rest of their profile when hunting. Thus, the round head and prominent shoulders of a human may be alarming to wild animals. This should tell us what needs to be camouflaged if we want to be undetected.
6. Test the hypothesis in #4 by presenting different cardboard profiles to animals in cages (such as birds) and recording their reaction. Does the profile of a human on all fours with head slunk below the rest of the profile reduce the alarm? Does the profile of just head and shoulders alarm the animal as much as the profile of a whole human? Are animals that have never seen a human alarmed by the profile? Does the placement of eyes on the front rather than the side of the head alarm an animal? Are non-human animals as alarmed by the image or profile of a non-human ape as they are by a human? If you cannot find a non-human animal to test these hypotheses, have another person sit in the woods at dusk and present profiles to them from a distance. Have the person report whether they think the profile presented is a human or not.
7. Compare the number of stars visible within a section of sky during a full Moon to those during a new Moon, or within a city compared to an area away from surrounding light. Take a

flashlight with you to a dark place. Cover the light with clear red plastic and determine whether the filtered light requires less time for your eyes to adjust than with white light.
8. The sky is a place that remains wild. Even in the largest cities some stars and planets are visible at night. Insects are usually present during the daylight hours. Birds and other objects pass in the wind. Precipitation of many kinds falls from the clouds. Dust and pollen floats through almost all air. Record five things new to you by inspecting the sky or atmospheric column above you.

REFERENCES

Biello, D. 2009. The origin of oxygen in Earth's atmosphere. *Scientific American.* August 19, 2009. scientificamerican.com.

Gooley, T. 2014. *The Lost Art of Reading Nature's Signs.* The Experiment, LLC, New York.

Grant, P.R. & B.R. Grant. 2006. Evolution of character displacement in Darwin's finches. *Science* 313:224–226.

Herron J.C. and S. Freeman. 2014. *Evolutionary Analysis* 5e. Pearson, Boston, MA.

Janke, P.R. 2004. *A Correlated History of the Universe.* Pan Terra Inc., Hill City, SD.

Kritcher, J. 2009. *The Balance of Nature: Ecology's Enduring Myth.* Princeton University Press, Princeton, NJ.

Kumar, S., A. Filipski, S.V. Walker, S.B. Hedges. 2005. Placing confidence limits on the molecular age of the human-chimpanzee divergence. *Proceedings of the National Academy of Sciences* 100:18842–18847.

Lomolino, M.V., B.R. Riddle, R.J. Whittaker, and J.H. Brown. 2016. *Biogeography.* 4e. Sinauer, Sunderland, MA.

Martin, P.S., and C.R. Szuter. 1999. War zones and game sinks in Lewis and Clark's west. *Conservation Biology* 13:36–45. https://doi.org/10.1046/j.1523-1739.1999.97417.x

Sadava, D., D.M. Hillis, H.C. Heller, and M.R. Berenbaum. *Life: The Science of Biology.* 10e. Sinauer, Sunderland, MA.

13

Introduction to statistics

INTRODUCTION

The first step toward understanding statistics is to have a firm grasp on graphing and sampling. If one can properly differentiate between the need for a bar graph (when comparing means), versus a best-fit line graph (when looking for a relationship between variables), versus having no graph but a tally table instead, one is on their way to understanding statistics. Statistics is using mathematical formulas, definitions, and computers to predict, define, and tell exactly how one treatment is different from another (Magnusson and Mourao 2004). Statistical inference uses standardized criteria for decision making to help ensure that decisions are not swayed by personal opinion or political pressure (Sinclair et al. 2006).

Statistics may not be easy for beginners (Magnusson and Mourao 2004). One problem is that many statistics courses are taught by mathematicians and their job is to emphasize theory. Things may get too deep, too fast, and never cover examples in one's field. Another problem is that practitioners, the biologists, chemists, and physicists who advise you, may have years of experience and tricks using the techniques they need, but their theory is absent and their language is inconsistent. How can you expect to learn anything from anybody?

A good place to start may be the first page of each chapter of a statistics book. With practice, one can peel through to the next layer with more complex graphs, experiments, and examples. The perfect class, book, or teacher will never emerge for something so personal. Instead, there are books at the library that must be read. Once the basic concepts are learned, the results section of science journals and experimental design books should be read to find specific examples that are of interest. Even if one has taken a statistics course, experimental design may not have been addressed as it is in this chapter.

THE NULL HYPOTHESIS

For every statistical test there is a null hypothesis. "**Null**" means no, nothing, none, nada, zip, or zilch. Depending on the type of experiment conducted, a **null hypothesis** means there is (Table 13.1):

- no difference between means being compared (t-test and ANOVA),
- no difference between observed frequencies and those expected by chance (chi-square),
- no relationship between two variables (correlation and regression).

Corresponding to each of these, the alternative hypothesis is the opposite:

- There is a significant difference between means being compared.
- There is a significant difference between observed frequencies and those expected by chance.
- There is a significant relationship between two variables.

THE PROBLEMS WITH NULL HYPOTHESES

Statistical tests are effective at ruling out null hypotheses, for example, "animals do not move." The trouble for most students is that the null

DOI: 10.1201/9781003271833-13

Table 13.1 Most commonly used statistical tests and their null hypotheses

	t-Test	ANOVA	Chi-squared
What is being compared	Two means	More than two means	Two or more frequencies
Null hypothesis	$H_o: \mu_1 = \mu_2$	$H_o: \mu_1 = \mu_2 = \mu_3$	$freq_1 = freq_2$
Type of variable	Continuous	Continuous	Categorical

hypothesis is the opposite of what we would expect. Additionally, the alternative hypothesis is what we are supposed to infer if we reject the null, but this is only inference. There may be more than one alternative hypothesis. Which one are we supposed to choose?

P value

When testing a null hypothesis, the result of the statistical test is a P value. It shows whether the null should be rejected. P has a complex definition, but beginners can think of it as the probability that the null is true.

- if $P < 0.05$, reject it. The data supports the alternative hypothesis and one can conclude the means are significantly different.
- if $P > 0.05$, fail to reject. There is not enough support for the alternative hypothesis.

When $P < 0.05$, beginners could think of the meaning as less than a 5% chance that the null is true. A more accurate way to express it is, "a difference as great as what we found between treatments would be expected less than 5 out of 100 times" (Gotelli and Ellison 2014).

WHY DO WE USE P=0.05 AS THE CUTOFF POINT?

This cutoff seems stringent if we reject only when we would expect it less than 5% of the time. If you used this rule in your everyday life, you would not take an umbrella unless the forecast for rain were at least 95% (Gotelli and Ellison 2014). It means that the evidence must be exceedingly strong for us to reject the null hypothesis. The jury does not issue a guilty verdict unless there is more than 94%

surety. We would certainly take precautions if we knew there was a 94% chance of a tornado. The reason the standard is so high is because:

- the convention is based on probability, not certainty. We do not measure whole populations, only samples. The estimate based on sampling is sometimes wrong and noisy. We need to be conservative, which means a high standard.
- two types of errors may occur as illustrated in Table 13.2. In a **type I** (alpha error) the null is rejected when it should not be. In a **type II** (beta error) we fail to reject when we should have rejected.
- A type I error is the worse type because it is a false positive. The researcher has rejected the null hypothesis when it was really true. More significant differences were declared than were actually there. It is like convicting an innocent person for a crime he or she did not commit.
- The problem is, the higher P you choose as your critical value, the more you increase your chance of making a **type I error**.
  - If you use a P value too low, you increase the chance of making a type II error. You let a guilty person go scot-free, perhaps to commit another crime.

Experience has shown that P=0.05 is usually the right balance between type I and II errors for most situations as long as at least 30 replicates in each treatment were taken.

The extent that a statistical test minimizes type II errors is called **power**. The power of the test increases with sample size. For almost all tests, there is sufficient power when there are at least 30 replicates in each treatment. In some situations, especially epidemiology, avoidance of type II may be more important. Ask your research adviser.

Table 13.2 Delineation of type I and II errors

	Analysis indicated that we should fail to reject H_0	Analysis indicated that we should fail to reject H_0
H_0 *true*=in reality there is no difference $\mu_1 = \mu_2$	Our analysis is correct	Type I error (alpha)
H_0 *false*=in reality there is a difference $\mu_1 \neq \mu_2$	Type II error (beta)	Our analysis is correct

USING CERTAIN WORDS

- Be careful about using "prove." Just because $P<0.05$ and you reject the null, it does not prove the null hypothesis is false.
- Be careful about using "accept the null hypothesis." It is equivalent to "proving." You may "reject" the null or "fail to reject," but you may not "accept the null" unless you do a power analysis to precisely calculate the probability of a type II error.

AN OBSESSION WITH REJECTION– STATISTICAL VERSUS SCIENTIFIC SIGNIFICANCE

A preoccupation with null rejection can overshadow more important concerns (Sinclair 2006). A data set with a small number of replicates, or a faulty null **hypothesis** in the first place, should always be treated with caution or suspicion. Statistical significance does not necessarily imply scientific significance.

MEANS COMPARISONS

A **t-test** is used to determine if there is a statistical difference between two means. Computer software from R, **Excel**, **Instat**, or others can be used to calculate t. A difference exists between independent and paired t-tests. Paired t-tests have more power to detect change and should be used whenever appropriate, but most comparisons are independent and not paired. **Paired t-tests** can be used when pairs of sampling units are correlated. In other words, a plot in a forest with a large value the first year is likely to have a large value the second year, which means they are correlated. Other examples include the right side of bilateral animals compared to the left, or comparisons between identical twins. If you are not sure whether a **t-test** should be paired, it

probably should not. Proceed as if it should not. A t-test can also be either one-tailed or two. If you are not sure, assume it is two-tailed.

PARAMETRIC VERSUS NON-PARAMETRIC TESTS

The most commonly used and most accurate statistical tests are based on specified distributions in their histograms (especially normal or Poisson). When they are based on these known distributions they are considered **parametric** tests. These include **t-test, chi-square**, analysis of **variance**, and others. They have certain rules (assumptions) that must be met, otherwise the results are invalid. The assumptions of a **t-test** are:

1. **data are normally distributed**: In other words, if a frequency histogram was drawn for each treatment, each histogram would form a bell-shaped curve (Figure 13.1). This will almost always be true if there is a sample size of 30 replicates for each treatment because of the Central Limit Theorem.

2. **the variance (or SD) of one treatment is approximately equal to the variance (or SD) of the other treatment**: This is the more important of the first two assumptions.

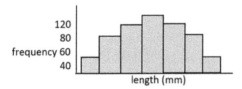

Figure 13.1   A bell-shaped curve in a frequency **histogram** showing a normal (Gaussian) curve. Each bar represents a range of values, with most values grouped tightly around the mean.

Variance refers to how much variation there is among the samples in one treatment.

3. **observations are independent**: One replicate does not in any way influence the value in another replicate and there is true replication, not pseudoreplication.

WHICH IS THE BEST CHOICE, PARAMETRIC OR NON-PARAMETRIC?

Despite the restrictions, you should always try to use parametric tests if possible. Non-parametric should be used only if a correction or transformation does not meet the assumptions. Parametric tests are far more accurate and do the most to reduce both type I and II errors If you use a non-parametric test you should justify your use of it, otherwise we may think poorly of your choice of statistical test.

WHAT IF THE FIRST TWO ASSUMPTIONS OF THE T-TEST ARE NOT MET?

There are two choices.

1. Some statistical packages make corrections for unequal variances such as using the Welch correction within Instat, but you have to say yes for this choice when prompted. Most statistical software does not provide the option.
2. Each number in the data sets can be transformed by ln x, sqr root of x, or ln (x+0.1). This is like changing units from miles to km. The natural log transformation is the most common and has the particular effect of making the standard deviations relatively smaller and therefore more equal. This seems like a magic

trick, but it is as valid as changing units from miles to km. Note that when dealing with percentages, the first two assumptions are almost always violated. The best transformation to use with percentages is to take the arcsin square root of each of the data points in each treatment. Although you may transform the data for analysis, you should report the results in the original units when making graphs or tables.

3. If the problem is not corrected when the test is run again on the transformed data, the researcher should use a **non-parametric test**. These have less stringent assumptions. **Parametric** means that the probability fits a specific distribution, almost always implying a bell-shaped (normal) curve. Non-parametric tests are usually based on ranks, a far less accurate way to assess differences. The non-parametric alternative for comparing two means is a Wilcoxon signed rank test or Mann-Whitney U test. The alternative for comparing more than two means is a Mann-Whitney U test or Kruskal-Wallis.

WHAT IF THE RESEARCHER IS COMPARING MORE THAN TWO MEANS?

An analysis of variance (ANOVA) using an *F* test is employed when comparing the means of more than two treatments (Figure 13.2). The basis of an ANOVA is different than of a t-test. In an F test, F is the ratio of the variance among groups over the variance within. The idea is that if the variance among is much greater than within, the treatments must be significantly different. This makes intuitive sense and is why this is called analysis of variance.

To do this in practice, the sum of the squared deviations among treatments $\sum(\bar{x}-x_1)^2$ is divided by the sum of the squared deviations within treatments $\sum(\bar{x}-x_1)^2$ to produce an F value. This F value is checked against a table of values to determine P. Simple.

The computer output for an ANOVA is in a format as in Table 13.3 (Dytham 2011). The business end of the table, of course, is the P value. In this case P>0.05, which indicates there is no significant difference among treatment means. We

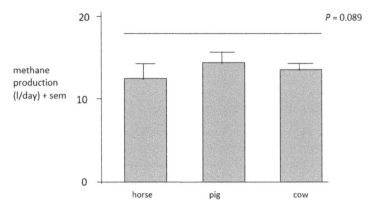

Figure 13.2 Methane production by type of farm animal (n=5) when fed standard diets administered for 6 weeks.

Table 13.3 An example of a standard ANOVA table produced in a standard computer output

Levene Test for Homogeneity of Variances

Statistic	df1	df2	2-tail Sig.
0.7616	1	8	0.408

Analysis of Variance

Source	D.F.	Sum of squares	Mean Squares	F Ratio	P
Between	1	16.3840	16.3840	15.3624	0.089
Within	8	8.5320	8.5320	1.0665	
Total	9	24.9160			

HOW TO TEST THE T-TEST ASSUMPTIONS

ASSUMPTION 1 (NORMALITY ASSUMPTION)

1. Some statistics packages check this automatically and will not let you proceed if the assumption has been violated. Some of the other statistical packages test it on request or provide results automatically with the final P value, but they do not stop you from proceeding to the end.
2. If you have 30 samples or more, assume the assumption is met.
3. Seat of the pants rule if you do not know another way: when there are less than 30 replicates per treatment, use a number line to examine the mean and individual values for each treatment. If the individual values are evenly distributed around the mean, assume the distribution is normal for that treatment.

ASSUMPTION 2 (EQUAL VARIANCE ASSUMPTION)

1. Some statistics packages check this automatically. In other software, check the results of the Levene test or other computerized test on the printout for results of the equal variance assumption.

2. **Seat of the pants rule if you do not know another way**: calculate the variance or standard deviation for each treatment. (Excel can do this.) If the standard deviations of the treatments are roughly equal (within 30%), consider the assumption to be met.

ASSUMPTION 3 (INDEPENDENCE ASSUMPTION)

There is no computer program that can tell you this. Use common sense and judgment. Was there true replication, or was it pseudoreplicated? Do the samples in one treatment or replicate influence the others? If this assumption is not met, the analysis must be abandoned. There is no choice.

fail to reject the null hypothesis. Note that when we refer to differences between two means we use "between." When we refer to differences among more than two means we use "among." We must check the assumptions of ANOVA just as we did in the t-test, and the assumptions are the same. For some statistics software, the assumptions are checked automatically. The computer output will now include something like Table 13.3 with a Levene's test:

The "2-tail Sig." value is the P value for the Levene's test. It tells us there is no significant difference in the variances among treatments, thus we have met the equal variance assumption. This is good and we can proceed to validly accept the information in the analysis of variance table. If the P value for the Levene's test were less than 0.05, we would have to take corrective action such as transforming the data. It would not be valid for us to accept the rest of the results in the analysis of variance table.

POST HOC TESTS TO COMPARE PAIRS OF MEANS

If the P value for our ANOVA results tells us that some treatment means are different from others, how do we know which mean is different from which when we have more than two means? To determine differences between means we need a **post hoc test**. Post hoc = after the fact. This is also sometimes called *a posteriori* = after the fact. This is also called a **means-comparison test**. These compare every mean to every other mean and provide P values for every pairwise comparison.

The one thing that is not valid is to complete multiple t-tests to compare every mean to every other mean. This compounds the probabilities and renders the tests invalid. In other words, we are no longer testing our null at the 0.05 level. If we conduct two t-tests, we are now testing the null at the $0.05 \times 0.05 = 0.025$ level. If we use three t-tests, it is at the $0.05 \times 0.05 \times 0.05$ level and so forth. Reducing our P standard will drive up our chance of making a type II error.

There are several **post hoc tests** available, Least Significant Difference (LSD), Student Newman Keuls (SNK), Sheffe, Tukey's, and Duncan Multiple Range. Mathematical research shows that the best one to use is Tukey's. Here is why: LSD uses multiple t-tests, which we just reported was invalid; it produces a result that is too conservative = too many type IIs. Duncan's Multiple Range is too liberal with too many type I = worst kind of error. Thus, Duncan's Multiple Range test should never be used. It means too many innocent people in jail. Tukey's has fewest type II or type I errors – use this.

The computer output with Levene's test, the ANOVA table, and the post hoc test is in Table 13.4.

According to our Tukey's test (Table 13.4) there are no significant differences between any of our pairwise comparisons among means because none of the P values were less than 0.05. Actually, we would probably not have run the post hoc tests on this analysis in the first place because we did not find a significant P value when the overall ANOVA was run.

HOW DO I SIGNIFY PAIRWISE SIGNIFICANT DIFFERENCES ON MY GRAPH?

Look again at the figure in our methane example. There is a horizontal line over all three bars. The conventional rule is that this horizontal line is placed over the top of all treatments that are not significantly different. In this case the original P = 0.089. Because it was not less than 0.05, there

Table 13.4 Example of standard ANOVA table with output for Levene test and post hoc test

Levene Test for Homogeneity of Variances

Statistic	df1	df2	2-tail Sig.
0.7616	1	8	0.408

Analysis of Variance

Source	D.F.	Sum of squares	Mean Squares	F Ratio	P
Between	1	16.3840	16.3840	15.3624	0.089
Within	8	8.5320	8.5320	1.0665	
Total	9	24.9160			

Tukey-Kramer Pairwise Comparisons 2-tail Sig.

	horse	pig	cow
horse	0.00		
pig	0.072	0.00	
cow	0.099	0.124	0.00

were no significant differences. A line can be drawn over all three treatments.

Consider an example when there is a significant difference among the means (Figure 13.4). The mean for horse is significantly different from the means for pig and cow, but pig and cow are not significantly different from one another. Notice that the graph has a line over the treatments that are not significantly different. Notice that to make this convention work, the means must be placed in order from lowest to highest, or highest to lowest on the graph, thus the order of the bars has been rearranged in this example.

The convention of placing a line above the treatments that are not significantly different can be used for some very sophisticated differences. In Figure 13.3, the lines tell us the mean for horse is significantly different from every other mean. The mean for snake is different from every other mean. Cow and pig are not different from each other. Pig, hamster, and lion are not different from each other. Remember, this convention will only work when you order your means from lowest to highest or highest to lowest. Altogether there are 11 significant pairwise differences depicted in Figure 13.3. Can you name them all?

TABULAR COMPARISONS: COMPARING FREQUENCIES THROUGH CHI-SQUARE TEST

It is appropriate to use a chi-square test when:

- frequencies are being compared, not means.
- categorical data are being used, not continuous data.
- the null is that the observed and expected frequencies are not different.

Frequencies are not means. They are the number of each organism or object and can only occur as whole numbers, not decimals as is possible for a mean. It is how many times a coin lands heads or tails, how many individual beetles reproduce or not, or whether organisms are present or not. It is the abundance of something, a unitless number, the number of times something occurs. You keep a tally when assessing frequency.

EXAMPLE HYPOTHESES TESTED IN CHI-SQUARE

- Is the observed and expected frequency the same?
- Are phenotypic ratios in a monohybrid cross the same as the expected 3:1 frequency?
- Are sex ratios the same as what we would expect?

Comparisons among frequencies are useful in genetics and some chemistry research, but with few exceptions they are rarely appropriate for answering field biology questions (Magnusson and Mourao 2004). Too many individuals or plots have to be sampled to record something like presence

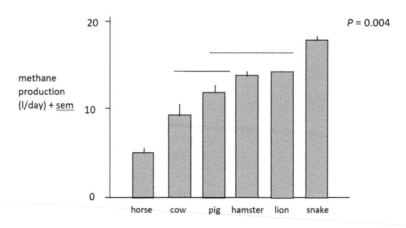

Figure 13.3   Methane production by type of farm animal (n=5) when fed standard diets adminis-tered for 6 weeks. Lines above bars indicate no significant difference in pairwise comparisons.

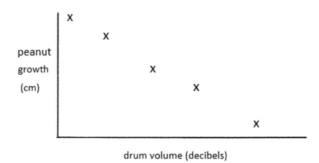

Figure 13.4   Peanut growth versus drum volume showing negative correlation.

and absence. It ends up being too expensive and time consuming to visit 1,000 Chestnut trees and determine whether they do or do not have repro-ductive structures. A thousand test tubes, however, may be reasonable.

Ecologists often find means and the varia-tion among samples to be more appropriate and more enlightening for small sample sizes. Chi-squared tests are very sensitive. They almost always show statistical significance to the point where the results become meaningless. If you are doing an ecological project and you find your-self using frequencies, see if you can turn your question into something using a mean. This is almost always possible. Leave the chi-squared tests for the laboratory researchers with many more replicates.

CORRELATION AND REGRESSION

For both correlation and regression we usually draw a graph with two axes, and plot points. This is called a scatter plot (Figure 13.4) or if it has a line with it, a line graph. If the two variables are cor-related, the data set will slope either one way or the other, positive correlation or negative correlation.

- In a graph plotting drum volume versus peanut growth, we interpret a positive correlation as "the greater the volume, the greater the peanut growth."
- We interpret a negative correlation as "the less the volume, the greater the peanut growth" or "the more the peanut growth, the less the volume."

Table 13.5 Statistical tests available for comparing samples

Purpose	Parametric test	Non-parametric test
Comparing Frequencies from Two or More Treatments, categorical, not continuous data		Chi-square test OR G tests OR Fisher's exact test OR Cochran-Mantel-Haenszel
Comparing Means from Two Treatments, Continuous, not categorical data		
• Samples independent	Independent sample t-test	Mann-Whitney U test
• Samples paired	Paired t-test	Wilcoxon's signed rank test
Comparing Means from More than Two Treatments, Continuous, not categorical data	Analysis of variance with Tukey's – could add Welch's correction for unequal variance	Kruskal-Wallis test OR Mann-Whitney U test with Bonferroni correction
	A post hoc means-comparison test should follow a significant ANOVA	A Post hoc means-comparison test should follow a significant procedure
Comparing Means from More than Two Treatments, one factor involved		
• Samples independent as in independent t-test	Analysis of variance for independent samples	
• Samples blocked as in paired t-test	Analysis of variance with blocking	
• Replicates nested at different spatial scales	Nested analysis of variance	
• Sampling is repeated on the same replicates over time	Repeated measures analysis of variance	

(Continued)

Table 13.5 (Continued) Statistical tests available for comparing samples

Purpose	Parametric test	Non-parametric test
• Only one replicate	Before-after controlled impact (BACI) design	
Comparing Means from More than Two Treatments, Two Factors Involved		
	Two-way analysis of variance	
• All treatments fully crossed	Factorial analysis of variance, also called orthogonal design	
• Only one replicate for each treatment	Experimental regression analyzed using regression, not ANOVA	
Correlation between Two or More Variables		
• Neither variable is clearly independent or dependent	Correlation analysis	
One variable is clearly the dependent variable	Regression analysis	
• Two or more independent variables	Multiple regression analysis	
• Two or more independent variables and two or more dependent variables	Multivariate analysis	
• Independent variable is categorical rather than continuous	Logistic regression	

HOW ARE REGRESSION AND CORRELATION DIFFERENT?

Correlation is plotting two variables and looking for a pattern. The researcher did not specify which variable is on the x axis and which is on the y. There is no predictor and response variable, no cause and effect. Regression does specify cause (independent (x) variable) and effect (dependent (y) variable) because the researcher knows which is causing which.

A best-fit line can be added to correlation or regression. This is done to minimize the average distance of the points from the line and can be done mathematically or by eyeball. For linear regression, the squared vertical distances from a line are generally minimized (Magnusson and Mourao 2004). For correlation, the horizontal and vertical distance of each point is minimized and this is called least-squares (also referred to as Model II regression) (Gotelli and Ellison 2014).

More advanced students may be interested to know that least-squares regression is logically and mathematically the very same as ANOVA. The distances to the best-fit line are "residuals." The variation about the line is the residual variation not explained by our model (the line). A plot of residuals after an ANOVA can be enlightening to establish how much variation exists within or between treatments. It is also used to calculate r^2.

SOME OTHER DIFFERENCES BETWEEN REGRESSION AND CORRELATION

In regression one can draw a line in the form $y = mx + b$ to predict y values based on x. A prediction is not appropriate in correlation. In regression one can obtain a P value and a significance test. It tests the null that one variable does not depend on the other. In other words, **P** is the probability that the best-fit line has a slope of zero. In correlation this is not appropriate.

THE STRENGTH OF THE CORRELATION CAN BE MEASURED

Strength is signified by **r** which stands for the Pearson's product-moment correlation. It varies from −1 (negative) to +1 (positive), with 0 indicating no correlation at all. It represents the percent variation in one variable explained by the other. In other words, it represents consistency in the data. We can ask, is there a high degree of error, or do the data points make a straight line? We can also calculate a P value that tests the H_0: one variable is not related with the other. It tells us something about the strength of relationship but not the slope.

WORDS OF CAUTION ABOUT CORRELATION

Pearson's correlation assumes that both variables are normally distributed. This assumption is almost never heeded. Correlations are frequently used, but do not report their statistical assumptions. Even if a correlation exists, it does not imply cause and effect (one variable causes the other). Consider "the higher the drum volume, the greater the peanut growth." All a correlation can do is establish a possible pattern but nothing else (Table 13.5).

REFERENCES

Dytham, C. 2011. *Choosing and Using Statistics*. Blackwell Science, Malden, MA.

Gotelli, N.J., and A.M. Ellison. 2014. *A Primer of Ecological Statistics*. Sinauer, Sunderland, MA.

Magnusson, W.E., and G. Mourao. 2004. *Statistics without Math*. Sinauer, Sunderland, MA.

Sinclair, A.R.E., J.M. Fryxell, and G. Caughley. 2006. *Wildlife Ecology, Conservation, and Management*. Wiley-Blackwell, Hoboken, NJ.

14

Population ecology basics

CLASS ACTIVITY: Population Study

Exercise 1: Map populations you observe. Take a walk in your neighborhood or in a park and make a map that includes at least ten populations you encounter. Check your yard, waterways, and gardens that host populations and monocultures. Consider taking binoculars and a hand lens on your walk. Find both wild organisms and those managed or introduced by humans.

Exercise 2: Complete math problems below as a warm-up for a mark-recapture study (from M. Brenner, pers. comm.).

a. The pond on a campus is filling with sediment because of erosion and it needs to be dredged. Doing so requires that the water be drained and the fish population removed. Because natural dispersal of fish to the pond would not occur easily, you want to restock it with fish after draining. You introduce 50 bluegill sunfish (*Lepomis macrochirus*). From the size of the pond, you estimate the carrying capacity is 600 bluegills. You get a per fish growth rate constant from the literature of R=0.4 fish/month.

b. You want to allow the population to be close to the carrying capacity before fishing is allowed, so you want to estimate how long it will take to reach 600 fish. Use simple algebraic iterations by month to do the calculations, then plot number of fish versus time (on graph paper, remember to label axes).

c. So far in our modeling for the restocked swim pond, we have assumed that no fish died each month, but of course this is unrealistic. Redo your calculations assuming that the mortality rate is 10% per month. Show your work.

Exercise 3: To estimate the number of fish in a real pond, mark/recapture methods can be used. To do so, animals are captured and marked during a precount period. The animals are then released and allowed to fully disperse. During the ensuing count period, as many specimens as possible are collected. The ratio of those marked (recaptured) to unmarked roughly represents the ratio of those originally marked to the unknown population size:

$$\frac{M}{N} = \frac{R}{C}$$

M=number marked in precount period,
N=total individuals present in precount period. This is the number we are trying to estimate.
R=number recaptured, that is, caught in count period with marks,
C=total individuals captured in count period, marked and unmarked.
Solve for N.

DOI: 10.1201/9781003271833-14

WHY FOCUS ON ONE SPECIES AT A TIME?

What if we could:

- manage disease epidemics,
- bring rare species back to high densities (Figure 14.1),
- control pest organisms and introduced species,
- sustainably harvest fish, game, agricultural products, range grasses, and trees needed for lumber.

Achieving these goals requires an understanding of populations and their dynamics. In fact, some of the most devastating problems of our time – biodiversity loss, world hunger, medical disasters, and economic injustice – are addressed through knowledge of population dynamics.

In short, population ecology applies to any situation in which there are too many or not enough of one species, or any case in which the changing size of a population is important. Thus, population ecology might be called the **busy-bee, blue-collar worker of ecology** because it can do many things.

Conservation biologists use population ecology to maintain rare species or control invasive species. Agriculture, fisheries, forestry, and range management specialists use population ecology for the production of economically favored species.

Mathematicians study populations to investigate intricate patterns of growth and decline. Theorists find that population ecology addresses questions common to all subdisciplines of ecology:

- Is there a balance of nature?
- Why are organisms where they are and in the relative numbers we observe?
- How did evolution shape the patterns we observe?
- How can we form and use theory to predict patterns?
- How can theory help us protect nature from humans, protect humans from nature, and obtain resources sustainably?

In fact, it is population ecology that most clearly lays out theoretical issues. In the summary list of principles outlined at the end of this book, it is population ecology that has the most consensus.

WHERE SHOULD A STUDENT BEGIN?

Chapters 14 and 15 are long because they teach some of the most important concepts in ecology. Once these concepts are learned, much of the rest of ecology theory falls into place. Step one is to learn the basic vocabulary of population ecology. Step two is to gain a thorough understanding of

Figure 14.1 Several generations of elephants can make up one population. These elephants are drinking from a waterhole in the Madikwe Game Reserve, Africa. (*Photo by Udo Kieslich, used with permission.*)

modeling for predicting population growth. Step three is to learn methods of measuring and managing growth. Learning to adequately predict growth is really the critical factor needed for management. All applied issues hang on this task, which is why this chapter and the next move progressively toward an understanding of mathematical modeling. We start with basic vocabulary associated with population ecology.

WHAT IS A POPULATION?

An ecological **population** consists of individuals from one species that occur together at the same time.

Populations consist of organisms, living or dead (as in fossil beds), but always organisms, not rocks, nuts, or bolts (Allee et al. 1949). Members of a population are whole organisms, not cells or organs (although cells and populations have interesting similarities). Both grow, die, **senesce** (die back for a time), and have a division of labor. For instance, bees and wasps divide labor at the population level. Groups of cells divide labor as tissues.

WHAT IS POPULATION ECOLOGY?

Population ecology has roots in **demography** (the study of populations), a science even before ecology. Demographers address plagues, diseases, insect infestations, and human population growth – age-old issues. Population biologists study population genetics, evolution, and **epidemiology** (the study of epidemics and disease) not necessarily in the wild. Population ecologists focus on wild-born species interacting within their natural environment.

WHEN DID POPULATION ECOLOGY ARISE?

After Charles Darwin, population ecology did not immediately get its start. It might have if there had been a natural progression from Darwin, to further development of evolutionary principles, to Darwin's ideas about population growth, as outlined in his books. History did not happen that way.

After Darwin, the first ecologists were European botanists in the tradition of world explorers investigating biomes (McIntosh 1985). Botanists, then and now, tend to study communities rather than populations because plants intermesh their roots and leaves with other species, making it difficult to sample one species at a time. Zoologists, then and now, gravitate toward population ecology because animals move and hide, making it difficult to study anything but one species at a time. Thus, botanists led community ecology efforts right after Darwin's time. Zoologists led population ecology, which started later (early 1900s).

It was not ecology, for the most part, that further developed evolution (McIntosh 1985). It was geneticists that formed the modern synthesis in the 1920s, incorporating Mendel's theory of inheritance with ideas about evolution. It was in the laboratory, not the field, that geneticist Theodosius Dobzhansky used fruit flies to apply theories. It was in evolution journals, not ecology, that field studies were published when they addressed evolution.

HOW IS A POPULATION "DELINEATED?"

It is the observer that defines both geographic and **temporal** (time) boundaries of a population. Sometimes boundaries are obvious, such as for a tight herd of antelope. Otherwise, **delineation** (defining by borders) is arbitrary. An ecologist might delineate the 2022 Vancouver Island cattail population, or the New Orleans population of Fire Ant. Although populations do not follow political boundaries, it might be convenient for an ecologist to designate them as such.

Ideally, a population would be defined by the boundaries of interbreeding within a species. The questions is, how would an ecologist know the bounds by which a group of salamanders find each other, breed, and have genes flowing among them? Should the population include individuals who are too old or young to breed? Delineation problems also emerge when phenotypic plasticity occurs. A species of wheat in a desert may flower during spring. The same species may flower during summer on the surrounding mountains. Are both groups in the same population?

HOW MANY INDIVIDUALS ARE IN A POPULATION?

No standard size or scale has been established for a "population." Populations can be on a grain of sand or throughout the world. Yet, once a population is delineated one of the most sought-after attributes is **abundance**, a unitless measure representing a count of individuals, a word used interchangeably with **population size**.

Abundance is different from **density**, which is organisms per unit space. Note that density can be described in two dimensions or three (volume). Density is sometimes measured in terms of weight per unit volume. This is applicable for crops, seeds, or plankton when the individuals are too small and numerous to count. While **density** refers to individuals per unit space, **growth rate** is individuals per unit time. Unlike density, which is always a positive number, growth rate can be positive or negative. **Rate** always refers to number per unit time.

HOW ARE THE NUMBER OF INDIVIDUALS COUNTED?

When the abundance of all individuals in a population is counted, it is a **census**. A census, however, is usually impossible to achieve, even in the human population census, conducted every 10 years in the U.S. (Mills 2013). In an ecological population, not every individual may get counted. There may be too many, or the individuals are too small to see, or some die while others are being counted. Animals may be afraid and evasive, too far away, hiding, camouflaged, active only in the dark, or too dangerous to encounter.

More realistic than a census is an estimation of population size through sampling. Plots or transects measure just a calculated portion of the population. Mark-recapture and distance sampling are methods used for estimating animal population size (Mills 2013).

WHAT IS RELATIVE ABUNDANCE?

Relative abundance or **relative density** is valuable when a census is hopeless. Crabs caught per day or birds heard per hour are compared in one area versus another. It is relative because numbers per sampling effort are compared, knowing that not all individuals are detected.

WHAT IS AN INDEX?

An **index** is a field count, not of live animals directly, but animal sign, such as tracks, calls, pellets, burrow entrances, or lodges (Mills 2013). An index comes from non-capture methods. Other examples are counts of roadkill, band returns in birds, antler or tooth returns at a check station, analysis of hair traps, or questionnaires about wildlife sightings. An index assesses relative abundance or relative density, the number of something in one area compared to another.

WHAT IS NOISE?

The problem with all estimation and indexing is **noise**, variability in the data because of sampling challenges. Noise always makes comparisons from one time or place to another seem more variable than they are. Techniques are available to estimate some of this noise (Mills 2013).

MORE DIFFICULTIES: WHAT ABOUT COUNTING MODULAR UNITS?

Whether a census or estimate is used, unitary organisms are more easily assessed. **Unitary organisms** grow to adulthood from a zygote and have a preprogrammed size and shape, as in the case of most animals. **Modular organisms** grow more units over time as in stems, leaves, and colonial animals like corals and polyps.

WHAT ABOUT COUNTING CLONES?

Clones present a problem for counting as in Duckweed (*Lemna* sp.) or Black Locust (*Robinia pseudoacacia*). Duckweed grows in colonies, and Black Locust grows in groves. Trees of Black Locust produce seeds, but they seldom germinate. Instead, new trees grow vegetatively by sprouting from a common root system. Each is genetically identical. It leaves us wondering whether a grove is an individual or a population.

Special terminology exists to help differentiate the situation. For plants, fungi, and bacteria, **genet** refers to a genetic individual, which may consist of several modules and it may extend over a large area, but the whole thing arose sexually from one zygote, like a grove. **Ramet** refers to an individual module of a genet (Figure 14.2). These ramets are **clones** of the parent and were produced asexually. Ramets may become separated if modules of the original genet have died away or if some float away from each other as in Duckweed. These ramets might get counted in a survey as "individuals."

Knowing these issues, plant ecologists need to be precise in their terms. Is it vegetative tillers, inflorescences, meristems, stolons, genets, or ramets that are counted when "abundance" of a

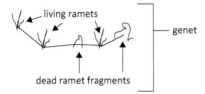

Figure 14.2 A collection of modules is a genet and individual modules are ramets. Dead fragments make it difficult to know the boundaries of a genet.

population is established. Compounding the problem, seeds may constitute most of the representatives of a plant population. If they are difficult to detect or in dormancy within the soil (**seed bank**), we might want to discount them entirely when assigning an abundance value to a population.

HOW ARE INDIVIDUALS DISTRIBUTED?

Global and regional distribution patterns of species are covered within macroecology and biogeography, while local distribution patterns of species are covered by population ecology. In doing so, population ecology differentiates between density and distribution in the following way.

Let's say we fly over our sampling space wearing special glasses and get a perfect density count per hectare of our population (Brown 1995). As helpful as this might be, it would not tell us what we commonly want to know, which is **dispersion**, the distribution pattern at a place. Without being aware of the dispersion concept, we may throw away valuable information by calculating density or abundance alone. Note that dispersion and dispersal are not equivalent. **Dispersal** is when an organism leaves the area of birth or activity for another area. Dispersion refers to the spatial pattern and has three types (Figure 14.3):

- **random**: lacking pattern or order and influenced by chance events.

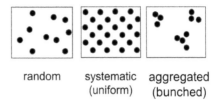

Figure 14.3 Dispersion patterns can occur in both volumes (3D) and areas (2D).

- **systematic** (uniform, regular): caused by territoriality or perhaps allelopathy.
- **aggregated** (bunched, clumped): caused by clonality, animals feeding together, or herding.

The concept of dispersion is valuable because we can match a dispersion pattern with a mechanism that could explain why an organism is where it is. A random distribution occurs when random factors are responsible for placement. Systematic spacing occurs when territoriality happens. Competition among individuals produces it, such as bullfrogs setting up territories along a lakeshore or plants exerting allelopathy.

An aggregated distribution might occur in nature for several reasons. For plants, the species might be clonal, propagated vegetatively as in Duckweed, Black Locust, goldenrods, sassafras, and aspens. In animals, a group might be hibernating together, aestivating, or mobbing the only water or food source. Individuals may form groupings for social reasons, such as staying warm at night. A clumped group might be a breeding aggregation. A congregation of chrysomelid beetles once formed a belt 15 feet thick and 100 yards wide that subsisted for 2 days (Allee et al. 1949).

Abiotic factors may be the cause of aggregations. Waves may sweep snails or crayfish into one location. Alternatively, a site good for one individual may be good for other individuals, such as a place in the shade or a site out of the wind or current. Note that some circumstances counteract clumping. Widely spaced plants are less likely to be infected by pathogens or eaten by herbivores (Barbour et al. 1999).

A handy application within field ecology is to quantify territory size based on dispersion pattern. If dispersion is systematic and the density of a population (organisms per unit area) is known, then territory size (unit area per organism) can be calculated. We do so by taking the reciprocal of density, assuming the territories are tightly packed. If eight pairs of Pileated Woodpecker occur per 1,100 acres, what is the territory size of a pair of woodpeckers, assuming the pattern is uniform and tightly packed?

CONCLUSIONS ABOUT THE POPULATION CONCEPT

If you get nothing else from this chapter, realize that the population concept is vague. Populations have no definitive size; the boundaries are set by the investigator; the individuals cannot always be defined, and even when defined the individuals cannot always be counted. Additionally, the genetic constitution and breeding behaviors of any population are difficult to assess. Still, the population concept is extremely useful, one of the most valuable in all of ecology.

VITAL RATES

After abundance, density, and dispersion are known, the next step within applications is assessment of population growth. For this we need a model, which can be in graph or equation form. Graphically for animals, we start with age structure diagrams, then move arithmetically to life tables. Life tables offer a wealth of insight for the eventual understanding of more sophisticated computer models used within wildlife management, fisheries, agriculture, and conservation biology. Life tables introduce elements of age and stage structure, time lags, and deterministic versus stochastic growth. We start with graphical models.

GRAPHICAL MODELING FOR ANIMALS

When envisioning a population, keep in mind that populations are not static. They continually increase or decrease. Particular age groups within the population grow or decline in abundance. Genders change in ratio. We must do our best to form a snapshot accounting for a population at one time.

For animals we can organize populations into age categories using **age structure diagrams**, the number or percent of the population in each age group stacked on top of each other as in Figure 14.4. By plotting males and females on separate sides of the axis, sex ratios can be compared by age group. Some age structure diagrams are further divided into preproductive, reproductive, and postproductive groups in what is known as **stage structure** (not shown in Figure 14.4). Age structure diagrams provide information about population growth to some extent. A growing population has a large number of young in the pre-reproductive group forming a broad base and narrow top in a triangle. An aging population has a narrow base and broad top, indicating that the population is not growing.

INCORPORATING VITAL RATES TO MAKE A LIFE TABLE

Age structure diagrams are useful for examining a population's history and its general growth and decline. For more precise assessments, models called life tables are used. Several **vital rates** (vital = "life") may be associated with life tables:

- **life expectancy**: average number of years to live for a member of a certain age class.
- **mortality**: rate of deaths within the population.
- **natality**: rate of living births per reproductively capable female. **Litter size** is measured for mammals and **clutch size** for birds.
- **fecundity/fertility**: in older literature these words were interchanged and used in reference

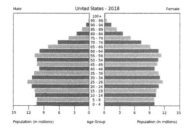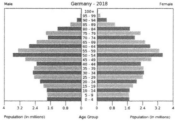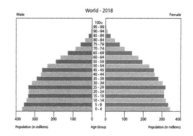

Figure 14.4 In age structure diagrams for humans, fast growing populations have a broad base, and declining populations have a narrow base. Populations sustaining their numbers have a base at same numbers as most groups above. (*CIA World Factbook 2019, Wiki Commons.*)

to the number of births/individual female of a given age in one time step. In newer literature, fertility may be used specifically within stage-structured models to refer to the proportion of the cohort that breeds.

- **recruitment**: net population production, also known as yield. These terms refer to the net number of new individuals in a population over a unit of time, characterized by dN/dt. It tells us whether a population is sustaining its numbers.
- **survivorship**: number alive at a given time.
- **sex ratio**: proportion of males to females. The number of males is usually given first as in 65:35, meaning 65 males, 35 females.

WHAT IS A LIFE TABLE?

A **life table** is a tabulation of mortality or reproduction data used to calculate population trends like life expectancy and population growth. Numbers are taken either from a cross section of a population at a given time, or from a cohort of organisms followed until their death. A **cohort** is a group of individuals in a population that are born during the same time interval, or in other words, a generation.

For calculating life expectancy, life tables mostly keep track of death, despite the name. Calculating life expectancy for humans is the basis of premiums derived by life insurance companies, the original developers of life tables. Particularly valuable for life insurance purposes is the average life expectancy at birth, the age at which 50% of the cohort is dead (Rockwood 2015). For modern humans this number is after 80 years, although it is only 57 for members of the population in Table 14.1, whose members spanned three centuries.

Table 14.1 Demography data for human females from 1750 to 1990 were collected from a cemetery by students at Newberry College, Newberry, South Carolina

Age	Raw number females alive at start of period	Number dying within age interval	Age-specific mortality rate	Average alive in each age category	Total years lived into the future	Life expectancy mean length of life remaining in years
x	l_x	d_x	q_x	L_x	T_x	e_x
0–9	2307	306	0.1326	2154	13173.5	57.102
10–19	2001	61	0.0304	1970.5	11019.5	55.069
20–29	1940	122	0.0628	1879	9049	46.644
30–39	1818	128	0.0704	1754	7170	39.438
40–49	1690	142	0.0840	1619	5416	32.047
50–59	1548	212	0.1369	1442	3797	24.528
60–69	1336	331	0.2477	1170.5	2355	17.627
70–79	1005	452	0.4497	779	1184.5	11.786
80–89	553	430	0.7775	338	405.5	7.3327
90–99	123	117	0.9512	64.5	67.5	5.4878
100–109	6	6	1	3	3	5

Only the l_x column includes raw data. The rest were derived from the raw data used to calculate life expectancy. The average life expectancy at birth for members of this population is 57.1 years. Raw data were acquired from academics.hamilton.edu.

HOW TO READ LIFE TABLES

For non-human species, life tables are usually compiled for animals, and life expectancy is usually not the most valuable entity. Instead, it is the first three columns and the resulting survivorship curves that allow important conclusions from instant inspection (Table 14.1). Life tables tell a story from the cells at the top left to the cells at the bottom right. Only the first two columns include information collected from the field. Other columns are calculated as various proportions from field data.

To read a life table start with the age groups listed in the left-most column for the cemetery data in Table 14.1:

- From the first column, we conclude that life span for this species stretches to 109 years.
- From the third column we determine that most mortality occurs after age 60.
- From the right-most column we see life expectancy for each age.

From the second column we can construct **survivorship curves,** plotting the l_x (survivorship) column against age (Figure 14.5). However, the number of survivors is standardized per 1,000 individuals. This is a graphical way of depicting the ages at which most individuals in the cohort died. Because the survivorship axis is standardized, the shape of curves across species can be compared.

The three main shapes for survivorship curves include Type I, which is convex and characteristic of populations that have high mortality in old age, typical of large mammals. Type II is a straight line, typical of organisms with constant mortality rates. They do not have a greater or lesser rate of survivorship by growing old. Type III is concave, typical of organisms with high

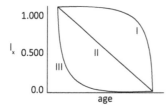

Figure 14.5 Survivorship curves have three main shapes.

mortality rates in early life, characteristics of most species that produce many offspring. Few survive to adulthood.

These categories represent extremes. In actuality, survivorship curves may vary from year to year and place to place for one species (Rockwood 2015). Often species such as humans with something close to a Type I curve have considerable mortality during the first year, but not as they would in a Type III curve.

LIFE TABLES IN ECOLOGY

One of the most important uses of life tables for non-human animals is to determine the major cause of death. The first step in that assessment is to determine at what age most mortality takes place. The second is to determine what threats occurred at that age. Key factor analysis is the name for this type of analysis.

For instance, interpret the life table of the soft-shelled crab (*Mya arenaria*) from the White Sea off the coast of Russia (Gerasimova et al. 2015). What is the maximum age for this population (Table 14.2)? Note something about the beginning age in this table. Usually, life tables begin with age 0. In this case, the offspring were planktonic and uncountable during their first year. The study began at age 1 when the clams settled into the sediment.

Next, look at the l_x column and notice the numbers are in proportions and not raw data in this case. To interpret this column, it helps to get rid of the decimal point in your head and think of these as the number out of 1,000 that were alive at the start of the age. What is the shape of the survivorship curve? Plot it.

Examine the d_x column, which provides the percent dying within the year. Between which birthdays did the greatest mortality take place? At what age did the second greatest mortality percentage take place?

Does the mortality have a pattern or do years of low mortality alternate with years of high mortality? What explains the pattern? According to Gerasimova et al. (2015), very little weather disturbance occurred at the study site. Only during the first years were predators a major factor. So, what could explain periods of mortality? Gerasimova et al. (2015) report that these organisms were sessile and therefore prone to stunting and self-thinning (see Chapter 15). This explains the mortality

Table 14.2 This is a cohort life table for the Soft-shell Clam *Mya arenaria* in the White Sea of Russia (Gerasimova et al. 2015)

Age in years	Proportion surviving at start of year l_x	Proportion dying within age interval d_x
1	1.000	[a]0.478
2	0.521	0.095
3	0.426	0.150
4	0.276	0.018
5	0.257	0.025
6	0.232	0.039
7	0.193	0.073
8	0.120	0.029
9	0.091	0.040
10	0.051	0.005
11	0.045	0.006
12	0.039	0.010
13	0.029	0.002
14	0.027	0.017
15	0.025	0.0004
16	0.008	0.0005
17	0.008	0.0004
18	0.007	0.0004
19	0.007	0.0004
20	0.006	0.0005
21	0.006	0.0004
22	0.005	0.0004
23	0.005	0.0028
24	0.002	0.0011
25	0.001	-

[a] Number derived by 1.000 − 0.521 = 0.478.

pattern in which periods of low mortality alternate with periods of higher mortality.

WHAT IS A FECUNDITY TABLE?

Rather than tracking mortality and calculating life expectancy, **fecundity tables** assess growth for the cohort, specifically as the rate of producing offspring. For monitoring rare and endangered species, life tables tell us whether a generation of the population is growing or in decline. Consider the Common Mud Turtle (*Kinosternon subrubrum*) (Figure 14.6). Measurements from the field were collected to obtain age (column 1), survivorship (column 3), and fecundity (column 4). What would a survivorship curve look like for this population of Common Mud Turtle? Plot it.

Other columns were calculated from these data. Instant inspection of the table can tell us the following (Frazer et al. 1991):

- life span was 37 years (column 1);
- most mortality occurred in the first year (column 2) when 73.9% of the population died;
- individuals reached breeding at age 4 (column 4);
- for the whole population, most offspring hatched between ages 4 and 5 (column 5) when 7.5% of the hatchlings were produced (think 75/1,000);

Age	proportion dying during age interval d_x	proportion surviving at start of year l_x	age specific fecundity m_x	annual fecundity for population $l_x m_x$
0	*0.739	1.000	0.00	0.000
1	0.125	0.261	0.00	0.000 total young produced
2	0.038	0.136	0.00	0.000 by each age group
3	0.019	0.098	0.00	0.000
4	0.010	0.079 _x_	0.96	⟶ 0.075
5	0.009	0.069 _x_	0.96	⟶ 0.066
6	0.007	0.060	0.96	⟶ 0.058
7	0.007	0.053	0.96	0.051
8	0.005	0.046	0.96	0.044
9	0.005	0.041	0.96	0.039
10	0.005	0.036	0.96	0.034
11	0.004	0.031	0.96	0.030
12	0.003	0.027	0.96	0.026
13	0.003	0.024	0.96	0.023
14	0.003	0.021	0.96	0.020
15	0.002	0.018	0.96	0.018
16	0.002	0.016	0.96	0.015
17	0.002	0.014	0.96	0.014
18	0.001	0.012	0.96	0.012
19	0.002	0.011	0.96	0.010
20	0.001	0.009	0.96	0.009
21	0.001	0.008	0.96	0.008
22	0.001	0.007	0.96	0.007
23	0.001	0.006	0.96	0.006
24	<0.001	0.006	0.96	0.005
25	0.001	0.005	0.96	0.005
26	<0.001	0.004	0.96	0.004
27	0.001	0.004	0.96	0.004
28	<0.001	0.003	0.96	0.003
29	<0.001	0.003	0.96	0.003
30	<0.001	0.003	0.96	0.002
31	0.001	0.003	0.96	0.002
32	<0.001	0.002	0.96	0.002
33	<0.001	0.002	0.96	0.002
34	0.001	0.002	0.96	0.001
35	<0.001	0.001	0.96	0.001
36	<0.001	0.001	0.96	0.001
37	-	0.001	0.96	<0.001

the number out of 1000 = 0.602 = R_0
that were reproduced

* number derived by 1.000 – 0.261 = 0.739

Figure 14.6 This is a fecundity table for the Common Mud Turtle in South Carolina (Frazer et al. 1991). Start with the idea that there are 1,000 young/year (not raw number hatched). This is a composite life table constructed from a 20-year mark-recapture study.

- females produced 0.96 hatchlings per female/year when they were at full breeding potential (column 4);

The most important thing to gain from a fecundity table is the R_0, **the generational replacement rate** (bottom right number). This is the number of hatchlings produced per female per generation. In this case the $l_x m_x$ sums to 0.602, which is less than 1.000 and therefore indicative of a declining population by 39.8% for this cohort. For this population, the pond was drying up and filling in over the course of two decades, causing the decline. Note that fecundity tables are often based on the life of females because it is usually easier to tell who the mother is than the father. Mothers can be observed laying eggs or giving birth.

WHAT ARE THE DIFFERENT TYPES OF LIFE TABLES?

Life tables that follow a cohort of individuals (**cohort life tables, horizontal life tables**) from birth until the last one dies can be extremely accurate if good assumptions are made when collecting data. However, they are difficult if not impossible to construct for long-lived species like sea turtles (that may outlive researchers) or any time researchers have only a short time to work.

Time-specific life tables (static life tables, vertical life tables) require less time. The population is sampled only once to determine the age of all individuals (see spotlight illustration in Figure 14.7). For instance, age rings could be counted on Mountain Goat horns at a hunting check station, or a cemetery could be visited once to collect age at death from all headstones, or turtles could be caught from a pond and their ages determined by counting rings on their shell.

A composite life table is a combination of cohort and time-specific life tables. It records growth for several years, but not for every year the cohort is alive. For the Common Mud Turtle, data were taken for 20 of the 37 years. For long-lived species, time-specific or composite life tables may be the best we can do or used when we want to model current populations only and project future growth. Sometimes we do not care what happened 100 years ago to the life of particular turtles. We are more interested in modeling all age groups alive now, facing current threats, with added study on particular groups like hatchlings. For the cemetery life table (Table 14.1) is this a cohort or time-specific life table? What about the *Mya* (Table 14.2) life table?

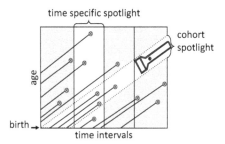

Figure 14.7 Think of time-specific and cohort life tables in terms of this spotlight diagram. Life span is represented by a diagonal solid black line. Black dots indicate death. (*Weber adapted from Begon 1996.*)

WHAT CAN WE CONCLUDE OVERALL ABOUT LIFE TABLES?

Life tables are extremely valuable for ecologists and natural historians, both for their value in modeling growth and for what they teach about natural history. They also take a great deal of devotion. In the case of the Common Mud Turtle (Frazer et al. 1991) a drift fence around the pond and pitfall traps were maintained for 10 years. The fence was patrolled by humans once or twice a day year-round (probably a muddy walk in a mosquito-prone area). Turtles were caught in the pitfall traps thousands of times, taken out of the traps, the data recorded, and each individual then placed on the other side of the fence. Females were taken back to the lab and X-rayed for a count of eggs. Some individuals were caught many times. With this much observation, researchers often learn fascinating natural history details about the species, its habitat, and animal behavior.

Only some species make good candidates for cohort life tables. The organisms must:

- be an animal (usually);
- have the ability to be marked for identification as an individual;
- have a countable number of young;
- have a large sample size (couple of 100 at least);
- have a stable age population. If age distributions are unstable, the populations change continuously and do not reflect the true nature of the population.

Plants are often not modeled well by life tables. We must be able to count individuals, which means modular organisms may not be good candidates unless we can make assumptions about our definition of an individual. Two different plants with the same genes may have different sized leaves just because of phenotypic plasticity, demonstrating different production, but not different abundance. Plants may also have several stages: dormant seeds, small rosettes, medium rosettes, or asexual reproduction early in life before they can produce seeds (Rockwood 2015).

Finally, life tables have been valuable because they have exposed shortcomings in projecting population growth. This has led to better models. Researchers methodically considered and reconsidered each assumption in data collection and analysis. Improvements led to key factor analysis and the stochastic models used now when natural history information is combined with simple mathematical equations for projecting growth, introduced next.

SIMPLE EQUATIONS FOR MODELING POPULATION GROWTH

We start by asking what it is that stops a population from growing indefinitely. Even Charles Darwin was interested in this question, asking how many offspring could theoretically arise from a single pair of a species that grew slowly, like elephants, over hundreds of years. The answer is a pair could give rise to 30 million after 1,000 years.

Yet, elephants have not taken over the Earth. Exponential growth may occur for a time, for instance when a species is introduced to a new territory, but something always stops it. It may be food shortage, weather, parasites, predators, competition, or the ivory trade, but something always stops it. Which species do you think would have the greatest chance of growing out of control and taking over the Earth? Which species do you think currently has the most individuals on Earth?

WHAT IS EXPONENTIAL AND ARITHMETIC GROWTH?

Darwin became interested in the elephant question by reading Thomas Malthus' 1798 book about human population growth. Malthus introduced the idea of exponential growth, the J-shaped curve when population growth (N) is plotted on the vertical axis and time (t) is plotted on the horizontal axis.

We can plot the shape of an exponential curve for ourselves with the following thought experiment. Suppose a bacterium divides every half hour, doubling the size of the population each time. Beginning with a single bacterium, what is the abundance of bacteria in the population at the end of 10 hours? Fill in the blank part of the function box in Table 14.3.

For the data in the function box, graph the population growth for the bacterium by plotting N (abundance) against t (time).

The J-shaped curve illustrates geometric or exponential growth, rising at an increasing rate on the Y (vertical) axis for each additional step on the X (horizontal) axis. The relationship is called **geometric growth** when growth occurs at discrete time intervals, and exponential when growth is continuous. Both types are called **exponential growth** because the growth is a function of an exponent:

$$N = 2^{20} = 1,048,576$$

The exponential equation allows us to calculate N at 10 hours (20 time steps) for our bacterium. It is known as an exponential function because it takes the form $y = q^x$ in which there is a constant base raised to a variable power.

The example provides a generalized mathematical equation for population growth:

$$N_x = (N_i)^x$$

where N_x is the abundance at the xth time step and N_i is the initial population size, which was 1 in our example.

Using this equation, what would be the N_x after 30 time steps?

Table 14.3 Fill in the blank rows of this function box for exponential growth

t (hours)	x (number of time steps)	N (pop.size)
0	0	1
0.5	1	2
1	2	4
1.5	3	8
2	4	16
2.5	5	32
3	6	64
3.5	7	128
4	8	256
10	20	1,048,576

Note that for exponential and geometric growth, the sequence of change on the Y (vertical) axis could be, but does not have to be doubling (2, 4, 8, 16, 32), for each single step on the X (horizontal) axis.

THE EQUATIONS FOR EXPONENTIAL GROWTH

Rather than counting time steps, we can write two generalized equations for the J-shaped curve (Summary Figures 14.a and b). One is exponential and uses calculus in the equation. It is appropriate for use when a population has continuous growth as in humans or microbes. The second is geometric and uses algebra in the equation. It is appropriate when a population has specific breeding seasons or intervals, thus the population grows in time steps. The two equations are:

$$\frac{dN}{dt} = rN \, (\text{calculus, exponential form}) \quad (14.1)$$

where $\frac{dN}{dt} = \frac{delta \, N}{delta \, t} = \frac{\Delta N}{\Delta t}$

with the final fraction described as "change in N over change in t."

The units are individuals/(individuals·time).

In equation (14.1) r **is instantaneous growth of the population**, also known as instantaneous rate of increase, or intrinsic rate of increase. If r is greater than zero, the population is growing. If less than zero, the population is declining. This is equal to the instantaneous birth rate minus the instantaneous death rate in very tiny time intervals. It is equivalent to the probability of a birth minus the probability of a death occurring in the population during a particular time interval,

$$= (b - d)$$

$$N_t = (1 + R)^t N_0 = \lambda^t N_0 \quad (14.2)$$

(algebraic, geometric form two terms that mean the same thing)

In equation (14.2) rate of increase for the whole population is symbolized by R or λ (lambda) (Rockwood 2015). More specifically **R is the net growth rate per generation**, also known as net reproductive rate, or discrete growth factor, or geometric growth factor. It is the constant proportion by which a population increases each time step in

a discrete model of population growth. The units for R are individuals/(individuals·time). If greater than zero, the population is growing. If less than zero, the population is declining.

It is equal to **per capita** (per individual) birth rate minus per capita death rate,

$$= (b - d)$$

In equation (14.2) λ **is the finite rate of increase**, the growth rate per time period. It is a ratio measuring the proportional change in population size from one time step to the next in a discrete model of population growth. Because it is a ratio, it is a unitless number. If greater than one, the population is growing. If less than one, the population is declining.

It is equal to per capita birth rate minus per capita death rate plus one,

$$= (b - d) + 1$$

Note that in the two equations λ and r and R are not equal, but can undergo approximate conversions through the following (and see summary box at end of chapter):

$$\lambda = e^r \quad r = \ln \lambda \quad R = e^r - 1 \quad R = \lambda - 1$$

HOW DO WE ASSIGN VALUES TO r OR R?

We could observe a population for several generations and take the mean growth rate. However, for many organisms it is difficult to record the growth for even one generation. If it is possible to complete a life table, we can calculate a growth rate. The product of a life table is R_0 (average number of female offspring produced per female over her lifetime). We could convert it as in the equation below (Donovan and Welden 2002):

$$r = \frac{\ln(R_0)}{G} \quad (14.3)$$

where G is the mean time for one generation, in other words, the mean age of reproduction (Case 2000). In humans, G is approximately 28. In 1989 the R_0 in the U.S. was 0.9096. Thus, through use of the above formula, r equals −0.0035, a decline of 0.35% per year. If a country is growing at 5%, its R_0 and λ would be 1.05 and the r would be 0 if we measured G in years.

Geometric growth

- Using *difference* equations, when growth occurs at **discreet** (regular) time intervals such as when organisms reproduce seasonally. This is the *algebraic* form of the equation.

Terms

R = net growth rate per generation
= net reproductive rate
= geometric rate of increase
= discreet growth factor

N_{t+1} = population size at time t + 1
N_t = population size at time t
N_0 = initial population size
λ = lambda, finite rate of increase, growth rate per time period
$\lambda = \dfrac{N_{t+1}}{N_t} = (1 + R)$
N_d = population size at doubling time
t_d = time to double population size

Equations

$N_{t+1} = (1 + R)N_t$
$\quad = \lambda N_t$

$\begin{cases} N_{t+2} = \lambda^2 N_t \\ = \text{growth for two intervals} \end{cases}$

This is the form of the equation that tells you only what the population size will be one step at a time into the future.

$N_t = \lambda^t N_0$

This is the form of the equation that tells you the population size after any number of time steps.

doubling time formula

$N_d = 2N_0$ and $t_d = \ln 2/\ln \lambda$

Growth occurs only during one discreet mating season each year, yet the curve still begins to take the shape of the J over time.

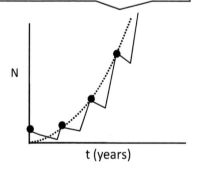

N

t (years)

Exponential growth

- Using *differential* equations when growth occurs continuously. This is the *calculus* form of the equation.

Terms

$\dfrac{dN}{dt}$ = instantaneous per capita rate of change
equivalent to "change in N over change in t"

r = instantaneous rate of increase
= intrinsic rate of increase
= per capita birth rate − per capita death rate

N_t = population size at time t
N_0 = initial population size
e = base of the natural logarithm
(e = approximately 2.718)
ln = natural log

N_d = population size at doubling time
t_d = time to double population size

Equations

$\dfrac{dN}{dt} = rN$

This is the differential form of the equation that tells you only what the instantaneous rate of change is at a given population value.

$N_t = N_0 e^{rt}$

This is the integrated form of the equation that allows you to plug in values for any time and get a population size.

doubling time formula

$t_d = \dfrac{\ln(2)}{r} = 0.69/r$

At any value of t, the slope of the line that is tangent to the curve at that point is the population growth.

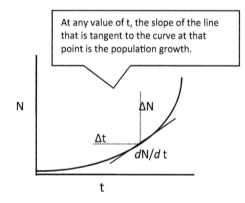

N

ΔN

Δt

dN/dt

t

Figure 14.a Summary for J-shaped curve, equations, terms, graphs. (*Adapted by Weber from Donovan and Weldon 2002, Case 2000.*)

Converting between λ and r

Note that one cannot interchange λ and r directly in the equations above as tempting as this might be. Think of it this way, if this were a bank, λ is discrete growth (annual yield) and it compounds only once a year (Case 2000). Continuous growth r (annual rate) compounds continuously. As in Figure 14-3, the λ yields a slower rate, thus we need to make it larger to be equivalent to the growth of r. The two variables must be converted from one to the other using one of the equations below. Failure to convert gives erroneous results:

$$\lambda = e^r \quad \text{or} \quad r = \ln \lambda$$

where ln stands for natural logarithm with the base e which is approximately 2.718.

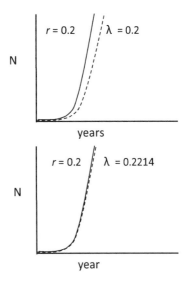

Figure 14.b (Upper right) Exponential growth and geometric growth are not equal when both r and λ=0.2 (lower panel). They are equal when λ has been converted to r. (*Adapted by Weber from Case 2000*.)

THINKING QUESTIONS

1. Consider the populations found in your neighborhood. Which ones consist of individuals of one species that just happen to live together, and which consist of individuals interacting with one another heavily?

2. Based on observations of populations where you live, are most species in dispersion patterns that are uniform, random, or clumped? Give an example of a species in each pattern.

3. From observations of a single-species bird flock in your area, identify which species it is and describe why the individuals flock. Is there a particular time of day when they are found in these flocks? If yes, what is their reason for being in that flock at that time?

4. What are the benefits and drawbacks for plants that form a monoculture? Choose a particular species of a non-native plant pest and describe its vulnerabilities.

5. Suppose that Caribou are moving south on migration. They are currently 145 kilometers from their most densely occupied summer home and are moving farther away at 8.3 kilometers per hour. Write an equation for their distance from home t hours from now, assuming their rate stays constant. Show your work. Put the answer in a box.

6. Suppose 18 pairs of Black-throated Blue Warblers per 33.4 hectares occur. What is the territory size of a pair of warblers assuming the pattern is uniform and territories are tightly packed? Show your work. Put the final answer in a box.

7. If $N_t = 400$ and $R_0 = 0.5$, what is the value of $N_t + 1$?

8. For invertebrates, fish, turtles, or small mammals, mark-recapture studies can provide important conservation insights. For instance, the size and approximate age of turtles can provide information on how well the turtle population is reproducing. If all the turtles are large and old, something must be stopping the egg laying, hatching, or survival of hatchlings.

9. Make a sound map (Shetland Nature Prescriptions Calendar 2018): Sit outside with blank paper on a clipboard. Place an X in the center to mark your location. Identify and draw on the map all sounds you hear, represented to scale by distance and volume. Determine how many populations of how many species occur.

10. Plot a species accumulation curve from data collected in 1 m² plots. If students keep a list of plant species found and compile data as a class they can keep a running total of plant species richness (total number of species, a measure of biodiversity). For instance, if plot 1 had eight species and plot 2 had a species not found in plot 1, species richness would be 9. If plot 3 found two additional new species, species richness would be 11. A species accumulation curve plots cumulative species richness on the Y axis against number of plots on the X axis. The point where the curve reaches an asymptote is the point at which species richness is estimated. It is also the point at which an adequate number of plots were taken to estimate species richness in nearby grasslands. For this method to work, students must agree on the names and ID of species, even if students need to make up names for hard to identify species.

11. Calculate a species diversity index: the Shannon Index or Simpson Index can be calculated if students record abundance of each species in their 1 m² plots. The index can be used to compare samples taken from two treatments (e.g., burned and unburned). A diversity index goes beyond species richness to incorporate species evenness (how equally abundant the species are). This is important because areas with several individuals of each species should be considered a higher quality of biodiversity than areas with the same number of species but only one of each species.

12. Volunteer at a local nature center: even if it is a one-time event, remember to reciprocate in your love of nature. Heal yourself by healing nature.

REFERENCES

Allee, W.C., A.E. Emerson, O. Park, T. Park, and K.P. Schmidt. 1949. *Principles of Animal Ecology*. W.B. Saunders Company, Philadelphia, PA.

Barbour, M.G., J.H. Burk, F.S. Gilliam, W.D. Pitts, and M.W. Schwartz. 1999. *Terrestrial Plant Ecology*. Benjamin Cummings, Menlo Park, CA.

Begon, M., M. Mortimer, and D.J. Thompson. 1996. *Population Biology*. Blackwell, Malden, MA.

Brown, J. 1995. *Macroecology*. University of Chicago Press, Chicago, IL.

Case, T.J. 2000. *An Illustrated Guide to Theoretical Ecology*. Oxford University Press, Oxford, UK.

Donovan, T.M., and C.W. Welden. 2002. *Spreadsheet Exercises in Conservation Biology and Landscape Ecology*. Sinauer, Sunderland, MA.

Frazer, N.B., J.W. Gibbons, and J.L. Greene. 1991. Life history and demography of the common mud turtle, *Kinosternon subrubrum*, in South Carolina. *Ecology* 72:2218–2231. https://doi.org/10.2307/1941572

Gerasimova, A.V., N.V. Maximovich, and N.A. Filippova. 2015. Cohort life tables for a population of Soft-shell Clam, *Mya arenaria* L. in the White Sea. *Helgoland Marine Research* 69:147–158. https://doi.org/10.1007/s10152-014-0423-2

McIntosh, R. 1985. *The Background of Ecology*. Cambridge University Press, Cambridge, UK.

Mills, S. 2013. *Conservation of Wildlife Populations*, 2e. Wiley-Blackwell, West Sussex, UK.

Rockwood, L.L. 2015. *Introduction to Population Ecology*. 2e. Blackwell Publishing, Oxford, UK.

Shetland Nature Prescriptions Calendar. 2018. https://www.healthyshetland.com/site/assets/files/1178/730-1309-17-18_nature_prescriptions_calendar_4sep.pdf

15

Population ecology's profound questions

CLASS ACTIVITY: Do Transects Really Measure Density Better Than Quadrats?

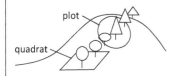

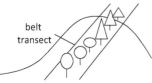

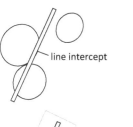

For measuring density of plants and sessile animals, standard methods use **plots, quadrats,** and **transects.**

Note that **plots** can be rectangular, square, or round. **Quadrats** are traditionally square.

Transects are long, narrow plots of three types as described below.

A **belt transect** is a long strip of terrain in which all organisms are counted and measured as if it were a long, narrow plot.

A **line intercept** is a straight, narrow line cutting across the community under study, usually dense vegetation. All organisms that intercept (touch) the line are counted.

A **line transect** is used by animal ecologists and involves traversing a line and stopping at designated intervals. Animal observations are recorded by sight, sound, or trap at each interval.

Your challenge: Elzinga et al. (2001) make the astonishing claim that measuring density of plants and sessile animals merely by using transects (long and narrow) rather than quadrats (always in a square) provides a truer mean (more accuracy) and greater precision (smaller standard deviations). True? Test the hypothesis using the following procedure.

Step 1: Choose a plant species that has a countable number of individuals and yet is somewhat abundant in a 20 m×20 m grassy field. Dandelions are appropriate, or maybe white clover flowers, or even cow pies in a pasture.

Step 2: Within the 20 m by 20 m area,

- establish **three** quadrats that are **3 m × 3 m** each, placing each in a truly random way and not overlapping.
- count the number of cow pies in each quadrat and record the density in pies/m².

Step 3: In the same 20 m×20 m area used for quadrats, lay out three **randomly** placed transects measuring
1 m × 20 m all fitting within the 20 m×20 m area and all running the same direction.

- count the number of pies in each transect and record the density in pies/m².

Step 4: Count the total number of pies in the entire 20 m×20 m area to get the real density.

Step 5: Using the same units throughout, compare the **transect** estimate of the pie density to the **quadrat** density and determine which is closer to the **real** density. Which has the smaller standard deviation? Share with classmates.

Step 6: Based on your results, make conclusions about whether your results are consistent with the Elzinga et al. (2001) hypothesis that transects are more accurate than quadrats. Include an assessment of the standard deviations. To follow up, either read the Elzinga et al. text or use other sources to explain why transects are thought to give better results.

Based on a rough eyeball analysis, do you think the individuals you counted were dispersed randomly, regularly, or clumped (aggregated)? _____ Do you think there was any kind of a gradient in how the individuals were dispersed? _____In light of the dispersion or gradient you observed, revisit your conclusion about the accuracy of transects versus quadrats. Can you now better explain why your results were or were not consistent with the Elzinga et al. hypothesis?

WHAT IS THIS CHAPTER ABOUT?

This is the story of the 100+ year effort to find mathematical models that accurately predict population growth. Population modeling may not seem like a profound topic, nor does it seem like it should have taken 100 years to develop results, yet this topic has been one of the most prominent and divisive in the history of ecology. For three decades (1920s, 1950s, 1980s), ecology careers were defined by which side individuals came down on the arguments. What provoked the acrimony?

Clearly, the need for predictive population models is vital with applications that include: controlling pests that cause crop failure, identifying epidemics caused by infectious disease, saving rare species whose existence after millions of years depends on subtle management practices, and setting fisheries regulations with billion-dollar ramifications.

Disagreement arose over the age-old question of whether a balance of nature exists, proposed in the form of the equilibrium theory of ecology. It was hoped that like natural selection, the equilibrium theory could explain how nature regulated itself in the absence of direct intervention by a deity. While natural selection had addressed regulation acting on individuals, equilibrium theory addressed regulation acting on populations and communities.

Note that ecologists do not like the phrase "balance of nature" precisely because it alludes to the pre-Darwinian idea of a benevolent deity directly controlling nature. The post-Darwinian understanding is that nature itself may have the mechanism for some level of regulation. Thus, within population ecology rather than "balance of nature" it has been "population regulation" and "equilibrium" that have been debated.

THE CHALLENGE FOR STUDENTS

The problem for students is that most chapters on population regulation in ecology texts focus so much on math that overarching questions get lost. By the end students are too bewildered to say whether populations are regulated or why it matters. For decades this has been a problem. In the 1930s, British ecologist Charles Elton wrote the following about mathematician Alfred Lotka while reviewing his book on populations:

"Like most mathematicians he takes the hopeful biologist to the edge of a pond, points out that a good swim will help his work, and then pushes him in and leaves him to drown" (Elton 1935).

THE REMEDY IS CLARITY AND CONTEXT

To be clear this chapter focuses on the following overarching questions:

1. What are the best current mathematical models for predicting population growth?
2. What factors keep populations from growing indefinitely, and do these factors actually regulate populations at an equilibrium?

Beyond that, each reader needs some context for really appreciating population ecology. Growing a little garden would help. Much can be learned by denuding a square meter of soil or even a flowerpot and watching new populations grow either from planted seed or nature's lottery. Charles Darwin himself denuded portions of his yard and watched the plots for 2 years as a crucial step in learning population ecology (Costa 2017).

Likewise, an ant farm is instructive, or watching a test tube of *Paramecium* for 24 days and observing the population drama. The resulting exponential growth seemingly comes to the brink of bursting its chamber, but alas, something always stops it. Dramatic population crashes often occur. Kids watching the ant farm assume they have done something wrong to cause the apocalypse. Parents use the opportunity to nag children about responsibility. Debates ensue over who or what is to blame, a century of acrimony in ecology in a microcosm.

In other words, applying this chapter's information is a good way to bring the subject to life. Press yourself to think of issues in your own region: why have eradication efforts for some weed or pest species been futile while populations of native species flounder even during ideal conditions? Why do certain fish species persist despite repeated heavy fishing?

SPOILER ALERT – ANSWERS TO THIS CHAPTER'S BIG QUESTIONS

What is profound about this chapter's questions? For millennia the balance idea persisted, and it is easy to see why – the biome level of nature seems to have an obvious order. Regions that host forests continue to host forests, boreal in some areas, tropical wet in others, temperate deciduous in others. Even a child can see that forests have similar characteristics across a region. Something must direct this order. What is it?

The profound answer seems to be that it is not a specific governing mechanism that is responsible. Species composition is partly a result of chance, randomness, patchiness, perhaps even chaos plays a part. These combine with factors familiar to us – symbiosis, competition, predation, weather, soil quality, and other biotic and abiotic factors. Dynamic combinations produce the result.

So now you know the spoiler, but the how and why remain for you to read. Your job as a student is to appreciate both the mathematics and the overarching principles described in this chapter. Do not give up on either. It has only been through the struggle to find accurately predictive models that the subtle truths about equilibrium, randomness, and chaos have been exposed. It is by knowing which factors move population growth off a smooth progression that population dynamics become clear. Your journey begins with understanding history from the early 20th century. The next section tells that story.

LOGISTIC GROWTH MODELS

In the 1920s, Raymond Pearl (1879–1940) at Johns Hopkins University and Hospital in Maryland recognized the shortcomings of the exponential equation for accurately predicting population growth. Pearl's interest had begun while working for Herbert Hoover after World War I (Kingsland 1995, McIntosh 1985). The main issues were invertebrate pest outbreaks that caused widespread crop damage in the U.S., and potential starvation in

Europe amidst a rapidly growing human population. The exponential equation did not have the sophistication to be of help. As an alternative Pearl put forth the logistic growth model, with its characteristic S-shaped curve. An Italian, P.F. Verhulst, had written about the logistic model in 1838, but Darwin and the early ecologists were apparently unaware, having been published in an obscure journal.

In the logistic model, rather than a J-shaped curve, populations are thought to take on **logistic growth**. At low densities, growth is exponential, starting as a J shape, then curving into a slanted S known as sigmoid (Figure 15.1). In the exponential stage, the birth rate is higher than the death rate, but as population numbers approach K, births decline, or deaths increase, or both.

According to theory, the logistic curve forms an asymptote at an **equilibrium** when the per capita death rate exactly balances the per capita birth rate. These birth and/or death rates are **density dependent**. When density is low, rapid growth occurs. When density is high, growth declines. In what became classic logistic growth theory, the equilibrium forms when N is at or near K, which came to be known as **carrying capacity**, the maximum abundance the environment can sustain over long periods. The population size may overshoot or die back while wavering around K, but net growth is zero. The population size returns to equilibrium following a disturbance. Equations for the logistic curve are more fully described in Table 15.1.

THE EQUILIBRIUM THEORY FORMALIZED IN THE 1920s

The logistic curve came to represent the equilibrium theory of population ecology (McIntosh 1985). Like laws of thermodynamics or nature organized in a great chain of being, biologists

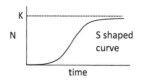

Figure 15.1 The logistic growth curve in the shape of an S. Population size is represented by N.

hoped to find an encyclopedia of explanations for order in nature. Pearl argued that the logistic curve was the next big law, extending self-organizing ideas beyond individuals to populations.

The balance of nature idea had never been quantifiable or testable, but the equilibrium theory was. Pearl hoped math could be used to show nature's way of regulating itself. This seemed like an easy prospect. No one anticipated the level of subtlety and complexity still challenging population ecologists to this day. Pearl hired Alfred Lotka in the 1920s to help. Lotka, a physicist, had worked as a demographer for an insurance company. Beyond the logistic equation, Lotka's hope was to produce mathematical equations that could predict the winner in predator-prey contests. Lotka defined r and derived the equations still used for exponential growth, as well as equations that predict predator-prey cycles.

THE EQUILIBRIUM DEBATE BEGINS IN THE 1930s

Debate over the equilibrium view began in the 1930s and lasted until at least the 1980s. One side represented the equilibrium school, led by vertebrate animal ecologists A.J. Nicholson and David Lack. Underlying their assumption was that nature had a strong ordering system. Proponents of classic logistic theory argued that certain factors, especially competition, held populations in equilibrium at carrying capacity. These factors were assumed to be density dependent. Then and now we know that density-dependent mortality must ultimately take place under crowded conditions because no population can grow indefinitely.

WHAT "REGULATES" POPULATION GROWTH?

The debate was whether populations regulated themselves around an equilibrium. **Regulation** took on a specific definition within the theory, meaning that density itself regulated growth above and below the equilibrium. The mechanisms accomplishing this came to be called **limiting factors**, pushing population levels away from exponential growth and toward carrying capacity (Figure 15.2).

Table 15.1 Equations for the logistic curve using either algebra for species that reproduce seasonally (left panel) or calculus for species that reproduce continuously (right panel)

Geometric growth	Exponential growth
• When growth occurs at **discrete** (regular) time intervals such as when organisms reproduce seasonally. Geometric growth uses an algebraic form of the equation.	• When growth occurs continuously. Exponential growth uses a calculus form of the equation.

Terms

R=geometric rate of increase
  = birth rate – death rate
N_{t+1}=population size at time t+1
N_t=population size at time t
N_0=initial population size

Terms

$\dfrac{dN}{dt}$=instantaneous rate of change in population size
r=instantaneous rate of increase
  = intrinsic rate of increase
N_t=population size at time t
N_0=initial population size
e=base of the natural logarithm
  (e=approximately 2.718)
ln=natural log

Equations

$$\Delta N_t = RN_t \; = \left(\dfrac{K - N_t}{K} \right)$$

This is the form of the equation that tells you population size after any number of time steps.

$$N_{t+1} = N_t + RN_t \left(\dfrac{K - N_t}{K} \right)$$

This is the form of the equation that provides the population size after only one time step.

Equations

$$\dfrac{dN}{dt} = rN \left(\dfrac{K - N}{K} \right)$$

• This equation is the same as for exponential growth except (K–N)/K is an additional factor that slows the growth rate.
• This additional factor bends the curve. As N gets bigger, (K–N) gets smaller, which means that (K–N)/K approaches 0.
• Thus, rN is multiplied by a smaller and smaller number as carrying capacity is approached. This means the population grows by smaller and smaller increments in the S-shaped curve.
• The above form of the equation is the differential form. It only tells you what the instantaneous rate of change is at a given population value.

$$N_t = \dfrac{K}{1 + [K - N_0]/e^{-rt}}$$

The equation just above is the integrated form, allowing you to plug in values for any time and get a population size.

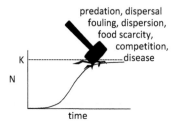

Figure 15.2 Density-dependent factors that limit population growth. The main question is whether these factors "regulate" growth at an equilibrium.

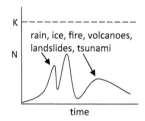

Figure 15.3 Density-independent factors affecting population growth.

From outside the population, biotic mechanisms can limit growth through interspecific competition, predation, parasitism, disease, or lack of symbiotic species. These may become more prevalent at high densities, causing growth to slow or waver near carrying capacity.

Several biotic factors are still thought to be limiting factors, driving N toward equilibrium. These include:

- **intraspecific competition**: competition for resources or aggression within a population prevents individuals from occupying a habitat or accessing a resource, thus limiting population size.
- **dispersal**: organisms may move out of their area of birth to escape crowded conditions, thus limiting population size.
- **dispersion**: distribution patterns may become uniform as territories are established in response to crowding. The uniformity leads to equilibrium. A **territory** is a space defended, with space itself as the limiting resource.
- **social behavior**: as density increases, alpha individuals may establish dominance and control the activity of the rest of the group. This occurs for wolves. Specifically, the alpha may sequester most of the mating, thus limiting population size. Instead of territories being defended, mates are defended. If a wolf pack is too large for the resources, subordinates may be driven off, all of which may maintain the population at carrying capacity.
- **self-regulation at the individual level**: an individual's own physiology may regulate population size. Queen bees under stress may signal messages to the rest of the colony through hormones. In other cases, chemical messages may suppress growth, sexual function, lactation, or affect the immune system.
- **waste and pollution**: crowded conditions may become so fouled that it poisons the population, thus limiting population size.

NON-EQUILIBRIUM SCHOOL OF THOUGHT

On the other side of the debate, Australian entomologists H.G. Andrewartha and L.C. Birch came to represent the non-equilibrium school beginning in the 1940s and 1950s (Zalucki 2015). They argued that abiotic effects (weather, geology, natural disasters, landform) were limiting factors and "determiners" of population density. "Determiner" was used instead of "regulator" because regulation implied feedback in an equilibrium.

By their argument, acts of nature like weather have an equal response in dense and sparse populations alike. For these **density-independent** factors, birth and death rates do not change in proportion to population density. They acknowledged that food limitation and other biological limiting effects exist, but rarely get a chance to operate because abiotic factors affect populations first (Figure 15.3).

According to theory, abiotic density-independent factors have the most influence on populations in temperate and arctic regions where weather is dramatic. Furthermore, biotic density-dependent factors are more influential in tropical regions where conditions are more stable.

SUPPORT AND EVIDENCE FOR DENSITY INDEPENDENCE

Andrewartha and Birch came to their conclusions by studying an insect pest in Australia called the Apple Blossom Thrip (*Thrips imaginis*). After extensive research, they determined that population density could best be predicted (with about 78% accuracy) using mainly abiotic variables in a mathematical equation (Andrewartha and Birch 1954). Rainfall, temperature, and size of the overwintering population were the main predictors.

WHICH SIDE WAS RIGHT IN THE EQUILIBRIUM DEBATE?

Both sides acknowledged that biotic and abiotic factors could influence population size. The question was whether an equilibrium tended to develop and whether density-dependent (mainly biotic) factors regulated population size around this equilibrium.

To settle the debate for ourselves, we could begin by plotting census data for familiar species to see if abundance wavers around an asymptote in a logistics curve. Take the Red-breasted Nuthatch, which occupies northern regions of N. America in coniferous forests. For the Long Island population, we could see the plot in Figure 15.4. The pattern seems more variable than we would expect, not smooth, and we do not see the expected J shape at the beginning of the curve. Upon reflection we would realize we should not expect the exponential part of the curve in this case. For an established

population like the Red-breasted Nuthatch, we would assume it were already at carrying capacity.

To see a full sigmoid growth curve, we would need to plot the growth of a newly colonized species. We could do this by studying species in laboratory or garden experiments. In the wild we could watch colonization after a disturbance or growth of an introduced species. A number of these studies were completed beginning in the 1930s.

DO WE SEE THE LOGISTIC PATTERN FOR NEWLY COLONIZED SPECIES?

Since the 1930s, essentially the sum of populations showing the logistic pattern include:

- laboratory examples: yeast (Figure 15.5 top), *Paramecium*, *Drosophila*, grain beetles, water fleas, diatoms (Allee et al. 1949), and several other examples of microbe species.
- field examples: Mute Swan population in England from 1823 to 1972 (Figure 15.5 middle), some populations of ants, bees, Tasmanian sheep, wildebeest (Figure 15.5 lower), willows, and barnacles (Rockwood 2015), elk in Yellowstone (Coughenour and Singer 1996), and Fur Seals (Kenyon and Wilke 1954).

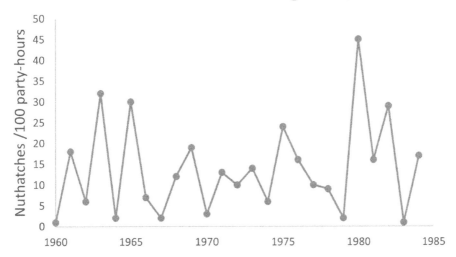

Figure 15.4 Abundance of Red-breasted Nuthatch from Long Island, NY, showing high variability. (*Data from Christmas Bird Counts 1960–1985, Yunick 1988.*)

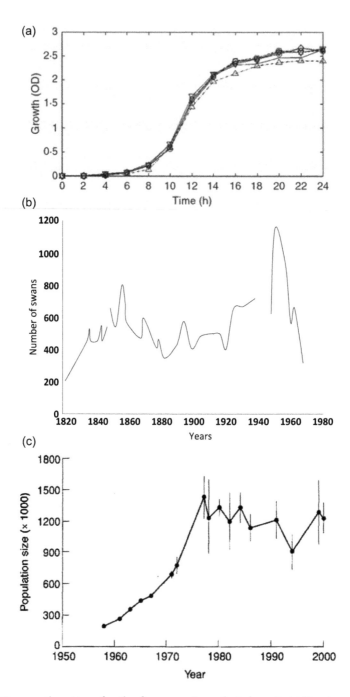

Figure 15.5 Logistic growth pattern for the few organisms that show it. (a) Yeast growth in optical density units on a spectrophotometer. (*Jomdecha and Prateepasen 2011 free access.*) (b) Mute Swan population of the River Thames, England. (*Redrawn by Weber from data in Cramp 1972.*) (c) Wildebeest of the Serengeti after rinderpest virus was eradicated. (*Mduma et al. 1999 free access.*)

For a 100-year period, this is a very short list. Despite decades of searching, our conclusion should be that the simple sigmoid curve of the logistic is rarely found except in microbes. For non-microbes we see more variability than we would expect. The sigmoid curve is the exception, not the rule. Why? What could explain this if at least part of the theory is indisputable?

Indeed, researchers commonly find evidence of density dependence, but population growth in the pattern of the logistic is rare (Fryxell et al. 2014). In other words, as conditions become crowded, it is common to find the number of offspring declining, or dispersal increasing, or mortality heightened. Still, this rarely results in an S-shaped growth curve. We can conclude that limiting factors affect population growth, but do not always lead to "regulation" of population growth at an equilibrium.

WHY NOT LOGISTIC GROWTH – ABIOTIC FACTORS

Over the rest of the 20th century, researchers sought the reasons for high variability in population growth. Some explanations include the influence of abiotic environmental factors (like weather) identified by the non-equilibrists. In the 1980s, these and other factors occurring with uncertainty were classified in a helpful way for modelers as stochasticity. Beginning in the 1940s certain other biotic factors were identified, including the Allee effect and stunting.

WHY NOT LOGISTIC GROWTH– BIOTIC FACTORS, THE ALLEE EFFECT

For some species, population growth does not follow the logistic pattern because the organisms need high densities to function. In the **Allee effect**, high population abundance enhances population growth because some aspect of group behavior is required for reproduction or survival (Allee et al. 1949). The Allee effect is an example of **positive density dependence**, which shows increasing growth as density increases. In contrast, the logistic equation is an example of **negative density dependence**, i.e., population growth diminishes as density increases.

- Examples of the Allee effect include: cooperative hunters such as lions, hyenas, pack dogs, and some predaceous fish species that are more successful at killing large-bodied organisms when in a group,
- terrestrial endotherm animals that stay warmer within a group,

- colonially nesting seabirds or shorebirds that guard against predation more effectively by nesting together,
- whale species that more readily find mates across the wide expanse of the ocean if more of them are present in a group.

The result for populations is either dense packing of individuals in one place or no individuals whatsoever. In other words, populations under the influence of the Allee effect are patchy, a sort of boom or bust existence in space.

WHY NOT LOGISTIC GROWTH – BIOTIC FACTORS, STUNTING

Another reason populations might not demonstrate logistic growth is because of stunting. In **stunting**, members of a population seemingly share resources instead of competing for them to death. Each member has less growth, but a better chance of survival.

Instead of a logistic curve the result can be high densities that occur straight from seed and stay that way for a long time, as in turf grass. Alternatively, the opposite of the logistic curve may occur – a population of stunted individuals may be dense at first, but then abundance declines exponentially and stays at low densities for a very long period with a few large individuals. (See section on self-thinning in trees later in the chapter.)

Stunting is common in annual plants when conditions are crowded, such as in a crop field. High-density treatments of potted plant experiments show it (Begon et al. 1996). The total biomass across treatments may be in equilibrium whether seeds are planted in high densities or low (Figure 15.6). However, the size of each individual is small in the crowded treatments. Only in cases of extreme crowding would competition cause mortality.

IF THE SIGMOID CURVE IS NOT TYPICAL, WHAT IS THE COMMON PATTERN?

For non-microbe organisms in natural settings, variability is common. If we had to classify typical patterns, we would identify (Begon et al. 1996, Fryxell et al. 2014):

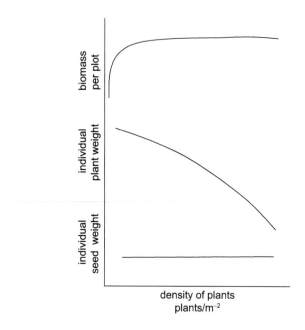

Figure 15.6 Biomass patterns for plants in increasingly crowded conditions. Note the axes labels. These are not abundance graphs. (*Summarized by Weber from Begon et al. 1996.*)

Short-eared Owl Statewide Totals

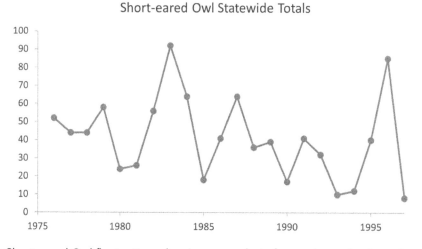

Figure 15.7 Short-eared Owl fluctuations showing somewhat of a regular cycle when observed from 1976 to 1997. (*Data from Christmas Bird Counts in New Jersey, Walsh et al. 1999.*)

- erratic growth without clear cycles as in Red-breasted Nuthatch (*Sitta canadensis*) (Figure 15.4),
- fluctuations that begin to show somewhat regular cycles as in Short-eared Owl (*Asio flammeus*) (Figure 15.7),
- boom and bust patterns, including irruptions as in Sea Lamprey, Goshawk, and Snowy Owl (*Petromyzon marinus, Accipiter gentilis, Bubo scandiacus*) (Figures 15.8–15.10), and mice species,
- exponential growth followed by steep decline with stunting of individuals and eventual self-thinning as in forested trees,
- oscillations and predator-prey cycles out of synchrony with one another as in Snowshoe Hare (*Lepus americanus*) and Canada Lynx (*Lynx canadensis*).

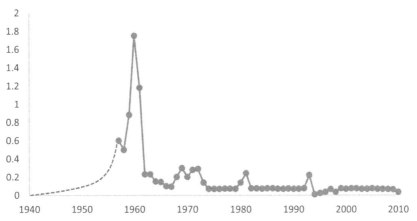

Figure 15.8 Sea Lamprey showing an irruption in abundance of adults caught in Lake Superior 1930–1966 after accidental introduction in the 1930s. (*Siefkes 2017, open access.*)

Figure 15.9 Goshawk showing 10-year irruptions between 1972 and 2001 in Duluth, Minnesota. (*Data from Hawk Ridge Nature Reserve 2002.*)

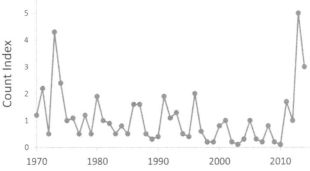

Figure 15.10 Snowy Owl showing irruptions. (*From Christmas Bird Counts and other surveys in N. America, Smith 2017, Robillard et al. 2016, Kerlinger et al. 1985.*)

EXAMPLES OF BOOM, BUST, AND IRRUPTIONS

Boom and bust patterns are one of the most common patterns, having been recorded for Sea Lamprey, Zebra Mussel (*Dreissena polymorpha*), Mute Swan (*Cygnus olar*), Snowy Owl, Goshawk, and many, many others (Figures 15.11 and 15.12). When the boom pattern is exaggerated this is called an irruption (not eruption). **Irruptions** are dramatic population explosions of large magnitude occurring occasionally, not regularly, and covering a wide geographic area. They occur only when conditions are perfect for a species that is usually at low densities.

WHAT EXPLAINS BOOM AND BUST– PHYTOPLANKTON EXAMPLE

Freshwater phytoplankton often show two abundance peaks a year, one in the spring and one in the fall (Figure 15.13). What explains this boom and bust? See if you can figure out the reason from the graph and other examples in this chapter. For

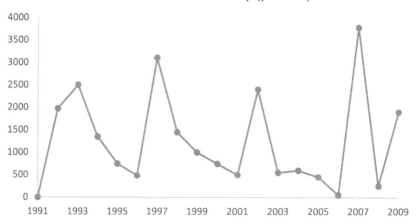

Figure 15.11 Zebra Mussel showing boom and bust in 5-year cycles (approximately) when observed on the Hudson River, New York. (*Data from Cary Institute 2018.*)

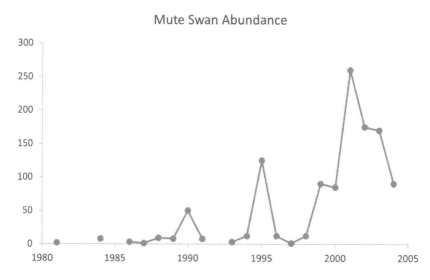

Figure 15.12 Mute Swan showing boom and bust abundance in the Detroit River. (*Data taken from Christmas Bird Counts 1981–2004, Craven 2008.*)

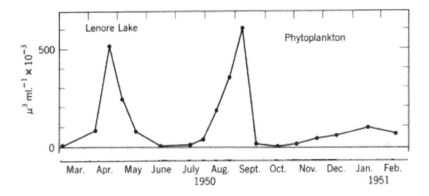

Figure 15.13 Phytoplankton showing two abundance peaks in 1 year (1950–1951) in Lenore Lake of Washington (*Anderson et al. 1955*).

phytoplankton, it may seem hard to determine at first. Phytoplankton consist of algae that drift in the water column, with populations that some-times cover vast areas of water. In that water is a soup of variables, both biotic and abiotic, that influence population growth.

Just from looking at the graph, do you have any initial hypotheses to explain the bimodal growth pattern of phytoplankton that is seasonal? Do you have an explanation for how populations can recover so effectively from extremely low abun-dances between the peaks? We will return to this example, but for possible answers, consider other situations that explain boom and bust patterns, including the very slow-motion action in trees of a temperate forest.

WHAT EXPLAINS BOOM AND BUST– SELF-THINNING IN TREES

Unlike fast growth in phytoplankton, slow popu-lation growth is characteristic of tree species that experience a winter. Seedlings establish just once a year and proliferate only when a major disturbance produces an opening in the canopy. Thus, within one stand or patch the dominant trees may be all the same age.

For this cohort of dominant trees, stunting may occur at first when the saplings are crowded (Oliver and Larson 1996). Trees in the stand grow taller and thinner than otherwise as they compete for sunlight, with few leaves in their lower trunks (Figure 15.14). Eventually the population under-goes **self-thinning** in which the weakest indi-viduals are selected out, leaving more sun for the survivors.

Figure 15.14 Self-thinning in action within a White Pine plantation, with three dead trees amidst living trees. The largest of the three dead trees had a bifurcated apical meristem, which limited its height.

For a population undergoing self-thinning, the growth pattern is not a logistic curve, but a parabola (Figure 15.15). If several generations occur within one forest as gaps in the canopy open, several parabolas can be plotted for just one tree species.

WHAT ELSE EXPLAINS BOOM AND BUST – BIOTIC EFFECTS, COMPENSATORY MORTALITY

One of the ideas that provided early insight was compensatory mortality (Ricker 1954). To under-stand this idea, think of a commonly hunted bird or mammal species, like a quail or rabbit species that has a summer breeding season followed by a cold winter. Typically, more individuals are born than the winter environment can support. Thus, a cer-tain number of young each year are **doomed sur-plus** (Errington 1956). Their death is inevitable by some means, although it does not matter whether it is in the form of starvation, disease, flood, drought, exposure, predation, or hunting, according to

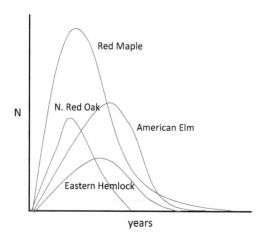

Figure 15.15 Individual cohorts of a tree species each undergoing self-thinning in a forest. Each cohort got its start as gaps in the canopy temporarily opened.

theory (Bolen and Robinson 2003). A rabbit population may have a winter carrying capacity of 30, but the population may grow to 50 during the summer. The surplus 20 die during the winter because the environment cannot support them.

a hunting mortality of 75% without decreasing the population size the next year. This is high even for an upland game species. Other studies projected that upland game species could sustain losses of 20%–40% by hunting.

While there is validity to the compensatory mortality idea, claims like the above have been generalized to include all hunted species in all places. In reality, when data sets are provided for specific populations, mortality rates change annually, among species, and among locations. The current reality for most marine and freshwater fish populations is that mortality is additive because of depletion by commercial or sport fishing. The average body size of the adults caught is often 50% less than catches 100 years ago.

The lesson to learn is that when setting harvest limits for hunting and fishing the percentage of doomed surplus must be estimated conservatively. The lesson for ecologists trying to explain boom and bust patterns is that many species overcompensate in birth rate, which makes up for high losses in death rate later in life. Hunted species, typically early succession herbivores, commonly

---

The doomed surplus strategy seems wasteful, but knowing what you do about rabbits, how might it evolve as adaptive? Provide an explanation.

In this case, when one form of mortality compensates for another, it is termed **compensatory mortality**. In contrast, **additive mortality** occurs when the additive effects of disease, exposure, predation, and other factors determine overall mortality. For instance, if food shortages and disease remove 40 individuals in a season and predation removes 40 more, the **additive mortality** is 80 individuals. This is not likely in a real population. If 40 individuals die of disease, it leaves more food for the other 40. Additive mortality tends to predominate when density-independent factors like adverse weather are responsible for mortality. Theoretically, compensatory mortality occurs when density-dependent factors dominate (Bolen and Robinson 2003).

Compensatory mortality has been used over the years to justify hunting, explaining why a certain percentage of some populations can be taken each year without causing population declines. One study found that Cottontail Rabbit could sustain

show this pattern. We will return to compensatory mortality later in this chapter.

WHY DO THE PHYTOPLANKTON POPULATIONS NOT GO LOCALLY EXTINCT – THE "LAW" OF DIMINISHING RETURNS IN HUNTING

Within the phytoplankton boom and bust growth cycles, at least some individuals continue to persist despite repeated busts. Why? How is the population able to rebound? Consider the so-called "law" of diminishing returns, another idea coming from wildlife management. It explains why populations of game species are often protected from local extinction (Connelly et al. 2012).

For example, some squirrel populations can sustain hunting losses of 40%, but populations hardly ever suffer losses that great. Hunters give up searching as population numbers decline. For species that excel at hiding and that live in expansive areas, the search time exceeds the reward value for each hunter. (This is the idea in theory. In today's reality, night vision goggles, deep sea trawl nets, and sonar make it possible for humans to catch every individual through modern methods. The idea has validity, however, when modern human methods are excluded.)

is simply a result of summer warmth and sunshine. Even in lakes without high zooplankton density, however, spring and fall spikes of phytoplankton are common. An alternative explanation is that the flush of nutrients coming from spring and fall turnover, or seasonally high inputs from streams cause phytoplankton population growth.

Lakes have herbivores like the crustacean *Daphnia* that eat the phytoplankton. Use the idea of diminishing returns as part of your hypothesis for explaining how phytoplankton can recover from frequent steep declines in abundance.

REVISITING THE PHYTOPLANKTON CASE, WHY THE BIMODAL CURVE?

To explain the spring and fall spikes in phytoplankton density in Lenore Lake of Washington, first note the abundance of zooplankton during the time phytoplankton density was low (Figure 15.16). Would phytoplankton density be high all summer if not for zooplankton that eat it? If yes, perhaps the phytoplankton growth

IN THE EQUILIBRIUM DEBATE, WHO WAS RIGHT?

Population growth curves from real species do not confirm the contention of equilibrists that populations stabilize around an equilibrium. At best, populations are weakly regulated (Ziebarth and Abbott 2010). Most non-microbe species have abundances with a range of variability, and growth tends to be boom and busty. The more interesting

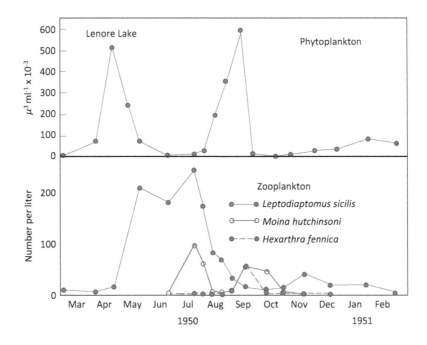

Figure 15.16 Phytoplankton showing two abundance peaks in 1 year (1950–1951) in Lenore Lake of Washington. Note the high zooplankton populations that limit phytoplankton growth in summer (*Anderson et al. 1955*).

question is, what causes the variability? Beyond abiotic factors like weather, biotic factors like the Allee effect, stunting, and self-thinning account for some of the deviations from the logistic.

There were other factors to discover beginning in the 1960s. One technique called key factor analysis was used to determine the greatest cause of death within each age group of a population (Begon et al. 1996). Using life tables, individuals in the population had to be laboriously tracked until death. For example, in a population of Colorado Potato Beetle (Begon et al. 1996, Harcourt 1971), the percentage of eggs that washed into a puddle during a hard rain was compared to the percentage of larvae preyed upon. These were compared to the percentage of adults that dispersed during migration and compared to the percentage that died during hibernation in an early frost.

The conclusion from many studies was that a complex combination of biotic and abiotic factors contributes to mortality overall in populations. Neither biotic nor abiotic factors are universally more prevalent. The highest source of mortality might change from year to year even for a single species or population (Begon et al. 1996). Still, patterns are not random. They are consistent with the r and K selected species pattern described next.

LIFE HISTORIES (K SELECTED, r SELECTED SPECIES) AND A TRUCE

A truce in the equilibrium argument was reached upon hearing one of the most important insights in the 50-year debate. This was the categorization of organisms into two life history strategies: r and K selected species. The terms came from the logistic equation, and the idea came from a tenet of MacArthur and Wilson's (1967) Island Biogeography Theory.

As further defined by Pianka (1970) and others, **r selected species** are early succession specialists. They proliferate after a disturbance (like in a patch of soil cleared for a garden, or after a fire in a forest) (Table 15.2). The rate of survival per individual is low, but reproduction and dispersal rates are high. Theoretically, abiotic factors limit their population growth (such as late season frosts). They are susceptible to early death, as in rabbits or dandelions, but breed profusely if they make it to adulthood. In contrast, **K selected species** have

Table 15.2 Characteristics of r and K species compared

r Selected species	K Selected species
Population growth explosive, in exponential pattern, below carrying capacity, not at equilibrium	Population growth slow, in logistic pattern, at carrying capacity, closer to equilibrium
Juvenile rate of development rapid	Juvenile rate of development slow
Mortality density independent	Mortality density dependent
Dispersal excellent	Dispersal fair to poor
Occurs in early stage of succession, exploit new habitat after disturbance	Occurs in late stage of succession
Short life span, less than a year	Long life span, more than a year
Body size small	Body size large
Reproductive rate high, many offspring, low survival	Reproductive rate low, few offspring, high survival
Pre-reproductive period short	Pre-reproductive period long
Parental care minimal	Parental care extensive, reproduction delayed and repeated
Mate selection not careful	Mate selection careful
Often communal, heavily preyed upon	Often territorial, competitive
Seeds do not have excess stored nutrients	Seeds have abundant excess stored nutrients, provide seedlings with head start
Tend to be semelparous, but much variation	Tend to be iteroparous, but much variation

Source: Summarized from Pianka (1970) and expounded upon by Rockwood (2015) and others.

high survival per individual, but low rates of reproduction and dispersal. They replace r selected species in late stages of succession when competition is high. Theoretically, biotic factors mainly limit their population size (such as predation). In other words, they live a long time, but breed slowly as in elephants or sequoias.

Additional characteristics of r selected animal species include **precocial young**, offspring that take care of themselves right after birth with little parental care. K selected species produce **altricial young**, which need abundant parental care after birth.

In plants, r selected species tend to have many small seeds that disperse widely. K selected tree species may not produce seeds until 6–12 years old. Once produced, the seeds of K selected species are large and heavy, falling near the parent tree. Although not dispersed widely, the seedlings have a nutrient-rich bed for growth under the parent plant.

In conclusion, r selected species have more of a boom and bust existence. K selected species theoretically have population growth closer to the logistic pattern. The r and K hypothesis vindicated both sides depending on which life history strategy were one's study organism. The equilibrists had mainly studied warm-blooded vertebrates (birds and mammals) in late succession situations. Density dependence and the effects of biotic influences in K selected species were the tendencies. The non-equilibrists had mainly studied insects, such as Apple Blossom Thrips and grasshopper species. They saw more tendencies toward boom periods after disturbances from abiotic events.

In essence, the r selected species concept combined with compensatory mortality explained why so many species have boom and bust growth; r selected species have overcompensation in birth rate, making up for high losses in death rate later in life.

THE ITEROPARITY/SEMELPARITY PHENOMENON AND ITS SIGNIFICANCE

Later studies showed that r selected animal species tend to have **semelparity** (one-time breeding toward the end of life), whereas K selected tend to show **iteroparity** (repeated breeding throughout life) (Figure 15.17) although many exceptions occur.

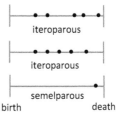

Figure 15.17 Iteroparous species breed several times per lifetime. Semelparous species breed once and then die.

Semelparity, "big bang" reproduction is exemplified in dragonflies, mayflies, and periodic cicadas (Gotelli 2008, Rockwood 2015). They have long lives for invertebrates but breed once just before death (Rockwood 2015). The juvenile stage is often underwater or underground for years but lasts only a few days or weeks during adulthood. Survivorship is high in juveniles, but as adults, breeding makes them vulnerable to predation. Adult numbers irrupt in swarms great enough to satiate (satisfy) the predators. Only some survive long enough to lay eggs. Those that do lay many eggs that hatch into juveniles. The juveniles once again begin a long stable period of growth until it is time to swarm as adults.

Species show semelparity when they have high mortality in adulthood, but low mortality for juveniles. Species show iteroparity when they spread out reproduction over several years – not putting all eggs in one basket. For example, turtle adults have long lives because they have few predators when in the water, even though turtle eggs suffer chronically high predation (Frazer et al. 1991).

PROBLEMS WITH THE r AND K HYPOTHESIS

The r and K selected species hypothesis was a major idea, leading to the development of many further concepts, and providing decades of testable hypotheses. Beyond ecology, animal behaviorists took up these ideas including the notion of budgeted energy within a trade-off. Despite these achievements, the concept has several problems. MacArthur and Wilson's original premise was that species showed r selected traits in uncrowded situations and K selected in crowded areas. In other words, one species could switch categories depending on population density. Studies in the mid-20th century on fruit flies, protozoa, and others showed that species do not switch based on crowding. For this reason and others (Gotelli 2008, Rockwood 2015), the theory behind the hypothesis has been abandoned even if the vocabulary is still used.

Another problem is that for plants, some species do not fit neatly in either category. Like turtles, tree species may have long lives and sequential reproduction periods like K species, but high juvenile mortality like r species. Think of long-lived tree species like oak that put out multitudes of acorns each year. Other plant species are adapted for chronic stress rather than early succession r or late succession K species. Phillip Grime (1977) proposed a three-pronged approach as an alternative. He added the additional category of stress, characteristic of species in poor environments such as arctic, alpine, arid, shaded, or nutrient-deficient areas. They have slow growth rates, long-lived leaves, and high rates of nutrient retention. Examples include mosses, lichens, succulents, and evergreen needles.

Inclusion of the additional category was not the only contribution of the Grime model. The r, K, and stress categories were to be envisioned as placements on three points of a triangle (Figure 15.18). Each plant species was meant to have a place on a spectrum within the triangle rather than being lumped into a single group. Grimes' model made predictions about the winners against other species competing in natural habitats, dependent on size of individuals competing.

ADVANCES IN MODELING: TIME LAGS, STAGE STRUCTURE, LESLIE MATRICES

By the late 1970s, it was evident that the simple logistic model was a highly convenient theoretical model, but not a law of population growth. Using the logistic equation as the base, modelers added a variety of algebraic features that greatly improved realism for predicting growth. A library of modifications to the logistic model is now available specific to circumstances, species type, and data collection method (see Dinsmore and Johnson 2012 for a summary).

Theoretical ecologist Robert May and others introduced modifications that accounted for delays and time lags (May and Oster 1976). Species could be accommodated like elephants or oak trees that require a decade to become reproductively mature. The Greek letter tau (τ) was used to represent the carrying capacity based on the population size Tau-time units in the past (Rockwood 2015). The product of r and T represented growth patterns with large deviations from equilibrium, in other words, boom and bust patterns that accurately depicted reality.

Recognition of time lags made it obvious that birth and death rates were dependent on age, sex, and genetic makeup. In other words, for accuracy in prediction, ages (and stages) should be considered separately. Individuals that are

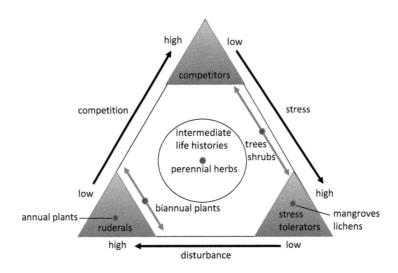

Figure 15.18 Grime's C-R-S triangle, a model depicting three life strategies commonly used to categorize plants based on the environment. C-R-S is competitors, ruderals, and stress tolerators.

too young or too old do not breed at all, thus a single continuous growth estimate for r that represents every individual is unrealistic unless it models single-celled bacteria or protista (Gotelli 2008).

WHAT DOES STAGE STRUCTURE MEAN?

It is not purely age that categorizes breeding groups. Individuals of several ages can make up a life-cycle stage. For instance, teenage humans between the ages of 11 and 18 might be capable of producing offspring, but not at the same rate as the group just older. Thus, the 11- to 18-year-old group would be placed in their own stage. Combining ages into breeding groups make up what are known as **stage-structured population models**.

MATRIX PROJECTION MODELS

To actually put age and stage structure into computer models, George Leslie introduced the idea of matrix algebra to calculate population growth for age groups or stage groups separately (Figure 15.19). In this system vital rates are kept in a **matrix**, a rectangular array of numbers, with rows and columns to keep straight the book-keeping of birth and survival. These **matrix projection methods** can start with any number of individuals in different stages or ages and keep track of relative numbers as well as population growth over time. The result is the computation of λ (lambda), the finite rate of increase for a population via the lambda for each stage or age (Donoman and Weldon 2002).

THE CURIOUS PHENOMENON OF CHAOS

In the 1970s Robert May introduced the idea of chaos and its possible effects within ecology (May 1974, Coulson and Godfray 2007). May discovered that even without biotic or abiotic influences, the equation for the logistic could produce oscillations around K if the population had a high enough r. In other words, the simple discrete form of the logistic model could show complex effects without biological causation. These patterns were simply a property of mathematics. With a high enough r, the growth trajectory could even fall into chaos.

Chaos refers to fluctuations that do not repeat themselves and are highly sensitive to initial conditions (Figure 15.20). Chaos ensues when initial conditions get amplified over time. The fluctuations in chaos are not random even though the dynamics may look random because of the non-repeating pattern. Unlike randomness, the same chaotic fluctuations occur each time if the same inputs go into the logistic equation.

This seemed like a very big discovery during the 1970s. Chaos had the potential to explain why population growth was erratic rather than moving in a smooth projection toward carrying capacity. It effectively tempered the equilibrium debate for many population biologists. If math alone could explain variation in growth, it seemed time to move to a new topic. Topics like metapopulations, patch dynamics, interspecific interactions, and the emergence of conservation biology came to the forefront as population ecology moved into the 1980s.

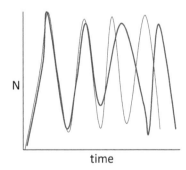

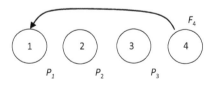

Figure 15.19 Four ages depicted in an age structure model. Only age four can reproduce. P values represent the probabilities of survival to the next age. F represents the fecundity (birth) rate for the fourth age group.

Figure 15.20 The equations for two lines may have slightly different numbers, resulting in the trajectory of the second line to veer off the path of the first and slide into chaos.

It now appears that chaotic dynamics are rare in ecology. Only organisms with very high R_O values would show it (Figure 15.21). Small-bodied organisms with high population growth after a disturbance might be capable of it, as shown in laboratory studies with insects (Coulson and Godfray 2007). Outside the laboratory only viruses and bacteria may be able to reproduce that fast.

FURTHER THEORETICAL IDEAS – METAPOPULATIONS

The premise of metapopulations was put forth by Richard Levins in 1969. This idea contributed to further tempering of the equilibrium debate and caused an entire paradigm shift in population ecology. Before these ideas, natural populations were assumed to be numerous, widespread, and contiguous, an unrealistic scenario even before humans fragmented the environment. In reality, wild populations are often small and somewhat isolated. Thus, a **metapopulation** is defined as a group of subpopulations linked by dispersal (Figure 15.22). We might think of these subgroups as colonies. For example, mice living in a collection of barns and outbuildings on one farm are separate populations but linked if mice regularly disperse among the structures. Patch models were a further idea that could be related to metapopulations. Any area where birth rates exceeded death rates was a source patch. Any area where death rates exceeded birth rates was a sink patch. The sink patch could become inhabited by immigrants from source patches. The idea of metapopulations has become one of the most insightful in ecology.

Once metapopulations were recognized, the concept fundamentally changed population ecology analysis and modeling, providing more accuracy and realism. The concept vastly changed the idea of equilibrium and carrying capacity. It had been assumed that carrying capacity and equilibrium were the same, and at carrying capacity, all habitable places were occupied.

After metapopulations were proposed it became obvious that habitable places were regularly not filled even when the overall population was at carrying capacity. Some previously occupied areas "blinked out" for a time as part of regular changes in dispersal, chance, weather, or predation. These areas filled later while others blinked out. This is another idea that explains why populations can sustain themselves despite repeated losses. They "rescue" themselves from low population numbers by emerging from small source subpopulations that escape epidemics and catastrophes. The metapopulation idea was so useful that it quickly merged with ideas from patch dynamics, island biogeography theory, and landscape ecology (Rockwood 2015).

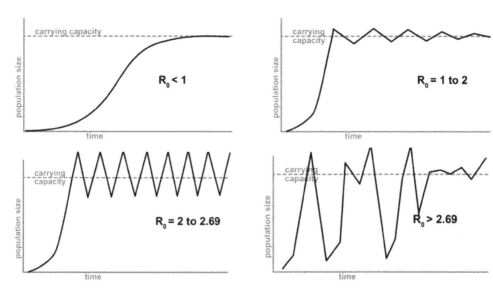

Figure 15.21 Population growth from the discrete form of the logistic equation eventually leading to chaos as R_0 grows (*Mackenzie et al. 2001*).

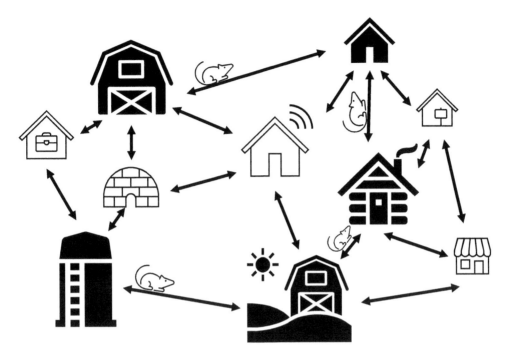

Figure 15.22 In Levins' metapopulation model, patches are either occupied or unoccupied, with subpopulations linked by dispersal. The dark structures are currently occupied by mice subpopulations. The white structures are currently unoccupied but have the capacity to be occupied.

WHAT ARE STOCHASTIC MODELS?

The idea of stochasticity was a further major advancement in population modeling. It was introduced to population ecology in the 1930s, then again in the 1970s, and went mainstream by the end of the 1980s. Stochasticity may have been the single most important idea adding reality to population models. Using the logistic equation as the base, modelers had already added time lags, age structure, and stage structure to improve models. Problematically, this still overestimated population abundance and drew a growth curve that was too smooth. With stochasticity, the models depicted variability more realistically, especially if they were under the influence of abiotic factors like weather, which could be incorporated.

Stochasticity means uncertainty or element of variation. The inclusion of stochasticity allows the inclusion of probability. The initial reaction by students to stochastic modeling is usually mistrust and skepticism, but use of stochasticity in weather forecasting is so common we do not recognize it as such. Weather forecasting reports the probability of rain for tomorrow, rather than a yes or no

prediction of rain. It tells us the degree of weather severity predicted, light showers or heavy thunderstorms. We understand that these forecasts are predictions, but we trust them because of their basis in historic records and real-time data gathering. Stochastic population models use these same principles and have the same advantages as in weather forecasting compared to deterministic models.

Deterministic models are equations that arrive at the same solution every time, without varying by chance. The base logistic equation is deterministic. Life tables are deterministic. If we input the same data each time, we get the same answer each time.

Stochastic models add probability of an outcome based on known events and historic records. We may input the same data each time, but we get a different answer each time. One of the ways stochastic models do this is by considering the fate of each individual in a hypothetical population. At each time step a random number between 0 and 1 can be assigned to each individual, which determines whether that individual survives or dies, for example, from environmental effects.

Say we know that the chance of dying during winter is 50% on average. Instead of applying 50% to the entire cohort as we did in life tables, we can program the computer to assign survival or death by individual. Survival is indicated if an assigned random number is greater than 0.5. Death is indicated if less than or equal to 0.5. For a stochastic model with a population of ten individuals, sometimes seven survive, sometimes three, two, nine, or ten, etc. We can run the projection 100 times and calculate a probability of extinction. Use of this technique has repeatedly shown that stochastic models more accurately predict the variability we see in real populations compared to the simple logistic.

Within stochastic computer models, a variation of the logistic equation is often included called the Ricker equation. Basically, this fixes the model so that any time abundance slips to zero, extinction is declared, which makes sense. In the simple logistic equation, abundance could go below zero, but then "recover" and continue growing, an impossible scenario.

STOCHASTIC MODELING FOR SMALL POPULATIONS – ENDANGERED SPECIES MANAGEMENT

To determine the probability of extinction for small populations of rare species (as in endangered species management), modelers developed a type of computer model called a population viability analysis (PVA). As developed by Shaffer (1990) the idea is that small populations tend to fall into an **extinction vortex** as their population declines (**vortex** referring to a positive feedback loop, a downward spiral, or snowballing effect).

Four factors were identified as the major problems bringing down small populations: environmental, demographic, genetic, and catastrophe problems (Figure 15.23) (Shaffer 1990). These could be modeled stochastically within population viability analyses using software by the trade name of "Vortex" or "Ramas."

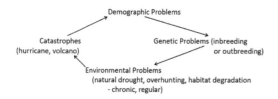

Figure 15.23 An extinction vortex with its four components.

Demographic (population) problems include skewed age or sex ratios. For instance, the last ten Sumatran Rhinoceros from a population in Malaysia were brought into captivity in a desperate attempt at breeding (Ahmad et al. 2013). Only one healthy male was found, and his capture was 10 years after the first female was caught. Breeding was not successful, and the population was extirpated.

Demographic problems can arise if all individuals are too old to breed. Alternatively, individuals could be victims of the Allee effect, for instance in the case of Whooping Crane. An attempt was made in the 1980s to introduce a population of Whooping Crane to the Rocky Mountains. The attempt failed because the adults were too spread out from one another to mate once they arrived in Idaho after spring migration.

Note that demographic problems are only an issue in very small populations. If 95% of 10,000 frogs are lost, 500 remain and the chance of them all being one gender is low. If 95% of 100 are lost, only five remain and the chance of all being one gender is high. As a rule, demographic problems are only an issue for abundances in the 10–100 range.

Environmental problems are chronic, regular issues rather than catastrophes. Natural weather variations, pathogens, parasites, competition, predation, and habitat degradation are examples. A large population can withstand this stress, but small populations cannot.

Catastrophes are acute, one-time disasters at a larger and more devastating scale than environmental problems. They are capable of destroying a population. Examples include disease epidemics, earthquakes, hurricanes, volcanoes, tsunami, and fires. Do these affect populations of all sizes? Yes, but problems can be alleviated by locating populations in more than one geographic area so that eggs are not in one basket.

Genetic problems include genetic drift, inbreeding, outbreeding, and the founder effect. One of the problems with inbreeding is that individuals can essentially become clones and therefore more susceptible to disease. Just because inbreeding occurs, it does not necessarily mean that **inbreeding depression** occurs (low fertility among offspring, high mortality among offspring, and physical deformities). Inbreeding depression only happens after repeated breeding among relatives.

In contrast, **outbreeding depression** is the mating of individuals from distant and different populations. The progeny have reduced "vigor," meaning they have less resistance to disease, slower growth, and less survival.

Despite the damaging effects that genetic problems can cause in some situations, conservation biologists have realized that the effects of weather and catastrophes are usually greater for small populations. Thus, the most effective conservation management is usually to protect against abiotic factors.

A SUMMARY OF CURRENT POPULATION MODELS – AND A WISH LIST

Modeling of population growth began with the idea of exponential and logistic equations for populations. The most popular models are still based on the logistic but have evermore realism because of added features. Some of the best have been worked out for endangered species management within an extinction vortex. We would like even better models. The following compilation includes both existing features and others the modelers would like to include (Mills 2012):

- time lags – growth occurs in spurts and sometimes complex cycles, and sometimes the reproductive rate depends not on the density of the same generation (t), but on the density in the last generation (t − 1),
- stochastic rather than deterministic models to incorporate weather, catastrophes, genetic problems, and demographic issues,
- stage and age structure,
- immigration and emigration,
- metapopulations and patchiness,
- special circumstances for disease and epidemics,
- multispecies interactions.

ARE THERE ANY GOOD MODELS NOT BASED ON THE LOGISTIC?

Occupancy models are based not on population abundance but on presence-absence data. Data collection is easier than for the logistic because each data point consists only of whether an individual is present rather than the more difficult task of counting individuals and determining per capita growth rates. For a rare species that is virtually unknown to science, completion of a population viability analysis is not possible. Information about its threats and basic biology may be unknown. Additionally, there may not be time for population growth to be measured. Occupancy models may provide a quick and easy alternative compared to PVAs.

For instance, if geographic range can be reliably assessed, such as from citizen science reports, and the range is shrinking, the rate of decline can be estimated. This is potentially an easier prospect than using a model based on probabilities and stochasticity (Fryxell et al. 2014).

WHAT IS THE CURRENT UNDERSTANDING IN THE EQUILIBRIUM/NON-EQUILIBRIUM DEBATE?

For non-microbe species, it now seems evident that:

- most populations have variability in their growth without a distinct carrying capacity. At best, carrying capacity falls within a range of values rather than a strict equilibrium;
- boom and bust growth patterns are common and explained by the Allee effect, stunting, compensatory mortality, abiotic effects, and life history characteristics categorized into r, K, and stress strategies;
- ideas about metapopulations and diminishing returns reveal how populations can sustain themselves despite dramatic declines in growth at times;
- for more accuracy in mathematical modeling, new features can be added to the logistic equation. These account for time lags, delays, growth rates defined by ages and stages, stochasticity, and better knowledge and modeling of what causes mortality in small populations.

By the late 1980s, most ecologists were ready to accept a mainly non-equilibrium understanding of population ecology. This is the idea that few macroscopic species are at a self-maintained equilibrium. It is no longer assumed there is a balance of nature, or that population growth necessarily results in a stable equilibrium. Most populations and communities are in a continuing state of change, caused by chance and the dynamics of weather, climate, and intra- and interspecific effects by organisms. Parasites and other organisms can cause radical changes, especially when affecting key species.

Further research in population ecology and a debate in the 1980s about interspecific competition eventually led to the neutral theory. More will be written about the competition debate and the neutral theory in Chapter 17. In brief, the neutral theory approaches any ecological situation with the premise of no equilibrium, then determines if the accompanying data show otherwise. This approach applies to all ecological interactions, not just population growth. For instance, instead of assuming that predators and prey affect one another because that is what past theory suggests, the neutralists advocate an expectation of no interaction, then make a determination based on whether this hypothesis is violated. Interestingly, equilibrium states have been found for populations, but in small patches for short periods. Biotic processes as hypothesized by the equilibrists may indeed be at work as predicted, but in a constantly shifting mosaic of small patches.

The equilibrium debate in population ecology diminished in the 1980s as practitioners began to turn toward conservation biology, the growth of rare species, and other practical applications. The equilibrium debate, however, migrated to community ecology, where it became an argument about interspecific interactions. Chapter 17 further describes this development of the equilibrium debate.

THINKING QUESTIONS

1. Why is the model named "logistic?" Use the internet to find out.
2. Find examples of population growth for other species. Data from Christmas Bird Counts are available via internet. How might data for bird species collected during annual Audubon Club counts be somewhat biased?
3. For the examples in this chapter that show boom and bust or erratic growth patterns, such as Red-breasted Nuthatch, examine the diet, life histories, and habitat requirements to speculate on the immediate causes of variation from year to year.
4. How can chance, randomness, patchiness, perhaps even chaos be responsible for patterns we see in nature like the formation of biomes?
5. Mayflies have an adult emergence period of only a few days (earning the apt name Order Ephemeroptera). So many adults swarm at once that snowplow equipment has been used to remove piles of cast juvenile skins and dead adults from the streets of waterfront towns. Likewise, 17-year cicada adults can be so numerous that dogs and other animals get sick from eating too many. Describe situations like this you have experienced. Does heavy predation on cicadas still leave enough adults to lay eggs and renew the cycle?
6. Fox (2011) has relegated the r and K selected species idea to the list of *Zombie Ideas in Ecology* – ideas now hard to kill. Widely believed and intuitively appealing, they tend to persist, according to Fox, in spite of repeated theoretical refutations and whole piles of contrary facts. In his words, they are undead. As a solution, Read and Harvey (1989) proposed "fast life histories" to describe early, fast reproduction and "slow life histories" to describe delayed reproduction. What has been given up, what is gained, and what is lost in these new descriptors?
7. What other zombie questions persist in ecology, whose untruths remain despite repeated demonstration of their fallacy? Why do people often prefer falsehood over truth?

8. Research the pros and cons of population viability analyses. What is the meaning of the GIGO principle when using mathematical models? What is sensitivity analysis and how does it relate to population viability analyses? In the end, when mathematical models are used for endangered species management, is this better than a wild guess? Do computer models have to be perfect to be helpful?
9. Find examples of microbe species that actually have logistic growth.
10. Provide an example of the Allee effect besides those described in this chapter, either an example you know or an example found on the internet.
11. What are other examples of irruptions in wild populations besides those in this chapter? Use the internet.
12. Have recent examples of chaos been observed in wild populations?
13. Briefly summarize the outcome of the population regulation debate.
14. Besides flowerpot experiments, garden or lawn denuding studies are insightful. Darwin found 20 plant species over the course of the growing season in his lawn plot. How many are in your lawn, even if not denuded?

REFERENCES

Ahmad, A.H., J. Payne, and Z.Z. Zainudin. 2013. Preventing the extinction of the Sumatran rhinoceros. *Journal of Indonesian Natural History* 1:11–22.

Allee, W.C., A.E. Emerson, O. Park, T. Park, and K.P. Schmidt. 1949. *Principles of Animal Ecology*. W.B. Saunders Company, Philadelphia and London.

Anderson, G.C., G.W. Comita, and V. Engstrom-Heg. 1955. A note on the phytoplankton-zooplankton relationships in lakes of Washington. *Ecology* 36:757–759.

Andrewartha, H.G., and C. Birch. 1954. *The Abundance of Animals*. University of Chicago Press, Chicago, IL.

Begon, M., M. Mortimer, and D.J. Thompson. 1996. *Population Biology*. Blackwell, Malden, MA.

Bolen, E.G., and W.L. Robinson. 2003. *Wildlife Ecology and Management*. 5e. Prentice Hall, Hoboken, NJ.

Cary Institute of Ecosystem Studies. 2018. Zebra Mussels and the Food Web, Lesson 2 Answer Key. Caryinstitute.org, Millbrook, NY.

Connelly, J.W., J.H. Gammonley, and T.W. Keegan. 2012. Harvest management. In: N.J. Silvy (ed.), *The Wildlife Techniques Manual*. 7e, Volume 2. The Johns Hopkins University Press, Baltimore, MD.

Costa, J.T. 2017. *Darwin's Backyard; How Small Experiments Let to a Big Theory*. W.W. Norton and Company, New York.

Coulson, T., and H.C.J. Godfray. 2007. Single-species dynamics. In: R.M. May and A.R. McLean (eds.), *Theoretical Ecology: Principles and Applications*. Oxford University Press, New York.

Coughenour, M.B., and F.J. Singer 1996. Elk population processes in Yellowstone National Park under the policy of natural regulation. *Ecological Applications* 6:573–593. https://doi.org/10.2307/2269393

Cramp, S. 1972. One hundred and fifty years of Mute Swans on the Thames. Wildfowl 23.

Craven, J. 2008. Malnourished waterfowl dying in Michigan-Ontario. *Bootstrap Analysis: Chronicles and Musings of an Urban Field Ecologist*. bootstrap-analysis.com/2008/03/malnourished-wa.html

Dinsmore, S.J., and D.H. Johnson. 2012. Population analysis in wildlife biology. In: N.J. Silvy (ed.), *The Wildlife Techniques Manual*. 7e, Volume 2. The Johns Hopkins University Press, Baltimore, MD.

Donovan, T.M., and C.W. Welden. 2002. *Spreadsheet Exercises in Conservation Biology and Landscape Ecology*. Sinauer, Sunderland, MA.

Elton, C. 1935. Eppur Si Muove. Review of theorie analytique des associations biologiques by A.J. Lotka. *Journal of Animal Ecology* 4:148–150.

Elzinga, C.L., D.W. Salzer, J.W. Willoughby, and J.P. Gibbs. 2001. *Monitoring Plant and Animal Populations*. Wiley-Blackwell, Malden, MA.

Errington, P.L. 1956. On the hazards of over-emphasizing numerical fluctuations in studies of "cyclic" phenomena in Muskrat populations. *The Journal of Wildlife Management* 18:66–90. https://doi.org/10.2307/3797617

Fox, J. 2011. Zombie ideas in ecology. *Oikos Blog June 17, 2011*. https://oikos-journal.wordpress.com/2011/06/17/zombie-ideas-in-ecology/

Frazer, N.B., J. Whitfield Gibbons, and J.L. Greene. 1991. Life history and demography of the Common Mud Turtle *Kinosternon subrubrum* in South Carolina, USA. *Ecology* 72:2218–2231.

Fryxell, J.M., A.R.E. Sinclair, and G. Caughley. 2014. *Wildlife Ecology, Conservation, and Management*. 3e. Wiley-Blackwell, West Sussex, UK.

Gotelli, N.J. 2008. *A Primer of Ecology*. Sinauer Associates, Sunderland, MA.

Grime, J.P. 1977. Evidence for the existence of three primary strategies in plants and its relevance to ecological and evolutionary theory. *American Naturalist* 111:1169–1194. https://doi.org/10.1086/283244

Harcourt, D.G. 1971. Population dynamics of *Leptinotarsa decemlineata* (Say) in eastern Ontario. III. Major population processes. *Canadian Entomologist* 103:1049–1061.

Hawk Ridge Nature Reserve. 2002. Annual report, License: Public Domain.

Jomdecha, C., and A. Prateepasen. 2011. Growth of yeast. *Letters in Applied Microbiology* 52:62–201.

Kenyon, K.W., and F. Wilke. 1954. Migration of the Northern Fur Seal *Callorhinus ursinus*. *Journal of Mammalogy* 34:86–98.

Kerlinger, P., M. Ross Lein, and B.J. Sevick. 1985. Distribution and population fluctuations of wintering Snowy Owls (*Nyctea scandiaca*) in North America. *Canadian Journal of Zoology* 63:1829–1834.

Kingsland, S.E. 1995. *Modeling Nature: Episodes in the History of Population Ecology*. 2e. The University of Chicago Press, Chicago, IL.

MacArthur, R., and E.O. Wilson. 1967. *The Theory of Island Biogeography*. Princeton University Press, Princeton, NJ.

MacKenzie, A., A.S. Ball, and S.R. Virdee. 2001. *Instant Notes Ecology*. 2e. Bios Scientific Publishers Limited, Oxford, UK.

May, R.M. 1974. Biological populations with nonoverlapping generations: stable points, stable cycles, and chaos. *Science* 186:645–647. doi: 10.1126/science.186.4164.645

May, R.M., and G.F. Oster. 1976. Bifurcations and dynamic complexity in simple ecological models. *The American Naturalist* 110:573–599. https://doi.org/10.1086/283092

McIntosh, R. 1985. *The Background of Ecology*. Cambridge University Press, New York and London.

Mduma, S.A.R., A.R.E. Sinclair, and R. Hilborn. 1999. Food regulates the Serengeti Wildebeest population: a 40 year record. *Journal of Animal Ecology* 68:1101–1122.

Mills, S. 2012. *Conservation of Wildlife Populations*. 2e. Wiley-Blackwell, West Sussex, UK.

Oliver, C.D., and B.C. Larson. 1996. *Forest Stand Dynamics*. John Wiley & Sons, Inc., Oxford.

Pianka, E. 1970. On r and K selection. *American Naturalist* 104:592–597.

Read, A.F. and P.H. Harvey. 1989. Life-history differences among the eutherian radiations. *Journal of Zoology*. 219:329–353.

Ricker, W.E. 1954. Effects of compensatory mortality upon population abundance. *Journal of Wildlife Management* 18:45–51.

Robillard, A., J.F. Therrien, G. Gauthier, K.M. Clark, and J. Bety. 2016. Pulsed resources at tundra breeding sites affect winter irruptions at temperate latitudes of a top predator, the Snowy Owl. *Oecologia* 181:423–433.

Rockwood, L.L. 2015. *Introduction to Population Ecology*. 2e. Blackwell Publishing, Oxford, UK.

Shaffer, M.L. 1990. Population viability analysis. *Conservation Biology* 4:39–40.

Siefkes, M.J. 2017. Use of physiological knowledge to control the invasive Sea Lamprey (*Petromyzon marinus*) in the Laurentian Great Lakes. *Conservation Physiology* 5:cox031. https://doi.org/10.1093/conphys/cox031

Smith, J. 2017. What northern bird species will show up at your feeder this year? Cool green science. *Science by Nature*. https://blog.nature.org/science/2017/01/18/

Walsh, J., V. Elia, R. Kane, and T. Halliwell. 1999. *Birds of New Jersey*. New Jersey Audubon Society, Bernardsville, NJ.

Yunick, R.P. 1988. An assessment of the White-breasted Nuthatch and Red-breasted Nuthatch on recent New York State Christmas Counts. *Kingbird* 38:95–104.

Zalucki, M.P. 2015. From natural history to continental scale perspectives: an overview of contributions by Australian entomologists to applied ecology – a play in three acts. *Austral Entomology* 54:231–245. https://doi.org/10.1111/aen.12156

Ziebarth, N.L., and K.C. Abbott. 2010. Weak population regulation in ecological time series. *Ecology Letters* 13:21–31. https://doi.org/10.1111/j.1461-0248.2009.01393.x

16

Community ecology basics

CLASS ACTIVITY: How to Measure Species Diversity, a Primary Application in Community Ecology

Goal: Use diversity indices to measure species diversity (and evenness, the degree of equitability in the distribution of individuals among species).

Diversity indices measure species diversity and evenness. **Species diversity** refers to species richness and endemism, with **species richness** referring to number of species. **Evenness** is the degree of equitability in the distribution of individuals among species. One of the oldest and most widely used diversity indices is the Shannon-Wiener Index:

$$H' = \sum_{i=1}^{S} \frac{n_i}{N} \ln\left\{\frac{n_i}{N}\right\}$$

H'=diversity index
S=number of species
ln=natural log (can also use log)
n_i=number of individuals of the ith species
N=total number of individuals of all species

Communities with the most unique species have a high index value. If the number of species is one (S=1), the value is 0 because there is no diversity. The index is based on information theory, a measure of uncertainty, where the higher the H', the greater the uncertainty.

Idea 1: Use a W-shaped walk to collect population counts of butterflies (Wheater et al. 2011). Census methods established in the U.K. call for searching in heterogeneous areas in partial shade and/or full sunshine. Air temperature should normally be between 13°C and 17°C. Wind speed should be less than when tree leaves begin to sway. Walks are generally taken from April to September between 10:45 a.m. and 1:45 p.m. Butterflies are counted up to 5 m on either side of the W-shaped transect, avoiding high-flyers or wayward animals. Every effort should be made to avoid double counting. Compare two grassland areas.

Idea 2: Compare pollinating insects in gardens, or plant species in lawns/pastures, or tree species in wooded areas.

WHAT CHARACTERIZES COMMUNITY ECOLOGY?

Community ecology is the study of interactions among species that live in the same place. An ecological community might include the species in a lake or those living together on a cliff. If population ecology is the busy-bee, blue-collar subdiscipline, community ecology is the **flashy movie star of the subdisciplines**. It has attracted some

DOI: 10.1201/9781003271833-16

of ecology's most notable human personalities, named the most theories, and inspired the most nature channel episodes.

QUESTIONS ASKED BY COMMUNITY ECOLOGISTS

- What is the species richness of a community?
- Which species are **dominant**, referring to the most abundant species, or the ones with the most impact on the rest of the community?
- Why do some species become invasive, the super dominant?
- Is community diversity linked with **stability**, the ability to bounce back to the original composition after a disturbance such as fire?
- What is the structure of a community, the study of roles within the community?
- How can prey co-exist with predators without depleting each other?
- Is prey abundance determined more by what they eat (a bottom-up explanation) or by who eats them (a top-down explanation)?

While sorting through the specifics, keep in mind the overarching questions in ecology. Is there a balance of nature? Why are organisms where they are? How did evolution shape the patterns we observe? How can we form and use theory to predict patterns? How can theory help us protect nature, obtain resources, and protect ourselves from nature?

WHAT ARE THE WAYS ORGANISMS CAN INTERACT?

Because ecology is the scientific study of interactions among organisms, precise terms describe common associations. **Intraspecific** interactions occur within species. **Interspecific** interactions occur between species. Note that any given individual may simultaneously be interacting with multiple species as in pollinators, seed dispersers, root symbionts, and competitors for several resources and conditions.

WHAT IS THE DIFFERENCE BETWEEN SYMBIOSIS, MUTUALISM, AND FACILITATION?

Symbiosis is two species interacting with one another in close association. The species must be physiologically dependent on one another for most or all their lifetimes (Bronstein 2009). Interactions may be positive or negative or, more commonly, run in a gradient between the two (Table 16.1).

Mutualism is two species providing a benefit to one another in growth, fitness, or survival (Bronstein 2009). The interactions are necessarily beneficial for both participants. Note that only a subset of mutualists are symbionts, and only a subset of symbionts are mutualists.

Note that symbiosis and mutualism are interspecific interactions. When *intra*specific individuals help each other, it is called cooperation, facilitation, or helping. Note too that mutualists

Table 16.1 Community-level associations between two species (A and B) can be characterized as positive, negative, or neutral (0). Each relationship can be symbiotic. Relationships in gray consist of those in which feeding is central.

	Species A	Species B
Mutualism	+	+
Commensalism	+	0
Detritivory/decomposition	+	0
Scavenging	+	0
Parasitism	+	− or 0 or +
Herbivory	+	− or 0 or +
Predation	+	−
Parasitoidism	+	−
Amensalism	−	0
Competition	−	−

are not necessarily altruistic. Each partner is attempting to maximize its benefits independent of the costs to its mutual partner, e.g., most pollinators visit flowers to obtain nectar, not pollinate the flower (Bronstein 2009).

Facilitation is in the class of mutualistic interactions that are not symbiotic. **Facilitation** is modification of the abiotic or biotic environment by one organism, enhancing colonization, recruitment, and establishment of another. For instance, during succession one plant may enhance colonization for another by providing shade, but these are not in close association.

Probably every species on Earth is involved mutualistically with another species at some time in its life. Over 80% of all flowering plants have **mycorrhizae**, fungal components that live on or in roots (Bronstein 2009). The benefit to the fungus is sugar provided by the plant. The benefit to the plant is better nutrient uptake with the association of the fungus. Seed dispersal, pollination, human food production, and cycles of carbon, nitrogen, and phosphorus are all heavily dependent on mutualisms. Mycorrhizae relationships may be obligate or facultative.

Within mutualism, an **obligate** relationship refers to a particular species that cannot live without another particular species. For example, a leafcutter ant species may grow a particular fungus species in a garden underground (Figure 16.1). Fungus gardening mostly occurs in the humid tropics, but occasionally it can occur outside the tropics (Rockwood 2015).

In contrast, a **facultative** relationship refers to dependency on another species, but not a particular species. For example, several insect and flower species may rely on each other as a group, but not require a particular species. Slave-making ant species may associate with several other ant species. Coral species may depend on several photosynthetic dinoflagellates (**zooxanthellae**). In some mutualistic relationships, one species may be obligate and the other facultative. For instance, humans depend on *Escherichia coli*, but *E. coli* can live independently of humans.

Lichen is fungal hyphae intertwined with algae or cyanobacteria in a symbiotic, assumedly mutualistic relationship. The algal colony plays an important role by photosynthesizing, thereby providing sugars and oxygen to the fungus. In return,

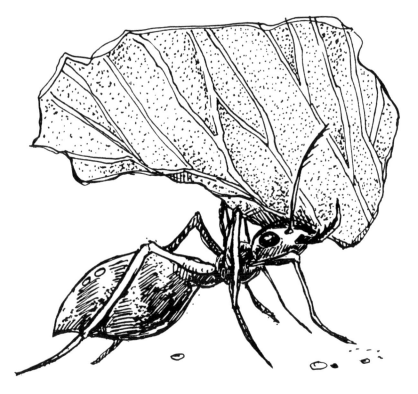

Figure 16.1 Leafcutter ant holding a leaf. (*Drawn by lisnycija, used with permission.*)

the fungus provides humidity, carbon dioxide, and perhaps protection from weather and UV rays. The relationship can be obligate or facultative.

Bacteria or protozoa living in the guts or cells of **ruminants** are symbiotic mutualists. This includes the organisms that ferment plant material within the multi-chambered stomachs of deer and cattle. Termites have bacteria or protozoans living within their hindgut. The microorganisms break down cellulose in wood eaten by the termites. These microorganisms are passed from one generation to the next by means of anal or oral **trophallaxis**, passing feces to another termite for reconsumption. Without this social behavior, termites would starve.

The ultimate mutualistic relationships at the cellular level include chloroplasts (where photosynthesis takes place) and mitochondria (where ATP is produced in eukaryotic cells). Both types of organelle are thought to have originated as free-living prokaryotes taking up life within eukaryotic cells of other organisms.

Other symbiotic mutualistic relationships fall into the category of **defensive mutualism**, organisms which defend the species against attack. These include grass species that host parasitic fungus. For instance, Tall Fescue (*Lolium arundinaceum*) may be infested with a fungus (*Neotyphodium coenophialum*) on its surface or living within the plant in what is known as **endosymbiosis**. These endosymbionts may produce alkaloid toxins that discourage grazers and seed eaters. Horses and cattle that graze the infected grass may have reduced productivity and fitness, even experiencing abortions and reduced milk output.

Defensive mutualism occurs in plant species with specialized glands called **extrafloral nectaries** occurring on leaves or stems. The glands secrete protein and sugar-rich fluids. The fluids attract ants, which attack herbivores or remove surrounding vegetation.

SUMMARY OF MUTUALISM

Mutualism is still largely in a stage of description. Despite its commonality, mutualism may be the most poorly understood of the interspecific interactions and is still in need of a theoretical framework (Hale and Valdovinos 2021).

Observations so far indicate:

- animal pollination of plants is more frequent in the tropics than in higher latitudes (Rockwood 2015). Almost all obligatory ant-plant mutualisms are found within the tropics.
- bats that provide pollination and fruit dispersal are found only in the tropics.
- every organism on Earth is probably involved in at least one mutualism in its life. Forest trees with mycorrhizal fungi are common in the boreal. The tundra is dominated by lichens. Deserts are inhabited by leguminous plants with nitrogen-fixing bacteria.
- the benefit in the vast majority of symbiotic mutualisms is nutritional, in either one or both directions, often with one species inhabiting the other (Bronstein 2009). Thus, co-parasitism may be a better term to describe the phenomenon than mutualism.
- even though mutualism is a net positive to both species, this has cost. Rather than considering species as purposeful cooperators, mutualists are probably self-serving, with a conflict of interest between the mutualistic partners.

COMMENSALISM AND AMENSALISM

Commensalism is when one species is positively affected, and the other receives no benefit. Some ecologists do not use these terms because of difficulty in determining if one species is truly unaffected. Purported examples of commensalism include:

- seed heads that stick to the fur of animals, thereby dispersing without cost to the disperser (Rockwood 2015),
- mites and bacteria living on the skin of animals at no cost,
- a bird's nest in a tree, doing no harm to the tree,
- Cattle Egrets eating more efficiently in the presence of a cow, but not disturbing the cow,
- mushrooms and other **saprophytic** organisms that consume dead or decaying tree roots and other decaying matter. The tree species does not suffer because it is already dead.

Other commensals include **epiphytes** (epi=upon, phyte=plant), organisms that live on the surface of plants, such as plants living on top of other plants instead of having roots in the soil. For example, bromeliad orchids live on tree branches in Latin American tropical forests. In some cases, the epiphyte might be considered a parasite rather than a commensalist, but if the plant provides a substrate only and does not gain or lose anything itself, it could qualify as commensalism.

Amensalism is when one species is negatively affected while the other is unaffected. Amensalism can result from:

- spite,
- unknowingly, when a human steps on an ant,
- some cases of **allelopathy**, in which a plant secretes a chemical substance from its roots that harms the growth and development of another nearby species. Allelopathy would qualify as amensalism if the allelopathic species produced the toxin even when the competing species was not present. However, it would be considered competition if the allelopathic species produced the toxin only in the presence of a competing species (Mackenzie et al. 2001).

PREDATION

In **predation**, one species kills and eats all or part of another, and this occurs several times in the predator's life. In predation, one species benefits and one species is negatively affected. A subsequent chapter further delineates the terms used in the predation literature.

HERBIVORY

Herbivory is the consumption of living plant material. If herbivores are not generalists, they can be classified by which plant parts they eat. **Browsers** consume woody material and bark, including that of shrubs, vines, and trees. **Grazers** consume leafy material. **Granivores** eat seeds. Frugivores eat fruit. Herbivores can also be classified by type of feeding such as chewing, sucking, boring, mining, and galling.

Herbivores are like parasites by not killing their victims, but unlike parasites they have many victims over their lifetime. Additionally, herbivores and plant species do not tend to cycle

dramatically as do parasites and their hosts. Herbivores are like predators in often playing the role of a keystone species. Herbivores that eat seeds are often keystone species more directly influencing plant populations than other herbivores.

There are circumstances in which herbivory inflicts dramatic impacts on plant populations (Morris 2009). Most notably, forests can be devastated by herbivorous insects. While mammals are more likely to have an immediate impact on the plants they eat, insects can be more lethal. Insects tend to be more specialized than mammals, and insects are more likely to be parasites forming a lifetime association.

PARASITISM

In **parasitism**, a species forms an association with an individual of another species (called a **host**) to obtain nutrients or energy and a habitat. Unlike grazers and predators, a parasite relates to only one or a few individuals over its life span. One species benefits and the other species has a neutral, positive, or negative reaction. The host is usually not killed except in parasitoidism. A **parasitoid** is parasitic only in the larval stage (Figure 16.2). It kills its host, then emerges to the adult phase in a free-living form. Examples occur mainly among the bees and wasps, but also among the flies, with larval stages known as maggots.

COMPETITION

In **competition**, adverse effects occur between two individuals as they try to use the same resources, either interspecifically or intraspecifically. Intraspecific competition is often the stronger of the two. Two mechanisms of operation occur. **Exploitation competition** is one organism reducing the resource for another, even though the individuals never come into contact. For example, ants and rodents may compete for seeds in the desert, but not interact directly because only one of them is active during the day. Aquatic plants tend to compete for phosphorus. Terrestrial plants tend to compete for nitrogen.

Figure 16.2 Leaf miner parasites leave a sign of their trail on the leaf.

The second mechanism of competition is **interference competition** in which one individual actively interferes with the other for a resource, preventing the other from occupying a habitat or accessing something. In animals this interference may include **aggression** – fighting, chasing, bluffing, or threatening with bared fangs. It includes posturing to establish dominance, vocalizing regularly, and marking an area with a chemical or physical signpost. It can include the establishment of a **pecking order** (dominance hierarchy). Interference competition is uncommon compared to exploitation competition, for good reason. Both participants must pay costs when engaging in interference, thus most organisms avoid it.

COMPETITION: THE CENTRAL CONCEPT IN ECOLOGY?

If nothing in biology makes sense except in the light of evolution, and natural selection is survival of the fittest, one might assume that competition is the central ecological interaction. It explains why competition, not evolution, has been the most studied concept in ecology, attracting some of the best minds over the years. Competition was at the forefront early in the 20th century, and again during the golden age of theory development in the two decades before Earth Day in 1970. It was at the center of a paradigm shift within ecology after that. Is competition still thought to play a universal role in the assembly of a community? Has it been the key to understanding evolution? The next chapter will step us through the evolution of ideas.

THINKING QUESTIONS

1. Which two community interactions have the greatest intimacy? Which have the greatest lethality?
2. Give an example of each: facultative and obligate symbiosis.
3. Give an example of how the biological interactions are like marketplace transactions among humans, especially for mutualisms.
4. Give an example of exploitation and interference competition not in the book.
5. Compare herbivory with predation and parasitism.
6. What is the difference between species richness and evenness? How can the two be combined in a quantitative way?
7. Do you think it is true that negative interactions tend to occur in pioneer communities, and positive interactions tend to occur in late successional stage communities? Explain.
8. Observe the community of pollinators for local native plants or in a domesticated garden. Determine which pollinator species come to which plant species. Determine how long they stay, what they are eating, how they collect pollen, and how far they fly to get to the next flower. If observing Honeybee determine where their hive is located. Watch them leave the flower area and go to a central location. Alternatively, watch the bees leave a known hive and determine where they go to feed. Do they fly farther in good weather?
9. Engage your senses to identify the community of birds (Monkman and Rodenburg 2016). Be patient. Nature study is not as depicted on TV. Concentrate, stay quiet, and do not make sudden movements. Close your eyes if you need to concentrate on hearing a particular bird. Cup your hand around your ear to amplify the sound. When searching visually for a bird use "soft eyes," scanning the landscape without focusing on any one object. Develop and trust peripheral vision and become aware of any movement within the vegetation. Marvel at your ability to see color, size, and some field marks in a very short time. "Something in blue" may be all the detail you get.

REFERENCES

Bronstein, J.L. 2009. Mutualism and symbiosis. In: S.A. Levin (ed.), *The Princeton Guide to Ecology*. Princeton University Press, Princeton, NJ.

Hale, K.R.S., and F.S. Valdovinos. 2021. Ecological theory of mutualism: robust patterns of stability and thresholds in two-species population models. *Ecology and Evolution* 11:17651–17671. https://doi.org/10.1002/ece3.8453

Mackenzie, A., A.S. Ball, and S.R. Virdee. 2001. *Ecology. Instant Notes*. 2e. BIOS Scientific Publications, Oxford, UK.

Monkman, D., and J. Rodenburg. 2016. *The Big Book of Nature Activities*. New Society Publishers, Gabriola Island, BC.

Morris, R.L. 2009. Interactions between plants and herbivores. In: S.A. Levin (ed.), *The Princeton Guide to Ecology*. Princeton University Press, Princeton, NJ.

Rockwood, L.L. 2015. *Introduction to Population Ecology*. 2e. Blackwell Publishing, Oxford, UK.

Wheater, C.P., J.R. Bell, and P.A. Cook. 2011. *Practical Field Ecology*. Wiley-Blackwell, Oxford, UK.

17

Theory in community ecology/ competition

ACTIVITY: Assessing Niches at a Bird Feeder

Goal: Learn the difference between niche and habitat. Know when to use chi-squared tests.
Objective: Determine whether bird species have different niches when they eat at a feeder.
   "Habitat" and "niche" (usually pronounced nitch in English) mean different things. Habitat refers to the type of environmental space occupied by a species. Niche refers to the unique abilities a species processes especially in reference to feeding capabilities. Think of habitat as an electrical outlet and niche as an electrical plug. The two fit tightly if they have evolved together over time.
   Watching birds at a feeder brings the niche idea to life, even allowing a test of a null hypothesis via a chi-squared test. Null means no or none. In a null hypothesis there is no difference between one thing and another. If you wanted to know whether one duck species had a longer beak than another, you would write the null hypothesis as "the beak length of a Pintail Duck and Mallard is the same." To test this null hypothesis you could measure the length of 30 Pintails and 30 Mallards and calculate the mean length in each treatment. You would use a t-test to compare the means of the two treatments and determine whether a significant difference occurred. A t-test compares two means. A chi-square test compares frequencies between treatments, not means. A chi-square tests the null hypothesis that there is no difference between frequencies. In the feeder study you will test the null hypothesis that "there is no difference in the frequency of how bird species feed at a bird feeder." In other words, there is no difference in the niches of birds at the feeder.

PROCEDURES

Step 1: Spend 10–15 minutes familiarizing yourself with the bird species you see at a feeder before you begin collecting data to test the null. Write the common names on row labels in a table.
Step 2: Notice different feeding strategies. Some birds sit on the feeder and feed directly. Some take seeds from the feeder and fly to a nearby perch to break open the seed. Some eat seeds on the ground under the feeder. Write labels on the columns in the data collection table such as "on feeder," "on other perch," and "on ground." These are the different niches at the feeder.

Step 3: Watch birds at the feeder until you observe 30 birds come and go. Make a tally mark for each bird in its strategy (niche) by species. You may count the same bird twice if it comes back to the feeder for another seed.

Step 4: Use chi-squared to test the null hypothesis. If you reject the null it indicates that species are in niches at the feeder. Statistics packages rarely calculate chi-square because it is easy to calculate values by hand. Find examples on the internet showing how to calculate chi-squared test values by hand.

EARLY ASSUMPTIONS ABOUT ECOLOGICAL COMMUNITIES

Early ecologists assumed communities had holistic properties. Like a super-organism, factory, or little town, species were assumed to have different roles within an ecological community working for the greater good in a self-sustaining, self-repairing, self-regulating group. Competition was thought to be the regulator that kept the species stable and in balance. Is this still thought to be true?

COMPETITION IN A GARDEN

Consider competition in the garden of Arthur Tansley, a British botanist, experimenting with two closely related species of bedstraw in 1917. When the species were apart, they grew in a variety of pH levels. When they were together each species grew best in the pH matching where it was most commonly found in the wild. Tansley inferred that competition was the driving force to produce the evolution of these different capabilities.

PREDICTING WINNERS THROUGH LOTKA-VOLTERRA EQUATIONS

Better than mere observation, early theorists sought mathematical predictive power in choosing the winners of ecological competition. In the 1920s an American, Alfred Lotka (1925), and an Italian, Vito Volterra (1926), formulated the same algebraic expressions independently to predict outcomes when two species competed for the same resource. Using derivatives of the logistics equation, population growth was assumed to be depressed by intraspecific and interspecific competition.

To test the Lotka-Volterra ideas, Russian biologist Georgy Gause used laboratory bottle experiments in the 1930s. Gause watched one species of *Paramecium* repeatedly outcompete a similar species. Other experiments with fruit flies, mice, flour beetles, and annual plants produced similar results, in some cases after as many as 30–70 generations. Sometimes the species coexisted indefinitely by living on opposite ends of the bottle in slightly different conditions or using different feeding modes.

Guase's experiments provided support for the Lotka-Volterra model. They helped establish what is now known as the **competitive exclusion principle** (Kingsland 1985). The principle is commonly worded as, "no two species can indefinitely occupy the same niche." Actually, the Lotka-Volterra model provides for cases of coexistence in competitive situations as found in Gause's bottle experiment. A better definition for the principle would be, two species with the same requirements for resources and conditions cannot ordinarily coexist.

HOW DO ENVIRONMENTAL CONDITIONS AFFECT COMPETITION?

In the 1940s, further analysis was conducted by microbiologist Thomas Park and associates at the University of Chicago who observed flour beetles in a 10-year series of laboratory experiments. Park found that each species' chance of winning was influenced by several factors including moisture, temperature, and parasitism, in other words, the particular environmental conditions. The implications were that environmental variation allowed competitors to coexist. In homogeneous conditions, one species tended to dominate. In patchy situations all competitors can be winners.

DOES THE LOTKA-VOLTERRA MODEL APPLY WELL TO NATURAL SITUATIONS?

No, and by now you will recognize problems just by considering the assumptions of the model. For the equations to predict accurately:

- resources must remain limited,
- carrying capacities must remain constant,
- intrinsic rates of growth must remain constant,
- no age structure must exist within the populations,
- no time lags must occur.

In wild situations, resources may not be limiting, and carrying capacities may change, or more than two species may be competing. One species may eat the other or cannibalize its own members. Use of the equations in their original form was confined to simple situations.

THE GOLDEN AGE OF ECOLOGICAL THEORY – HUTCHINSON, MACARTHUR, AND WILSON

In the 1950s, George Evelyn Hutchinson at Yale and his student Robert MacArthur took up theoretical ecology. They assumed there was a balance, an equilibrium among species that could be described by the logistic and the Lotka-Volterra models via competition.

MacArthur asked how five closely related species of warblers could co-occur in coniferous trees of the northeast U.S. Had the warblers reached a stable equilibrium as in the Lotka-Volterra model? Likewise, Hutchinson in 1961 presented what he called the **paradox of the plankton**, asking how multiple phytoplankton species could coexist in water bodies near the surface if they were in the same niche.

WHAT IS A NICHE?

Hutchinson and MacArthur assumed that no two species could occupy the same niche. Problematically, niche had a variety of definitions. In the 1917 definition by Joseph Grinnell, **niche** referred to the role (occupation) of a species.

By this definition a Little Brown Bat (*Myotis lucifugus*) would be labeled a nocturnal, aerial insectivore. In contrast, **habitat** referred to the type of living place (address) of a species, the home qualities of an organism. The habitat for the Little Brown Bat would be an open field next to a deciduous forest. Niche referred to the species' status within the community in relation to food and predators, but with labels like a profession. "There goes a badger" was equivalent to "There goes the vicar (preacher)" in human terms (Elton 1958).

In other words, the niche is a quality of the species. It is the species' role and set of requirements. The habitat is a quality of the environment. The habitat changes as a result of hosting the species. In theory the niche and habitat come to have a tight fit as the species adapts to its environment. The niche-habitat model is exemplified by situations in which one species of insect has evolved to pollinate one species of orchid with perfectly fitting mouthparts that provide a reward for the insect and pollination for the flower.

Be careful with your terms. "Niche" should only be used when referring to (Figure 17.1):

- a species' unique set of feeding conditions,
- perhaps in reference to an "empty niche" but with careful words as in the following example.

The American Dipper (*Cinclus mexicanus*) is a bird species occurring in mountainous rocky rivers of the western U.S. and Central America. It is absent in the eastern U.S. even in streams with flow rates the same as in the west. Dippers can fly, but they also swim and walk along the bottom sediments, sticking their heads underwater to prey on invertebrates. The habitat of the American Dipper is

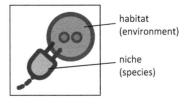

Figure 17.1 The simple difference between habitat and niche. The niche is a quality of the species, which you might think of as an electrical plug. The habitat is a quality of the environment, like an outlet. Together the two develop a tight fit.

rocky, fast-flowing rivers. The niche is an ability to feed on underwater invertebrates by swimming and walking in fast-flowing streams. We might say that no bird species fills this niche in the east, although it would be wrong to point to the eastern rivers themselves and say they are an empty niche. Niche does not refer to places, only to species.

THE HUTCHINSONIAN DEFINITION OF NICHE

In 1957 Hutchinson argued that to make competition predictive in the wild, the niche needed to be measurable. Thus, Hutchinson defined the niche as a **multi-dimensional hyperspace**, a volume with more than three dimensions, a theoretical space on a graph. Each axis represented one dimension. Each dimension represented an essential environmental state for a species. Just as humans need 26 vitamins (macro- and micronutrients), the essential needs of any species can be plotted as tolerance limits. The volume defined within the hyperspace by all the tolerance limits would be the niche. In further clarification, Hutchinson conceded that plots of more than three dimensions are difficult if not impossible to conceive. He suggested the niche could be represented by the zone of tolerance for the **limiting factor**. To make the idea quantitative, Hutchinson explained that the degree of **niche overlap** between two similar species could

predict the degree to which two species compete (Figure 17.2). The width of the tolerance zone was the **niche breadth**.

For these graphs to be created, Hutchinson had to describe the **fundamental niche** as "all the states of the environment" that allow a species to exist when in isolation from other species (Figure 17.3). The **realized niche** represented the state of the environment when the species was in the presence of a competing species.

For example, the Red-winged Blackbird (*Agelaius phoeniceus*) and Yellow-headed Blackbird (*Xanthocephalus xanthocephalus*) inhabit wet areas of the North American prairie (Orians and Willson 1964). Both use deep-water marsh for nesting, gaining maximum protection from mammalian predators in the deepest areas. Where the two competitors coexist, Yellow-headed Blackbird excludes Red-winged Blackbird from the deep areas. The shallow area becomes the realized area for the Red-winged Blackbird.

Today, ecologists have expanded the idea of a realized niche (Callaway 2009). It is not always smaller than the fundamental niche. Through facilitation or mutualism, other species may enhance the conditions and resources for other species, making the realized niche larger than the fundamental niche (Figure 17.3). Hutchinson's point was that competing species could develop a

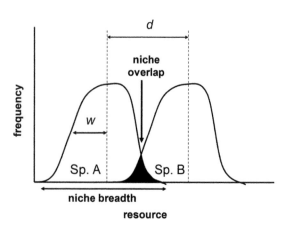

Figure 17.2 Overlapping zones of tolerance for a limiting factor show the degree of niche overlap. The variable *d* is the distance between optima, and *w* is the standard deviation within one species.

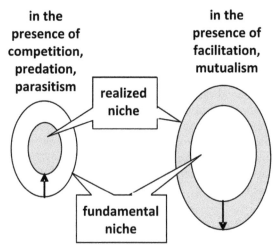

Figure 17.3 The realized niche can either be larger or smaller than the fundamental niche. It would be smaller in the presence of negative effects like competition, and larger in the presence of cooperative interactions like facilitation.

way to coexist, grow, and reproduce in the same place, partitioning the niche without actually competing. This became the explanation regarding MacArthur's warblers. The species occupied different heights and portions of spruce trees when they were together.

WHAT HAPPENS IF TWO SPECIES TRY TO OCCUPY THE SAME NICHE?

With the niche redefined, niche questions dominated the ecology literature from 1955 to 1980, relying heavily on the Lotka-Volterra equations and building upon them. Classic competition theory developed from field-based studies matched with theoretical concepts. The theory proposed that if two species tried to occupy the same niche, one of the following scenarios would occur:

- **competitive exclusion**: local extinction (extirpation) of a species because of interspecific competition.
- **niche partitioning**: two or more species coexist by sharing the resource at slightly different times or in different areas. The species could snap back to their fundamental niches when the competition ceased.
- **character displacement**: two species may have nearly identical niches when living separately. When living together, competition causes a separation by species through morphology, physiology, behavior, or breeding period. This begins to take on a genetic basis. Hence, evolution takes place.
- **speciation**: the eventual result of character displacement may be new species.

Beyond the above situations, species with similar ecological requirements could co-occur in:

- temporary situations, as in shorebirds probing for the same type of prey on mudflats during migration,
- situations with a patchy, frequently disturbed, and abundant resource base as in Hutchinson's plankton. The species never get a chance to grow to carrying capacity because of frequent disturbance even though they were together for a lifetime.

THE EQUILIBRIUM DEBATE

While asking questions about niches, two sides emerged over the existence of an equilibrium just as it had in population ecology. Within communities, **equilibrium** refers to both a consistent abundance within a population, and consistent number of species over long periods within a community. With time, equilibrists conceded that predation, not just competition, could be a regulator of species composition (Levin 1970). (Mutualism had largely gone unstudied up to that point and so was not included.) Eventually, equilibrists were further willing to concede subdivisions by patches having their own distinct stable equilibria (Levin 1974). In other words, the community could be modeled through metapopulations.

Non-equilibrium proponents continued to argue that communities rarely reach equilibrium. Density-independent (abiotic) factors through disturbance and chance have too much effect for biotic factors to take hold. In trying to settle the debate, new insight came from the theory of island biogeography.

THE THEORY OF ISLAND BIOGEOGRAPHY

In 1967, R. MacArthur and E.O. Wilson published *The Theory of Island Biogeography*, a short book outlining a series of hypotheses developed from the study of birds and ants on Pacific Islands. MacArthur became a professor at Princeton. Wilson was already a young professor at Harvard. Like Hutchinson and MacArthur, Wilson thought about the big questions in ecology and wrote 25 books. He was considered the world authority on ants and wrote groundbreaking works on sociobiology, biophilia, and biodiversity conservation.

The core idea of the theory was that the number of species within a taxonomic group on islands represented a balance between the addition of new species by colonization and the loss of established species by extinction. When one species was lost, a space for another similar species opened, providing equilibrium. The carrying capacity of an island depended partly on island size and partly on distance to the island. Since 1967 the following hypotheses from the Island Biogeography Theory have received the most research:

1. Larger islands have more species, as demonstrated through species-area curves (**species-area hypothesis**).

   Research since 1967 generally supports this hypothesis for vertebrates, invertebrates, and plants, and on mountaintop islands and in caves (reviewed in Stiling 2012, Schoener 2010, Lomolino et al. 2010a). There are exceptions, most notably for oceanic islands of vastly different ages as reviewed in Chapter 9. The much older geologic age of the Hawaiian Islands, the above-average variety of environments and altitudes, and a diverse climate explain its high biodiversity for its size. Species richness increases with area for several reasons (Colwell 2009), primarily because increasingly larger areas have more habitat types.

2. The number of species on an island is influenced by the distance from other islands and continents (**species-distance hypothesis**). Islands closer to other land have more species. This hypothesis has been generally supported since 1967, with research examining both oceanic and terrestrial situations (reviewed in Stiling 2012, Schoener 2010, Lomolino et al. 2010a). The relationship is less tidy than the species-area relationship because islands tend to be clumped in archipelagoes (Lomolino et al. 2010a) and again, the geologic age and additional factors such as nearby ocean currents can create exceptions, despite great distance from a continent.

3. Species diversity produces an "equilibrium" number of species (**equilibrium hypothesis**). This is the idea that has been the most highly debated. Two of the first researchers to test the hypothesis were young associates of MacArthur and Wilson. In 1969 Jared Diamond published a study of birds of the Channel Islands off the coast of southern California, comparing the bird composition to a survey done 50 years before. His conclusion was that a similar number of species remained and 20%–60% of the species had turned over since 1917. This appeared to provide support for the hypothesis but was widely challenged because some species were non-natives introduced through human influence. Some birds of prey were absent because of pesticides (Lomolino et al. 2010b).

To test the equilibrium hypothesis Daniel Simberloff, a graduate student of E.O. Wilson at Harvard, focused his dissertation on ant species richness of small mangrove islands off the coast of Florida. Contrasting with Diamond's observational study, Simberloff used manipulative experiments to exterminate species from a subset of islands, then recorded the recolonization rate. Within a year all but the most distant islands had recovered their initial species richness with a great deal of turnover, supporting the hypothesis (Lomolino et al. 2010b).

In the 1970s and 1980s, heated debate ensued between Diamond and Simberloff, which was not solely between two people. Allies supported one proponent or the other, publishing their responses in ecological literature. Simberloff eventually became a professor at Florida State University in Tallahassee. His group became known as the Tallahassee mafia as they took aim at their proponents in the Ivy League (McIntosh 1985). Diamond was a professor at Princeton, an Ivy League university.

At stake was an argument about competition as the source of community structure and the existence of an equilibrium. Simberloff's side won the most support. The importance of chance and random processes essentially killed the balance of nature idea and competition regulating species at equilibrium. Through Simberloff's arguments we have gained more appreciation for predation, succession, disease, parasitism, disturbance, and the ability to disperse as factors affecting community structure. They keep populations below carrying capacity, preventing the need for competition.

We now know that species occurrence could be the result of several factors for which we go back to the constraint hierarchy of Chapter 9. Establishment depends on the suitability of the habitat, dispersal capabilities, chance, geographic isolation, and cooperation among species. It depends on the size of the colonizing population, competition within its own species and with others. It depends on how much predation the colonists incur, and it depends on whether the organisms survive long enough to produce viable offspring. Competition, as the major organizing factor at a large spatial scale within an equilibrium model, can occur but should be called into question knowing that many other factors are at work. Organisms try to avoid competition. There are high costs for participation in interference competition.

SIMBERLOFF-DIAMOND DEBATE: HOW TO DO SCIENCE

Within the debate was the question of how to specifically demonstrate that competition was responsible for the number of species rather than any other factor. Diamond's work was merely observational, thus the best he could do was invoke the **ghost of competition past** argument. In other words, he inferred that past competition produced present-day patterns, but he could not demonstrate it specifically. To Simberloff and his allies, Diamond was merely verifying hypotheses rather than falsifying them, a weak way to do science. Simberloff pushed Diamond to at least supplement his analysis with a null model.

The emergence of **null models** (sometimes called **rule-based models**) in the 1970s was a breakthrough in the debate. Null models compare a scenario based on chance to an observable pattern. Any deviations from the chance situation are assumed to be evidence for specific interactions, like competition. Null models are different from null hypotheses. For example, Skelly and Meir (1997) evaluated the possible causes of distribution for 14 species of amphibians across a landscape of 32 ponds in Michigan. The researchers proposed three models that might explain the distribution pattern:

1. An isolation model (distribution was driven by distance between ponds);
2. A succession model (distribution was driven by sequential changes in vegetation of ponds);
3. A null model (changes were random events).

For this example, the succession model made the fewest mistakes in predicting species richness. Thus, the researchers were able to suggest management prescriptions based on the vegetation inhabited by the greatest number of amphibian species.

In the end, it became clear that experiments, null hypotheses, the proposal of alternative hypotheses, and the use of null models were generally superior to observational study when trying to advance ideas in science. At the same time, ecologists gained a more informed appreciation for natural experiments as practiced by Diamond. These influences are still with us, and use of the null has led to the neutral model, explained later in this chapter.

SIMBERLOFF-DIAMOND DEBATE: SLOSS

A further source of debate emerged in 1975, when Diamond related the theory of island biogeography to reserve design, the idea that small fragments of habitat are equivalent to oceanic islands. Diamond proposed guidelines to produce maximum species richness within terrestrial reserves. He proposed that:

1. A large reserve is preferable to small.
2. A single large reserve is preferable to several small reserves of equal area of the same habitat.
3. If there are multiple small reserves they should be close to one another.
4. If there are multiple small reserves they should be in a cluster as opposed to a line.
5. Connecting the reserves with corridors makes dispersal easier.
6. Reserves should be circular to minimize negative effects of edges, which have more invasions from introduced species.

This part of the debate was given its own name (single large or several small) with an acronym, SLOSS. In opposition to Diamond's scheme, the Simberloff side argued that large patches were vulnerable to catastrophic events by placing all their eggs in one basket. The Diamond side countered that small patches were more susceptible to population crashes.

One of the breakthroughs that eventually mollified the debate was the emergence of metapopulation ideas introduced in 1969 by Richard Levins. Although the metapopulation theory applied to populations within a single species and the theory of island biogeography applied to species richness in communities, implications were still clear. Through the lens of metapopulations, abundance is greatest in high-quality patches. It is not based on area and distance exclusively. If the habitat between patches is safe to travel, and if the quality of patches is high, several small patches forming a network connected by corridors may be superior.

In the end, Diamond's rules for reserve design, while important for stimulating further discussion especially about fragmentation, patchiness, and connectance, are now overshadowed by a more sophisticated understanding. Patch quality ideas within conservation and wildlife biology regarding edge, extinction vortices, dispersal dynamics, corridor ecology, disturbance frequency, and introduced species now subsume any argument based on size and shape alone.

WHAT IS THE NEUTRAL MODEL IN ECOLOGY?

Growing out of null models was the neutral model in ecology, related by theory to the neutral model in genetics. Proposed by Motto Kimura in 1964, the neutral model in genetics argued that genetic change was primarily the result of chance through mutation and genetic drift rather than selection.

The neutral model in ecology was developed in 1976 by Hal Caswell and used by Stephen Hubbell (2001) to examine tree species diversity on Barro Colorado Island. This is a tropical island in a lake created when the Panama Canal was built in the early 20th century. Like null models, the **neutral model in ecology** omits entirely the effects of biological interactions. The output of these models can be compared with the structure of natural communities, and from these comparisons, the effects of biological interactions can be evaluated (Caswell 1976).

To understand how neutral theory works, Harpole (2010) asks us to imagine a field as a local community consisting of many plant species. Surrounding this are more plant species, holding the regional species pool. The field and regional collection make up the **metacommunity**. Individuals in the local community die at random and the only way a new individual can establish is to use the vacant space. Individuals from the meta and local community randomly disperse seeds into the field. Individuals of one species may outnumber another, contributing more seeds than rare species; however, each individual seed has an equal chance of establishing. If dispersal from the regional community is strong, the local community will look like a small version of the region. If dispersal from the outside is weak, the local community composition will drift from random deaths and extinctions combined with random mutations and speciation. The local will differ more and more from other communities over time, but in an unpredictable way.

The strategy is to change different factors until the model predicts what is observed in the field. By building the model piece by piece, a clear picture emerges of how the pattern may have come about.

In short, the neutral theory is a set of null models against which the real situation is compared. For Hubbell, rather than assuming that intense competition must be going on among tree species in a highly diverse place, he assumed all species had identical birth and death rates, thus neutral in their interactions. In Hubbell's interpretation, biodiversity remained high because individual organisms only interacted with their local environment, and only with the individual competitors and predators nearby. Competitive exclusion may drive a species out of a given patch, but this only makes it locally extinct. A predator may drive a prey species locally extinct, but it takes up residence in a nearby empty patch.

Hubbell and his colleagues found that individual tree saplings compete with only 6.3 neighbors on average. Thus, weak competitors may subsist if chance puts them near other weak competitors. Think about trees in a trade-off between competitively dominant species and early succession colonizers. Competitively dominant species stand their ground. Weaker competitors simply vacate and exploit newly disturbed areas. Because competitive dominance is linked with low dispersal in late succession species, and poor competitors are linked with rapid dispersal, the process can lead to almost unlimited species diversity within a community.

IDEAS THAT EMERGED SINCE 1975 IN UNDERSTANDING COMMUNITY STRUCTURE

The keystone species idea developed from within predation theory (see predation chapter). Ideas about introduced species and how some of them become super dominant originated in succession theory (see succession chapter). Currently, it is clear that sometimes a single introduced species that may be diminutive on its home turf can cascade and change an entire community in its new home. Community structure has broken. What was it that broke and why did it break? We need more understanding.

Two-species interactions have received the majority of studies in the past history of community ecology. When there are more than two, indirect effects become prominent, including dramatic trophic level cascades (Loreau 2009). These ideas are better studied in an ecosystem context and will be examined in a later chapter. Many studies on competition have taken place since the 1970s. Ideas still thought to be true include:

- at low densities of individuals and for r-selected species, competition tends not to occur,
- competition is important:
  - for species high on the trophic ladder that have no predators,
  - when there is little physical disturbance,
  - in birds and mammals because they have multiple constraints as homeotherms (and flyers),
  - at high concentrations of herbivores,
  - when animals establish territories in high concentrations,
  - when plants have allelopathic chemicals and are in high concentrations.

When competition does occur, it is likely to be exploitation competition, and more likely intraspecific rather than interspecific. Individual encounters are most likely happening at small spatial scales in small patches, often with species put together by chance, within the context of a set of constraints regarding resources and dispersal at larger spatial and temporal scales.

ENDNOTES, UPDATES, AND CONCLUSIONS

A major paradigm shift occurred in the 1970s within community ecology because of the Simberloff-Diamond debate. The balance of nature/equilibrium/self-regulation idea was relegated to the trash heap and the new idea was that even in the absence of humans, nature is more redundant, patchy, random, opportunistic, disordered, chaotic, frequently disturbed, and frequently changed than formally appreciated. Abundance and distribution of organisms may be a matter of chance. What happens in one small patch may differ from a neighboring patch. Parasites and other organisms can cause radical changes, especially when they affect key species.

None of this is the end of the story. Recently, there has been a surge in studies on symbiosis, facilitation, and mutualism and their ability to provide cohesion and stability in communities (Leigh 2020). Ecological theory concentrated most of its energy for much of the last 100 years on competition as a structuring force, largely neglecting the immense power of cooperation among species within evolution and community building.

The next frontier is getting a handle on the theory of mutualistic networks in ecology (Bascompte and Jordano 2014). The theory of biological networks does not just apply to ecology. It is also important for understanding complex networks of gene sharing, protein interactions, neurons, and other applications – as well as species interactions, food webs, and community structure. The imperative for ecology is knowing how many links can break (from environmental destruction) before communities collapse. More about mutualism is covered in Chapter 19 on succession.

THINKING QUESTIONS

1. What is the difference between habitat and niche in contemporary thought? Other than the example in the book, think of an example of an empty niche and list it. Use the internet if necessary.
2. What part did metapopulation ideas, null models, and neutral theory play in the competition debate?
3. Explain the neutral theory in your own words. How does the neutral theory clash with niche theory?
4. Repeat MacArthur's warbler study, either investigating the species he observed, or other species that nest together in a forest.
5. Needham (1916) reminds us that the month before winter sets in (November) is a season of plenty, which provides a ready-made way to think of communities. Crop production is done, and living is easy for creatures. Animals can afford to be wasteful, which spreads the seed for growth next year.

a. Investigate and draw the skulls of seed-eating mammals and birds, including rodents with chisel-like teeth, and birds with seed-cracking and husk-opening beaks.
b. Investigate and draw the seeds that make up most of our diets, such as corn, wheat, and barley.
c. Make note of the ways seeds are dispersed. Microscopically examine the hooks and barbs of hitchhiker species and draw what you see. Is it true that hitchhiker seeds grow low to the ground, no higher than the backs of likely mammalian carriers? Examine seeds that disperse by wind, water, or mere gravity. Draw them. Examine those that have ejection capabilities like Jewelweed in N. America.

6. In the photograph, does the smaller tree next to the larger one compete or is it deriving a benefit from being so close? Both trees are well over 100 years old. How could the smaller one persist if it were under intense competition? This was in the old growth forest on the Boogerman Trail, Cataloochee section of Great Smoky Mountains National Park, North Carolina.

REFERENCES

Bascompte, J. and P. Jordano 2014. Mutualistic Networks. *Monographs in Population Ecology*, Princeton University Press, Princeton, NJ.

Caswell, H. 1976. Community structure: a neutral model analysis. *Ecological Monographs* 46:327–354. https://doi.org/10.2307/1942257

Callaway, R.M. 2009. Facilitation and organization of plant communities. In: S.A. Levin (ed.), *The Princeton Guide to Ecology*. Princeton University Press, Princeton and Oxford.

Colwell, R.K. 2009. Biodiversity: concepts, patterns, and measurement. In: S.A. Levin (ed.), *The Princeton Guide to Ecology*. Princeton University Press, Princeton, NJ.

Elton, C. 1958. *The Ecology of Invasions by Animals and Plants*. Springer, Methuen, London.

Harpole, W. 2010. Neutral theory of species diversity. *Nature Education Knowledge* 3:60.

Hubbell, S.P. 2001. *The Unified Theory of Biodiversity and Biogeography*. Princeton University Press, Princeton, NJ.

Kingsland, S. 1985. *Modeling Nature; Episodes in the History of Population Ecology.* University of Chicago Press, Chicago, IL.

Leigh, E.G. 2020. Neglected problems in ecology: interdependence and mutualism. In: A. Dobson, D. Tilman, and R.D. Holt (eds.), *Unsolved Problems in Ecology.* Princeton University Press, Princeton, NJ.

Levin, S.A. 1970. Community equilibria and stability, and an extension of the competitive exclusion principle. *American Naturalist* 104:413–423.

Levin, S.A. 1974. Dispersion and population interactions. *American Naturalist* 108:207–228.

Lomolino, M.V., B.R. Riddle, R.J. Whittaker, and J.H. Brown. 2010a. *Biogeography.* 4e. Sinauer, Sunderland, MA.

Lomolino, M.V., J.H. Brown, and D.F. Sax. 2010b. Island biogeography theory, reticulations and reintegration of "A Biogeography of the Species." In: J.B. Losos and R.E. Ricklefs (eds.), *The Theory of Island Biogeography Revisited.* Princeton University Press, Princeton, NJ.

Loreau, M. 2009. Communities and ecosystems. In: S.A. Levin (ed.), *The Princeton Guide to Ecology.* Princeton University Press, Princeton, NJ.

Lotka, A.J. 1925. *Elements of Physical Biology.* Dover Publications, New York, NY.

McIntosh. R. 1985. *The Background of Ecology.* Cambridge University Press. New York and London.

Needham, J.G. 1916. *The Natural History of the Farm.* Comstock Publishing, New York.

Orians, G.H., and M.F. Willson. 1964. Interspecific territories of birds. *Ecology* 45:736–745. https://doi.org/10.2307/1934921

Schoener, T.W. 2010. The MacArthur-Wilson equilibrium model, a chronicle of what it said and how it was tested. In: J.B. Losos and R.E. Ricklefs (eds.), *The Theory of Island Biogeography Revisited.*

Skelly, D.K. and E. Meir. 1997. Rule-based models for evaluating mechanisms of distributional change. *Conservation Biology* 11:531–538. https://doi.org/10.1046/j.1523-1739.1997.95415.x

Stiling, P.D. 2012. *Ecology: Global Insights and Investigations.* McGraw Hill, New York.

Volterra, V. 1926. Fluctuations in the abundance of a species considered mathematically. *Nature* 118:558–560.

18

Predation

CLASS ACTIVITY: Test the Hypothesis That Shorebirds on Migration Deplete the Prey Resource

Goal: Plot shorebird and prey dynamics during spring migration at a site, then use a t-test to determine whether a difference occurs in prey density between areas open to shorebird feeding and within exclosures.

At a coastal site in South Carolina, tens of thousands of migrating shorebirds take advantage of rich prey resources in the sediments of managed wetlands to fuel the next leg of their journey to the tundra (Weber and Haig 1997). Does this predation deplete the prey base, or do the invertebrates in the shallow wetlands reproduce at a pace that can keep up? Use the following data to make a graph that plots prey density on the vertical axis against calendar date on the horizontal. The graph should have two lines in which you connect the dots, one that tracks the invertebrates in exclosures, and one that tracks them in areas open to shorebird feeding. Exclosures were 1 m × 1 m areas that had thin nets stretched across four stakes to keep shorebirds from feeding within. For the final date only, complete a t-test between prey density inside and outside exclosures. Did the shorebirds significantly impact prey density? Note: use ordinal dates rather than months for the horizontal axis. Ordinal dates are the number of days since January 1. For instance, March 22 is the 81st day of the year.

How might prey biomass be a better variable than prey density for this comparison? How might the size of prey be affected by shorebird predation?

Table 18.a Practice results for shorebird exclosure experiment.

Date	Prey density in exclosures (inverts/400 cm³)	Prey density in areas open to shorebirds (inverts/400 cm³)
March 22	81	102
May 2	180	168
May 16	174	142
May 25	Results from 10 of the exclosures were: 200, 179, 167, 140, 176, 170, 182, 164, 183, 163	Results from 10 of the open areas were: 60, 51, 36, 31, 20, 47, 22, 34, 41, 29

TRUTH, MYTH, AND CONTROVERSY

The subject of predation is one of the most myth-laden and emotional topics in ecology. Beyond culturally and politically sensitive issues that make the headlines, each of us must at least occasionally worry about predators attacking us. As professionals in fisheries, conservation biology, or other

DOI: 10.1201/9781003271833-18

ecology fields, predation-related concerns often prevail over every other issue.

Some of the most emotional questions:

- Are predators dangerous pests of fable and fairy tales, or charismatic megafauna whose presence is essential for community stability (Rockwood 2015)?
- Will the cohesion of an ecological community collapse if a predator is removed?
- What can be done about deer overabundance that plagues parts of the U.S.? Is wolf introduction the best or only answer?
- Hunting? Ethical?
- Can cats, dogs, and predators released as pets harm the environment, help the environment?
- If there are mice in my house should I get a cat or use pesticides or what? What about bats or flying squirrels or snakes in my attic?
- Are human hunters the equivalent of a top predator and if so are human hunters in competition with other top predators?
- Why are wolves, lions, and tigers revered, and coyotes considered dirty, mangy, creepy pests that no one would pay to see in a zoo?
- How can I encourage game fish like bass in my favorite lake? Or understand dynamics in my fish tank?
- Pit bulls in my neighborhood? Sharks at the beach? Black Widow spiders in my garden? Poisonous snakes in the woods? Snapping turtles in the lake? Bears in my garbage? Should I eradicate?

THE ANSWERS START WITH DEFINITIONS

Can you describe an example of predation? Of course you can. Even children can. Unlike competition, predation is not hard to observe, but defining it is difficult. The concept is fraught with subtle distinctions. Ecology textbooks do not agree on definitions. For instance, some ecologists regard herbivory as a form of predation. The safe strategy is to choose words as precisely as possible to avoid misrepresentation. If you mean herbivory, use herbivory and not predation.

In that spirit, let us define predation in its purest form; **predation** is one organism killing then eating all or part of another animal or **protozoan** (an animal-like protist). Furthermore, a predator consumes several prey in its lifetime, in contrast to parasites that usually infect only one host. If predation were merely one organism eating another, it would not be distinct from:

- **carnivory**: consumption of animal tissue in general,
- **cannibalism**: when an animal eats a member of its own species,
- **scavenging**: when an animal eats a dead animal, recently killed by another source, or eats animal products like milk, dung, or blood,
- **detritivory**: when an animal eats decaying animal matter,
- **herbivory**: when an organism eats living plants as its primary diet,
- **parasitism**: when a species forms an association with an individual of another species (called a host), which it does to obtain nutrients or energy and a habitat.

Note that plants can be carnivores as in sundews, Venus Flytrap, pitcher plants, and bladderworts. Thus, predation can include carnivory (even by plants) and cannibalism, but not scavenging, detritivory, herbivory, parasitism, or parasitoidism. Keep in mind, too, that humans can be predators, and for most of human history, we were daily predators. It is only in recent times that a human can go a lifetime without killing then eating an animal, or at least helping in the preparation of recently killed animals.

MORE DEFINITIONS: NOT ALL PREDATORS KILL THE SAME WAY

Solitary predators such as Mountain Lion take down and kill their prey when alone, even though they are sometimes smaller than their victims (Rockwood 2015). In contrast, **pack predators** such as most wolf species kill as a group and are sometimes smaller than their prey.

What about the case of a gull harassing another seabird species to make it drop its prey, then stealing the prey in mid-air? This is not really predation because the prey is already dead. Parasitism by theft is **kleptoparasitism** and is extremely common even though most people are not familiar with the term. For instance, spiders sometimes steal and eat not only prey from another spider's web, but the web too. So too, it is common for birds to steal prey from spider webs. Why would they not when birds are hungry most of the time and spider webs are all around. The main deterrent for birds is that webs are usually close to the ground where predators might be hiding. It explains why birds are most commonly seen hunting in spider webs only during times of stress and desperation, like after a long leg of migration, or a long bout of adverse weather.

Sometimes pet dogs steal the breakfast of humans. Why would they not? Competing predators almost always try to hone in on the booty once another predator has made a kill. Even human children have to be taught to avoid taking food from another person's plate. Yes, our households are often infiltrated and surrounded by kleptoparasites.

Nest parasitism is when birds, insects, and bedbugs lay eggs in the nest of another species, and either the young are on their own, or the other species raises them. The Brown-headed Cowbird is a famous example.

WHAT IS HYPERPREDATION?

In communities with a single predator species and single prey, we would expect predator abundance to decline if prey abundance dropped dramatically. In **hyperpredation**, predator abundance remains high even when abundance of a prey species crashes. The predator either has a supplemental food supply, or some of the prey have a special circumstance keeping their abundance high (Mills 2007). For example, food placed outside for the benefit of stray domestic cats produces hyperpredation. The cats can subsist at higher densities than native carnivores. The same applies for predators that feed off garbage, as in gulls, crows, vultures, raccoons, rats, mice, and bears. Alternatively, hyperpredation can occur when both a new predator and a new prey species are simultaneously introduced to an area that already has similar predator and prey species (Mills 2007).

For example, think about a farm in a rural area that already has Bobcat (*Felis rufus*) and a native

mouse species. If feral cats and the House Mouse (*Mus musculis*) are introduced around the farm buildings, the area may now sustain a much higher density of feral cats than was possible with only the native mice species present. The feral cats prey on both the House Mouse and the native mouse species. Even if the native mice are only side snacks, they cannot sustain their population abundance.

The native mouse species probably has fewer offspring compared to the House Mouse because the native species is less likely to use a barn as its habitat. The native prey species is outside in harsh conditions, contending with ditching of wet areas, cutting of nearby forest, and cutting of all weedy areas to make the farm look tidy. The House Mouse population booms within the confines of a relatively dry barn full of grain. The Bobcat disappear because they are afraid of people and their prey base outside the barn decreases.

Hyperpredation can also occur when a new predator is introduced to an area without similar species (Figure 18.1). The Brown Tree Snake (*Boiga irregularis*) was introduced to Guam, probably as a stowaway in military planes from Papua New Guinea after World War II (Fritts 1988). This introduction led to the extirpation of several species of ground-nesting birds and mammals. Additional prey for the snake species includes lizards with a much higher reproductive capacity than birds and mammals. This explains why the snake species has not gone extinct. The lizard populations have been affected, but not at the same rate of decline as the bird and mammal species.

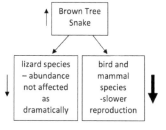

Figure 18.1 In hyperpredation one of several prey species may have a special circumstance keeping its abundance high, as in the case of lizards on Guam, a Pacific Island. An upward arrow indicates growing population numbers, a downward arrow indicates the opposite. The heavier the arrow, the more the growth or decline.

ECOLOGICAL THEORY REGARDING PREDATION

To address the important cultural and political questions regarding predation ecologists must be familiar with existing ecological theory. Predation theorists have framed their top questions as:

- Do predator-prey relations really cycle periodically?
- How can predators and prey coexist?
- Do predators hold prey abundance at an equilibrium stabilized around the carrying capacity, leading to a kind of balance?
- Is prey abundance controlled more by what prey eat (a bottom-up explanation), or by predators (a top-down explanation)?
- Are young and weak prey taken preferentially by predators?
- Do predators benefit prey populations by lowering the risk of disease and eliminating genetically weak individuals?

DO PREDATOR-PREY DYNAMICS CYCLE?

One of the best-known examples of a predator-prey cycle is the 10-year cycle between the Canada Lynx and Snowshoe Hare in the arctic. Data were collected for several decades from the Hudson Fur Company in the 1800s. For the lynx (Figure 18.2) and in other cases, there is often a delay in the response of one species compared to the other (Mackenzie et al. 2001). At first it was assumed the predator-prey dynamics themselves caused it: predation by lynx caused a decline in hare; decline in hare caused a decline in lynx; decline in lynx allowed the hare to recover, the pattern repeated.

However, observations of hare in areas without lynx showed that they cycled on their own (Krebs 2009). In fact, all the world's Snowshoe Hare are on the same cycle within 1 year. Krebs hypothesizes that dense hare populations cause a reduction in vegetation. The vegetation responds with high levels of toxins. This makes the hare population more susceptible to predation, causing a reduction in hare abundance. This creates a rise and fall in lynx abundance. Once the vegetation recovers, the hare density increases and the cycle repeats.

Additionally, Krebs (2009) explains that lynx is not the only predator of the hare. Radio collar studies show that Coyote, Great Horned Owl, and Goshawk prey on hares, with no predator predominating. During low hare periods, the predators turn on each other. The worldwide synchrony may be provided by an 11-year sunspot cycle, which produces periods of low snow accumulation.

Overall, regular population cycles are rare in nature except in viruses like measles. For all other

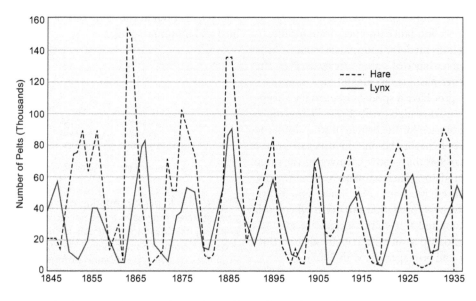

Figure 18.2 Cycles between lynx and hare plotted here from The Hudson Bay Company's records may now be explained by factors outside of the lynx and hare interactions (*G. Giacomin*).

organisms, regular cycles between two species tend to occur only in the following circumstances:

- A specialist predator preys upon one species,
- Both species reside in the north where fewer species of both predator and prey occur.

Cycles become dampened in the presence of a generalist predator, which is the situation generally found in the temperate and tropical areas.

HOW CAN PREDATORS AND PREY COEXIST?

Predators can but do not always cause species loss or permanent declines in prey abundance. In bottle experiments, G.F. Gause found that predators ate all the prey, then went extinct except when the prey were able to hide behind obstacles. In other studies, prey abundance declined when there were refuges because the prey became overcrowded in the enemy-free space leading to severe competition (Holt 2009).

How is it possible for predators and prey to coexist in the wild? On the Serengeti Plains of east Africa, many large predators (Lion, Leopard, Cheetah, wild dogs, Spotted Hyena) have had little impact on overall prey numbers, despite daily predation. Several explanations for predator-prey coexistence occur. First, most individuals taken by predators are **doomed surplus**, meaning that too many are born into the situation for the available resources. Through means of **compensatory mortality**, the surplus must die by one form of mortality or another. It might as well be predation as starvation.

Second, on the Serengeti, many prey species are migratory, yet the predators are resident (Smith and Smith 2001). For any one prey species, the effect on their population is minor because predation is seasonal. The ungulate (deer-like) prey species travel to higher elevations during the wet season.

Third, predators sometimes gain the upper hand because of better-evolved adaptations than their prey. In a hypothesis known as **co-evolutionary arms race**, predators develop adaptations to make them better predators, like better eyes, better chemoreception, or better behavior, as prey develop better defenses.

So, which has greater selective pressure, predators or prey? In thinking about the answer, keep in mind that if a predator is unlucky it misses a meal. If a prey is unlucky it loses its life. Indeed, it would seem that prey species have more selection pressure on them. However, prey must adapt to the capabilities of several predators at once, diluting adaptations of a particular specialty. At the same time, most predators are generalists and need to adapt to several prey species.

THE INSIGHTS OF HOLLING

Further insight into why prey can maintain their numbers in the presence of predators was provided in 1959 by C.S. Holling in a famous paper on functional and numerical response curves (summarized by Denno and Lewis 2009). In short, prey sometimes get a reprieve because predators are eating as fast as they can but cannot eat all the prey (eating is a "function" of mouth part morphology=functional response; Figure 18.3a). Some prey survive to reproduce. In other cases, predators become satiated (satisfied) before all the prey are gone. Most invertebrate predators (e.g., hunting spiders, praying mantis, and ladybug beetles) show this pattern.

In other cases, when a particular prey species is at low density, the prey are safe because the predator does not form a **search image**. At higher prey densities, a rapid rise in consumption rate occurs as predators become more familiar with and more efficient at eating an abundant prey type. At highest densities, predators become satiated or limited as above, and the functional response produces an S-shaped curve (Figure 18.5b). The explanation for this phenomenon is **switching**. Predators tend to eat whatever prey species is most abundant. As the most abundant form of prey becomes rare, the next abundant form becomes the target and the predator switches to the new species.

In a **numerical response**, the number of predators rises exponentially with more prey if predators congregate when they see other predators eating well. Eventually an asymptote forms when there is no more room for any more predators, or when the predators become satiated. Examples of this kind of crowding occur when shorebirds prey on a mudflat, or fish feed in schools, or herbivores feed in herds and no more room for any more individuals is available. The numerical response is also known as an **aggregative response** because predators tend to clump (aggregate) in the high-density prey patches.

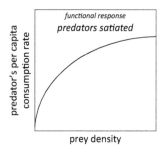
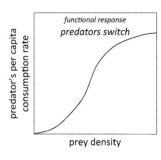
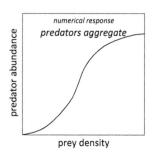

Figure 18.3 The leftmost panel is a functional response showing that as predators face higher prey densities their rate of consumption slows and does not keep up because they are eating as fast as they can. The middle panel is a functional response showing that predators switch to the most abundant and obvious type of prey as one prey type gets eaten away. The rightmost panel is a numerical response (notice the vertical axis has a different label). Predators congregate where the most prey are found.

PLANT DEFENSES AGAINST HERBIVORES

Just as prey can protect themselves from predators, plants can protect themselves from herbivores and thus coexist with their enemies. The plants evolve hair-like structures on leaves, spines on stems, or thorns on stems. Toxins develop in some plant species including the cardiac glycosides in milkweed species and alkaloids in cacti. Some plant species have evolved silica bodies in leaves and calcium deposits in stems that grind down the teeth in vertebrate herbivores and the mandibles in invertebrates.

MODELING PREDATOR-PREY INTERACTIONS

Questions about predator control, the reintroduction of large predators, and the introduction of non-native predators are extremely contentious in geographic areas where native predator species like lions and bears occur. Thus, the need for predictive models is great.

The difficulty is that while predation is more visible than competition, the dynamics of predation are more complex than competition. Multiple predator species may consume multiple prey species in any given place. For example, Red Fox is a major predator of duck eggs in the prairie pothole region of North America, but nest success can be higher in areas with coyotes because they kill or drive off Red Fox (Sovada and Carbyn 2003).

Alfed Lotka and Vito Volterra devised equations based on the logistic model to predict the dynamics of predator-prey interactions. Just as in the original logistic model, equations for both continuous and discrete generations were devised. Just as in the debates regarding population regulation and competition, the Lotka-Volterra equations needed modifications and more features to more accurately predict predator-prey dynamics in the wild. Modifications by Rosenzweig-MacArthur, Nicholson-Bailey, and others are commonly used (Rockwood 2015, Krebs 2009, Gotelli 2008).

At first it was assumed that predator abundances always cycled out of synchrony with their prey as observed in the lynx-hare example. However, these models were very sensitive to restrictive assumptions just as in competition theory (Gotelli 2008). In particular, they did not account for the functional response of predators, the phenomenon of a predator becoming satiated, or having a limit to how fast it could search for and handle prey.

Another insight concerns patchiness. Initially, it was assumed that foraging for prey occurred at random. We now understand predation to take place mainly in patches of high prey density. Optimal foraging theory and patch dynamics ideas, such as the marginal value theorem and ideal free distribution hypothesis, provide predictive contemporary models.

Additionally, the first models did not take into account prey hiding places. The visible prey may quickly be consumed by aggressive predators, but once the visible prey are gone, the predators begin to starve. Once predator numbers are low, the prey abundance increases and the cycle starts over. Other unrealistic assumptions were that predators are always specialists on particular prey species, and there is no immigration.

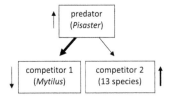

Figure 18.4 In Robert Paine's keystone species experiment *Pisaster* was the top predator and when its numbers were high, the abundance of the mussel *Mytilus* remained low.

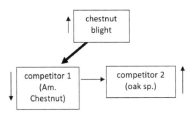

Figure 18.5 The Chestnut Blight (a fungus) can act as a keystone species. In its presence, oak species become dominant.

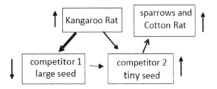

Figure 18.6 The Kangaroo Rat (a granivore) can act as a keystone species. In the presence of the Kangaroo Rat, plants with large seeds decline in abundance. Those with tiny seeds proliferate. Sparrows and Cotton Rats benefit because they prefer small seeds.

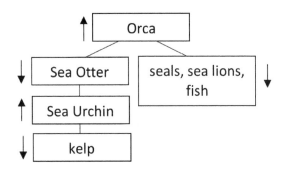

Figure 18.7 When Orcas are present, their effect as top predator cascades down the food chain.

KEYSTONE PREDATORS

Predation in aquatic situations brought new insight to ecology through a landmark paper by Robert Paine in 1966. Paine worked in the rocky intertidal zone (Figure 18.4) investigating **sessile** (very little if any motion) creatures of Washington state on Tatoosh Island.

The starfish, *Pisaster*, was fairly uncommon, considered unimportant as an ecological community member, but functioned in the role of what Paine called a **keystone species**. A keystone is the stone in the middle of the arch that holds the rest of the arch together. In the same way, a keystone species plays a critical role in determining community membership. In this case *Pisaster* kept population abundance low for *Mytilus*, the dominant competitor among the prey species. If the starfish was removed, the bully prey species *Mytilus* reduced species diversity among the prey by outcompeting most of them.

KEYSTONE SPECIES THAT ARE NOT PREDATORS

An array of mechanisms can facilitate coexistence among predators and prey. Top predators like a Grizzly Bear or Lion can be a keystone species, but an organism does not have to be large or predaceous to be a keystone. A beaver is a keystone species because it builds a dam, which dictates whether the habitat is a small shallow lake or a terrestrial forest with a small stream flowing through it. Once the area becomes a lake, a very different community inhabits it. Humans are a keystone species through use of fire (Bowman and Murphy 2010). Parasites can be keystone species, like the fungus causing the American Chestnut blight (Figure 18.5).

Grazers and granivores can be keystone species. Elephants push over small trees and destroy shrubs, transforming an area into a grassland for all species around it. Kangaroo Rats in the Chihuahuan Desert are keystone species because they prefer large seeds (Figure 18.6). Plants with tiny seeds thrive, promoting ground-feeding birds and mammals that would otherwise not be present. These examples demonstrate the multiple effects of competition and predation work synergistically to promote species richness.

The effects of a keystone species can cascade down the food chain, indirectly promoting population numbers of prey eaten by prey as in Prince William Sound, Alaska (Figure 18.7).

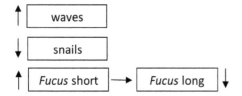

Figure 18.8 In the rocky intertidal zone of New England, the effect of crashing waves cascades down the food chain.

JANE LUBCHENCO

Based on Paine's work, Jane Lubchenco focused her dissertation on the rocky intertidal zone of New England. At her study site, short and **ephemeral** (temporary or seasonal) algal species grew, which were grazed by snails. A competing algal species (*Fucus*) was long and ungrazed. When Lubchenco removed the snails, short algae colonized to the exclusion of *Fucus* (Figure 18.8). *Fucus* only occurred in areas where snails were present. Furthermore, abiotic factors had control over snail distribution. Where crashing waves were present, snail abundance was low. The waves controlled snail abundance, and snails controlled the type of algae. It demonstrates that predation, competition, and disturbance can work together as the organizing factors in a community.

MENGE AND SUTHERLAND

Robert Menge and John Sutherland (1987) produced a conceptual model to capture the combined results of Lubchenco, Paine, and others. Physical disturbance, predation, and competition play important roles in community structure in the following ways:

1. In stressful environments herbivores have little effect because they are rare or absent. Plants tend to be regulated by environmental stress, thus predation and competition do not play a large role, as in the arctic tundra or desert.
2. In moderately stressful environments, herbivores are present, but tend to be ineffective at controlling plant growth. Consequently, plants attain high densities. Competition between plants is the dominant interaction.
3. In benign (low stress) environments, herbivores control plant numbers and thus plant competition is rare. Predation tends to be the dominant biological interaction.

In all cases, environmental stress is the key process in community organization.

THINKING QUESTIONS

1. Compare and contrast predation with herbivory and parasitism.
2. List three main reasons why predators and prey can coexist in nature.
3. Why is it that prey on the Serengeti do not become depleted?
4. Why do non-herding animals often have horns or antlers?
5. If an evolutionary arms race occurs, which is likely to have more selective pressure, predators or prey? Explain.
6. Why are aggregative response curves S-shaped?
7. Can predation structure communities? If so, how and in what type of situation?

REFERENCES

Bowman, M.J.S., and B.P. Murphy. 2010. Fire and biodiversity. In: N.S. Sodhi and P.R. Ehrlich (eds.), *Conservation Biology for All*. Oxford University Press, Oxford, UK.

Denno, R.F., and D. Lewis. 2009. Predator-prey interactions. In: S.A. Levin (ed.), *The Princeton Guide to Ecology*. Princeton University Press, Princeton, NJ.

Fritts, T.H. 1988. The Brown Tree Snake, *Boiga irregularis*, a threat to Pacific Islands. *US Fish and Wildlife Service Biological Report* 88(31):36.

Gotelli, N.J. 2008. *A Primer of Ecology*. Sinauer Associates, Sunderland, MA.

Holt, R.D. 2009. Predation and community organization. In: S.A. Levin (ed.), *The Princeton Guide to Ecology*. Princeton University Press, Princeton, NJ.

Krebs, C.J. 2009. *Ecology: The Experimental Analysis of Distribution and Abundance*. 6e. Pearson, New York, NY.

Mackenzie, A., A.S. Ball, and S.R. Virdee. 2001. *Ecology. Instant Notes*. 2e. BIOS Scientific Publications, Oxford, UK.

Menge, B.A., and J.P. Sutherland. 1987. Community regulation: variation in disturbance, competition, and predation in relation to environmental stress and recruitment. *American Naturalist* 130:730–757.

Mills, S. 2007. *Conservation of Wildlife Populations*. Blackwell, Malden, MA.

Rockwood, L.L. 2015. *Introduction to Population Ecology*. 2e. Blackwell Publishing. Oxford, UK.

Smith, R.L., and T.M. Smith. 2001. *Ecology and Field Biology*. Addison Wesley, San Francisco, CA.

Sovada, M.A., and L. Carbyn (eds.). 2003. *The Swift Fox: Ecology and Conservation of Swift Foxes in a Changing World*. Canadian Plains Research Center, Regina, Saskatchewan.

Weber, L.M., and S.M. Haig. 1997. Shorebird-prey interactions in South Carolina coastal soft sediments. *Canadian Journal of Zoology* 75:245–252. https://doi.org/10.1139/z97-031

Wiki Commons: File:Milliers fourrures vendues 90 ans odum 1953.jpg, "Wikimedia Commons, the free media repository." https://commons.wikimedia.org/w/index.php?title=File:Milliers_fourrures_vendues_90_ans_odum_1953.jpg&oldid=546279991 (accessed July 14, 2022).

19

Succession

CLASS ACTIVITY: Investigating Succession

Goal: Quantifying aspects of succession in four seral stages.

What would it look like if a pasture in your biogeographic region were left undisturbed for 1, 3, 35, or 60 years? You would observe old-field succession (so called because it happens when an old farm field is abandoned). The species present in the early stages would depend on previous history of the field, nearby seed sources, and time of year the field was abandoned. Eventually, the area would be dominated by late succession vegetation characteristic of your region. Your assignment is to use several measurements to record secondary successional changes by observing four stages.

Step 1: Take a coverboard (see activity in Chapter 5), tape measure, and four pin flags. At each successional stage, set up a randomly placed 20×20 m quadrat. Use random numbers from a random numbers table or internet random number generator to select a random number of paces from the trailhead or other landmark to set up the plot.

Step 2: At each stage, record
- coverboard index=(% covering top panel×4)+(% covering third panel×3)+(% covering second panel×2)+(% covering lowest panel×1),
- number of layers (ground, fern/herb, shrub, understory trees, canopy),
- an estimate of the total number of plant individuals in the plot.

Step 3: For each variable you measured in Step 2, draw a bar graph of results and determine which seral stage had the greatest quantity. Speculate why. Incorporating succession theory write a summary paragraph with your results from all seral stages.

Step 4: Suppose an additional measurement was percentage of each dominant vegetative species. In the third seral stage, Tulip Poplar trees dominated with 40% coverage followed by Red Maple with 22%. In the fourth seral stage, White Oak dominated by 42% and Shagbark Hickory by 12%. Write you own model of succession to explain how this change in ratio occurred.

DOI: 10.1201/9781003271833-19

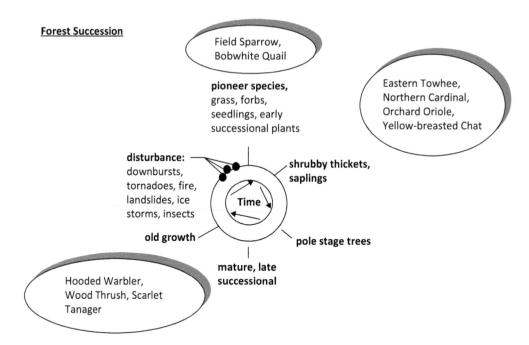

Forest Succession

Field Sparrow, Bobwhite Quail

pioneer species, grass, forbs, seedlings, early successional plants

Eastern Towhee, Northern Cardinal, Orchard Oriole, Yellow-breasted Chat

disturbance: downbursts, tornadoes, fire, landslides, ice storms, insects

Time

shrubby thickets, saplings

old growth

pole stage trees

mature, late successional

Hooded Warbler, Wood Thrush, Scarlet Tanager

Figure 19.1 Forest succession, showing pioneer species just after a disturbance succeeding to old growth. Disturbance sets it back. Species of birds are indicative of different forest stages in the U.S. eastern deciduous forest. (*Adapted from Yahner 2000 by Weber.*)

SUCCESSION OVERVIEW

Succession is the sequence of changes in the biota following a disturbance in which one kind of community is replaced by another. For example, a farm field in the eastern U.S. that was once tilled but now left undisturbed may be transitioning from grass or shrubs to forest over several years. This is **old-field succession** (Figure 19.1), named after abandoned farm fields. Succession can also happen in lakes or decomposition. The first colonizers are **pioneer species**. Each major community type in the successional sequence is a **seral stage**.

TYPES OF SUCCESSION

Degradative succession is when organic material is degraded by detritivores in a series of communities. An example is cow droppings decomposed by a series of microbes and invertebrates. A second example is the decomposition of a corpse, the successional sequence so predictable it gives forensic clues, including the host's time of death.

In contrast, **non-degradative succession** is the community sequence that develops in an empty environment following disturbance. **Disturbance** is a weather or geologic discrete event that removes organisms and opens space for recolonization (Smith and Smith 2001). **Primary succession** occurs after the environment has been sterilized or buried for a long time such as during glaciation, volcanic activity, or a meteor strike. **Secondary succession** occurs after a disturbance when life and seeds are still present from past communities. This could follow fire, windthrow, hurricane, flood, thunderstorm, ice storm, tornado, grazing, lawn mowing, or pesticide use.

Secondary succession occurs in old fields, dunes, and bogs. Bog succession happens in coniferous forest and is characterized by changes from open pond to quaking bog to nearly solid ground, which can then be set back to pond again by fire. Secondary succession may or may not follow the same course as primary succession, such as in mature tropical moist forests. Most of the nutrients are held in biomass, not soils, which means remaining nutrients leach from the soils rapidly once tropical forests are cut. Forest regeneration may not take place as a result, thus taking a different course than primary succession.

THEORETICAL QUESTIONS ABOUT SUCCESSION

Succession is analogous to aging, proceeding in one direction until disturbance occurs. Two factors characterize succession (Gotelli 2008):

- Species occurring immediately after a disturbance are not the same as those present later.
- Species occurring late in succession are the same as those in surrounding late succession areas, even if the early succession species were different.

To early ecologists, these patterns suggested a deterministically controlled cause, not the result of random processes. Climate and latitude or a combination of the two were proposed as the controller. In traditional thinking, a climax community characteristic of each region occurred, a regional community type that could be classified as organized and repeatable.

THEORY OF SUCCESSION

Theory regarding succession is usually framed in terms of the debate between **Fredric Clements** (1874–1945) representing the holistic view of communities, and Henry Gleason representing the individualistic view. Clements and his colleagues categorized groups of plants in what he termed *climax communities*. By **climax**, he meant, "a more or less permanent and final stage of a particular succession" that was "inseparably connected with its climate" (Clements 1936).

Clements, the developer of the biome idea with co-author Victor Shelford, regarded climax and biome to be "complete synonyms," but he used biome to emphasize that animals were part of this climax. Clements wrote that the climax constituted "the major unit of vegetation and as such forms the basis for the natural classification of plant communities." Clements and his associates put tremendous effort into devising a classification system for the U.S., although a finished product was never produced. They ran into more and more exceptions, needing to divide the environment into smaller and smaller sections, which became unwieldy.

Clements came to epitomize the **holistic view** of communities (Golley 1993):

- A single ultimate climax community (sometimes called **monoclimax**) was determined by the climate and characteristic of each region.
- Each climax community had a dominant plant species.
- Succession stopped when it arrived at an equilibrium or steady-state point.
- An ecological community could be studied like an organism either for its single parts as in an anatomy class, or as a whole individual.

Fredric Clements is credited with a great deal of work and effort over his professional career. It should be noted that his wife, Edith, contributed as his aide, but is often not recognized (Figure 19.2). She was the first woman to gain a doctorate (botany) from the U. of Nebraska (Bonta 1995). She wrote ten books mainly illustrating and photographing plants and describing their natural history. She was a scholar of Germanic languages and translated her own books. She served as Fredric's chief field technician, professor, stenographer, chauffeur, car mechanic, and cook during their early 20th-century field seasons consisting of long, remote trips in the western U.S. deserts and prairies. In college she was a Phi Beta Kappa, captain of the basketball team, president of the junior class, and participant

Figure 19.2 Dr. Edith Clements was a natural history botanist, artist, author, conservationist, mechanic, athlete, and college professor. Fredrick Clements said of her that she would have been ranked among the world's top ecologists had she spent less time assisting his career.

in exhibition fencing and tennis. In her autobiography published when she was 86, she reports that Frederic was a chronic diabetic and workaholic, forgetting to eat or sleep unless reminded by Edith (Clements 1960). She was never out of Fredric's sight for more than 24 hours in 40 years. She was a full participant in Fredric's work. On their honeymoon they conceived the idea of an alpine botany laboratory on Pikes Peak. She helped Fredric raise money for it, saw it through to fruition, and taught there during its 40-year existence.

THE INDIVIDUALISTIC VIEW OF SUCCESSION

H.A. Gleason came to epitomize the opposition to Clements by putting forth the **individualist view** published in 1926 (summarized by Hagen 1992):

- Communities should not be thought of as superorganisms because a whole community cannot reproduce itself.
- Community membership is a result of chance. Each community is a haphazard collection of species with no obligate interactions that tie all the species together. Colonization depends on seed bank, nutrient levels, grazing and disturbance frequency, and the time of year for the disturbance. Each situation is unique and arises through selection.
- No two communities are alike and individual species respond independently. As one walks along a gradient, communities gradually change. No sharp boundaries arise between them.
- Communities cannot be classified specifically because they are not repeatable. It is not possible to develop a precise categorization.

RESOLUTION OF THE DEBATES

Ecologists of the 20th century tended to believe Gleason was closer to the truth in the sense that species membership is more haphazard than determined in successional communities. Most ecologists eventually abandoned the idea of a simple climax community. When looking at old growth forests in New Hampshire at Hubbard Brook, Bormann and Likens (1979) observed that old growth forests are uneven-aged and gappy. Trees fall at random and what ensues is competition among the younger plants for space and light. The same is true for tropical rain forests and heathlands in Scotland. Bormann and Likens argued that forest succession is actually more like a **shifting mosaic steady state**. Rather than achieving equilibrium at a large scale, an ecosystem consists of patches in various seral stages, known as the **patch dynamics model**. At a metapopulation level, the community may be stable, but at the level of patches, it is not. This is a non-equilibrium model in which populations wink in and out with disturbance, competition, and climatic changes.

Even within the patch dynamics model, most forest ecologists recognize patches of late succession old growth but acknowledge it can take many years to reach that point. Tropical rain forests take approximately 100 years to become old growth, temperate oak-hickory forests in the eastern U.S. take 80–200 years, beech-maple forests on sand dunes take 1,000 years. A bog community can be stable for as long as 10,000 years once it is in the late succession phase.

In the end, the Clements were not successful in establishing a classification system for climax communities based on particular species. Their attempts were continually under change and a final classification was never published. The Clements group kept defining smaller and smaller entities, becoming too complicated and burdened with multisyllabic vocabulary (McIntosh 1985). Their mistake was trying to classify vegetation by species, not by life forms as was done with biomes. Their efforts to identify subregions by two or more dominant species were too specific for small areas because one community graded into another. The biome concept was more successful (Clements and Shelford 1939). A vegetation classification now based on formations has been developed for the U.S. (Faber-Langendoen et al. 2016).

MODELS THAT ORGANIZE IDEAS ABOUT WHAT CONTROLS COMMUNITY CHANGES DURING SUCCESSION

While the Clements-Gleason debate may have brought more appreciation for the contribution of chance within successional change, questions remained about why a predictable change occurs during succession. Think about the process of replacement for plant communities with time. If a forest is 40% oaks in one generation, it might be tempting to predict the next generation would be 40% oaks. This rarely occurs for the following reasons:

- Some individuals of some species live longer than others. The species with the longest individual tree ages tend to become dominant;
- Species that do well in shade for many years under the canopy will become dominant;
- Individual plants that compete well for light, water, and nitrogen sometimes drive growth;
- Biotic factors such as shading sometimes drive succession, but abiotic factors like soil type, aspect, and precipitation drive it in others;
- Vegetation becomes patchy from disturbance events. This sets back disturbance at random throughout the environment.

DISTURBANCE

One of the most important studies on disturbance was completed by J. Connell in 1979. Studying boulders off the coast of California, he counted the number of species that inhabited the rock surface after disturbance. The boulder studies allowed observation of the complete process of succession, but on a short sere (any one phase in a seral stage). With this quickened process, disturbance was less likely to interrupt the complete successional cycle. From his work, Connell introduced the idea of the intermediate disturbance hypothesis (Figure 19.3) used to explain the number of species coexisting in a community.

At high frequencies of disturbance, species richness is low because few species can tolerate the devastation. At low frequencies species richness is

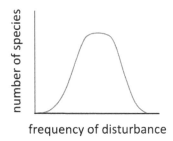

Figure 19.3 Intermediate disturbance hypothesis as proposed by J. Connell.

low because competition or predation is at work. It is only at moderate levels of disturbance that species richness is high because the short period between disturbance events does not allow competition to proceed to monocultures. While helpful, the hypothesis has been criticized for its lack of quantification.

ANALYSIS OF THE SUCCESSION AND COMMUNITY ECOLOGY PARADIGM – FACILITATION

It is easy to understand why early community ecologists put forward a superorganism model to describe succession and community ecology in general. Predation, competition, parasitism, and mutualism provide an astounding set of interdependencies. After a century of concentrating mainly on competition and predation, ecologists are now giving more attention to mutualistic interactions within communities.

To begin, consider facilitation as mutualism. **Facilitation** is modification of some component of the environment by one organism that enhances colonization, recruitment, and establishment of another, inter- or intraspecifically. This is not a symbiotic relationship. The species are not interacting for most or all of their lifetimes and are not physiologically dependent on one another.

For example, shade is one of nature's most important facilitative mechanisms, keeping plants below lethal temperatures, decreasing respiration costs and transpiration loss, reducing ultraviolet irradiation, and increasing soil moisture (Callaway 2009). In many cases shade is vital, as in the case of Saguaro Cactus in the Sonoran Desert of the American southwest. Nurse plants (offspring) of the cactus grow only in the shade (where they also have a wind screen) even when sufficient water is applied.

In general, facilitators can build up litter, decrease soil bulk density, intercept rain or fog, maintain soil moisture, and oxygenate soil (Callaway 2009). Deeply rooted trees may take up nutrients not available to shallowly rooted plants, depositing the nutrients on the soil surface via litter fall.

In the stress gradient hypothesis proposed by Bertness and Callaway (1994), facilitation is most likely to occur in very high abiotic stress. In less stressful situations, competition is more likely. The explanation is that without stress, there is nothing for a neighbor to facilitate (Callaway 2009).

In general, short-lived species tolerate stress well and can colonize or germinate quickly after a disturbance. We can think of them as specialists of disturbed sites that are only capable of colonizing disturbed sites. Early species are better dispersers than later species. They tend to be annuals and biennials. Once there, they often facilitate conditions for longer lived species that can tolerate conditions better, outcompete the early species, or inhibit them. These late species tend to facilitate their own offspring, unlike the early species. Late species are more resistant to disturbance. They tend to be perennials.

FURTHER ADVANCES IN UNDERSTANDING AND MODELING MUTUALISM

Darwin himself was fascinated by mutualist relationships, especially those between insect pollinators and specific plant species because they demonstrated the process of evolution so well. With the discovery of mycorrhizal relationships (symbiotic relationships between fungi and roots) in the 1890s, the phenomenon of mutualism became known to ecologists. Yet mutualism largely escaped the attention of ecology theorists during the 20th century, especially if more than a pair of species was involved (Leigh 2020). Perhaps this oversight was because of the complexity involved. Most participating species are unrelated and generalists rather than specialists, mutual benefits are present, and mutualistic relationships are ubiquitous. All terrestrial vertebrates, plants, and arthropods have mutualisms (Bascompte and Jordano 2014).

In actuality, mutualisms among species are one of the main wireframes of ecosystems, and ecosystems would collapse in the absence of animal-mediated pollination or seed dispersal of higher plants (Bascompte and Jordano 2014). Both processes depend on a reward to an animal while foraging (nectar, pollen, pulp, seeds, oil), which maintains animal diversity through keystone life histories. Five major types of mutualism exist (Bascompte and Jordano 2014):

1. Pollination;
2. Seed dispersal by animals for plants;
3. Protective measures by ants (and other arthropods) protecting plants from sucking insects like aphids, cicadas, and mealybugs;
4. Harvest mutualisms that include the gut flora and fauna of animals and the root rhizosphere inhabited by decomposers, lichens, and some epiphytes;
5. The interaction between humans and domesticated plants and animals.

The next frontier in ecological theory may be to embrace whole megadiverse assemblages in networks of mutualism and to infer consequences for biodiversity conservation if the mutualisms collapse (Bascompte and Jordano 2014). The theory of complex networks involves more than just ecology. Topics studied via "complex networks" are the internet, the animal brain, social behavior, sharing of genes, and protein regulation. Insights for techniques in these areas may help unlock the complexities of species diversity in, say, communities of tropical forests.

In short, the study of complex networks involves sophisticated graphs that consists of a set of nodes connected by "edges" (Figure 19.4). Food webs can be modeled this way, with modelers removing one species at a time and determining if the network collapses or adapts. Competitive and cooperative interactions can be studied together. Because plant-animal interactions tend to occur toward the base of the food webs (pollination, seed dispersal) the modeling is especially valuable for conservation efforts.

Three things influence network stability. Having these properties may buffer the effects of anthropomorphic changes somewhat. They include:

- Nestedness (specialists interacting with some species that are generalists),
- Redundancy (pollination by many pollinator species),
- Modularity (subclusters with dense nodes that do not have dense nodes with other subclusters).

Complex network computer models of this type might be able to answer the question of whether complexity leads to stability, a research area with ambiguous conclusions after decades of field studies. The study of animal behavior within the networks can reveal disease transmission, properties of personality (Hunt et al. 2018), and traits that reveal evolutionary pathways, among other things (Estrada 2011, Krause et al. 2015).

(a)   (b)

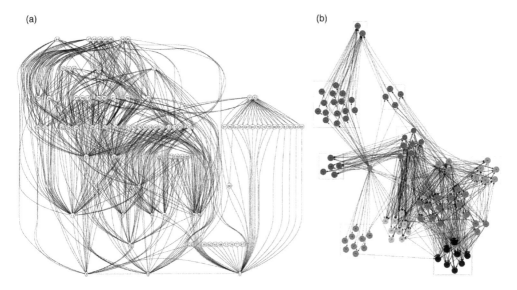

Figure 19.4 The structure of the food web at Secaucus High School Marsh with species sorted according to group affinity. The lines connect a consumer with a consumed species; the nodes represent species. The grouping configuration is determined by Allesina and Pascual (2009), assessed using AIC, a configuration containing 15 groups with the best fit of the data. (*From Anderson and Sukhdeo 2011, Creative Commons.*)

From here, we cannot go much further with theoretical conclusions until we add the insight of ecosystem and landscape ecology. Incorporation of abiotic processes benefits by thinking of nature as a machine with energy flow and matter recycling as in ecosystem ecology. Although one of the problems in ecosystem ecology has been the failure to appreciate patchiness in nature, landscape ecology can identify feedback-driven patterns at several spatial and temporal scales simultaneously. In short, we need to recognize feedback loops and cycles especially for resources like carbon and nitrogen in biological systems. Observation at several scales will require both ecosystem and landscape ecology covered in the next set of chapters.

THINKING QUESTIONS

Now that you have covered two of the major subdisciplines in ecology (populations and communities) take this intermediate self-test.

1. Community ecology has a particularly rich list of ideas it considers as theories. For each of the ideas below be able to summarize the main conclusions and describe the contribution to our overall understanding of ecology.
    Competitive exclusion principle –
    Niche theory –
    Succession theory –
    Keystone species theory –
    Theory of island biogeography –
    Shifting mosaic steady-state theory –
    Equilibrium theory –
    Non-equilibrium theory –
    Metapopulation theory –
    Intermediate disturbance hypothesis –

2. What would a metacommunity theory be?
3. Is succession logical and orderly? Can species composition be predicted accurately?
4. Consider the spatial scale of forest succession. Compare it to microbial succession. How might they differ in theoretical principles?
5. What is the scale of degradative succession?
6. Does succession have an end point?
7. What role do non-native species play in succession?
8. At what successional stage does carbon accumulate most rapidly?
9. From a contemporary ecology viewpoint, what structures (or organizes) most communities?
10. Vellend (2016) presents a comprehensive theory of ecological communities based simply on the four familiar concepts of speciation, dispersal, drift, and selection. Look back at the constraint hierarchy in Chapter 9. Could the constraint hierarchy be reduced to these four concepts?
11. Identify all the plant species on a nearby lawn. If you are in the eastern U.S. a familiar weed is Gill-over-ground (Ground Ivy, Creeping Charlie), a member of the mint species. It purportedly has a pineapple smell when crushed through the fingers. Try it and make your judgment. Find Broadleaf and Narrow Leaf Plantain, sometimes known as White Man's Footprint. Why does it have this name? Do these species have medicinal value? What other species are present in your lawn? What would happen if you did not mow for 2 years?
12. Examine the photo below. Is a landslide primary or secondary succession? What do you think this site will look like 15 years after the disturbance that created it? The slide occurred in Jones Gap State Park in South Carolina.

REFERENCES

Anderson, T.K., and M.V.K. Sukhdeo. 2011. Host centrality in food web networks determines parasite diversity. *PLoS One* 6:e26798. https://doi.org/10.1371/journal.pone.0026798

Allesina, S., and M. Pascual. 2009. Food web models: a plea for groups. *Ecological Letters* 12:652–662.

Bascompte, J., and P. Jordano. 2014. *Mutualistic Networks*. Princeton University Press, Princeton, NJ.

Bertness, M.D., and R.M. Callaway. 1994. Positive interactions in communities. *Trends in Ecology and Evolution* 9:191–193.

Bonta, M.M. 1995. *American Women Afield*. Texan A&M University Press, College Station, TX.

Bormann, F.H., and G.E. Likens. 1979. Catastrophic disturbance and the steady state in northern hardwood forests: a new look at the role of disturbance in the development of forest ecosystems suggests important implications for land-use policies. *American Scientist* 67:660–669.

Callaway, R.M. 2009. Facilitation and organization of plant communities. In: S.A. Levin (ed.), *The Princeton Guide to Ecology*. Princeton University Press, Princeton, NJ.

Clements, E.S. 1960. *Adventures in Ecology: Half a Million Miles: From Mud to Macadam*. Pageant Press, now open source.

Clements, F.E. 1936. Nature and structure of the climax. *Journal of Ecology* 24:252–284.

Clements, F.E., and V.E. Shelford. 1939. *Bio-Ecology*. John Wiley & Sons, New York.

Estrada, E. 2011. *The Structure of Complex Networks: Theory and Applications*. Oxford University Press, Oxford, UK.

Faber-Langendoen, D., T. Keeler-Wolf, D. Meidinger, C. Josse, A. Weakley, D. Tart, G. Navarro, B. Hoagland, S. Ponomarenko, G. Fults, and E. Helmer. 2016. Classification and description of world formation types. General Technical Report RMS-GTR-346. U.S. Department of Agriculture, Forest Service, Rocky Mountain Research Station, Fort Collins, CO.

Golley, F.B. 1993. *A History of the Ecosystem Concept in Ecology*. Yale University Press, New Haven and London.

Gotelli, N.J. 2008. *A Primer of Ecology*. Sinauer Associates, Sunderland, MA.

Hagen, J. 1992. *An Entangled Bank: The Origins of Ecosystem Ecology*. Rutgers University Press, New Brunswick, NJ.

Hunt, E.R., B. Mi, C. Fernandez, B.M. Wong, J.N. Pruitt, and N. Pinter-Wollman. 2018. Social interactions shape individual and collective personality in social spiders. *Proceedings of the Royal Society B: Biological Science*. Sep 5;285:20181366. doi:10.1098/rspb.2018.1366

Krause, J., R. James, D. Frank, and D. Croft. 2015. *Animal Social Networks*. Oxford University Press, Oxford, UK.

Leigh, E.G. 2020. Neglected problems in ecology: interdependence and mutualism. In: A. Dobson, D. Tilman, and R.D. Holt (eds.), *Unsolved Problems in Ecology*. Princeton University Press, Princeton, NJ.

McIntosh, R. 1985. *The Background of Ecology*. Cambridge University Press, Cambridge, UK.

Smith, R.L., and T.M. Smith. 2001. *Ecology and Field Biology*. Addison Wesley, San Francisco, CA.

Vellend, M. 2016. *The Theory of Ecological Communities*. Princeton University Press, Princeton, NJ.

Ecosystem ecology basics

WHAT CHARACTERIZES ECOSYSTEM ECOLOGY?

Ecosystem ecology is the study of the flow of energy and matter through organisms and the environment. In 1945, "ecosystem" was coined by Arthur Tansley (1871–1955), a British ecologist. "Ecosystem" and the idea behind it were in opposition to the Clementian superorganism view. Rather than a self-directed organism, Tansley emphasized the scientific interaction of the biota with the environment (Golley 1993).

Despite the long association ecology had with biogeography and its inclusion of the physical environment, community ecology had tended to emphasize biotic interactions over abiotic.

Tansley's work opened the door for the wider use of energy theory and matter cycling in ecology. Ecosystem ecology evolved to have the characteristics defined in the rest of this chapter.

CHARACTERISTICS OF ECOSYSTEM ECOLOGY – MACHINE THEORY

Matter and energy cross the boundaries of the living and non-living as if living beings were just bags of chemicals – machine theory applied to nature (Golley 1993). Ecosystem ecology is not the same as community ecology merely with added emphasis on the abiotic or larger spatial scales. Ecosystem ecology focuses on matter cycling and energy flow and can occur at any spatial scale, even the head of a pin.

CHARACTERISTICS OF ECOSYSTEM ECOLOGY – HOLISM

Ecosystem ecology is based on **holism**, an antireductionist whole-before-the-parts philosophy (Hagen 1992). Individual parts may be studied, but the primary focus is the overall structure and function of the ecosystem. A change in any one component may bring about change in other parts and in the operation of the whole. Consider the decline in ocean fisheries throughout the world. After years of species-based management, a more holistic approach considers the resource levels upon which all commercial fish depend.

CHARACTERISTICS OF ECOSYSTEM ECOLOGY – EMERGENT PROPERTIES

Emergent properties is a phrase that reflects the idea not just of holism, but that the whole is greater than the sum of the parts. Like a family, ecosystems take on their own personalities demonstrating emergent properties. Higher-level laws govern or change the behavior of the parts. Emergent properties of ecosystems may include characteristics of productivity, stability, or food web connectance (Vellend and Orrock 2010).

CHARACTERISTICS OF ECOSYSTEM ECOLOGY – SYSTEMS APPROACH

Ecosystem ecology takes a **systems approach**, an idea attributed to Ludwig von Bertalanffy, a biologist born in 1901. One distinction is between open and closed systems. **Closed systems** have no interaction with the outside. As physicists think of it, the solar system, an atom, and a pendulum are all closed. All biological systems are **open systems**, meaning inputs and outputs of energy and materials flow through. Although all ecosystems are open, the concept is useful as a relative term – some ecosystems are more closed than others.

In Bertalanffy's approach, components of a system include inputs, throughputs, and outputs. Think of a cat. The inputs are food, water, and oxygen. The outputs are urine, feces, carbon dioxide, and hairballs. In the cat, the boundaries of the system are the cat's fur.

In ecosystems, the boundaries are more arbitrary. Lakes may seem to have established upper, side, and lower boundaries. What about gas exchange? If a fish jumps out of the lake then back in, did it leave the ecosystem? Is a loon part of an ecosystem if it spends part of its time in water and part in air? Where does the stream begin and the lake end? Is groundwater included? Is the sediment?

It might be tempting to define ecosystems geographically with a common fixed unit, such as a watershed. The problem is that as much as 20% of the U.S. does not have defined drainage areas, for example, in the desert, in glaciated prairie pothole regions, and in wetlands of Florida and the Great Lakes (Bailey 2006). Even where the surface water can be defined, ground water may not correspond. In the end, ecosystem boundaries are arbitrary, but to a degree this is still minimal for the concept to be valuable.

Other components of a system include a **processor**, the central unit responsible for the functioning of the system. The processor can be described as a black or white box. A **white box** is transparent. We can see the inside parts as if it were under an x-ray machine and we know how the parts function.

A **black box** is opaque; we do not know the contents. Either we do not understand how it works, or we choose to ignore these reductionist underpinnings. For example, if too much pollution is pouring into the rivers of a city, we may not care which factory puts out the most. The goal is to limit pollution for this city perhaps by putting in a new sewer system and maintaining a vegetative buffer alongside the river.

Is Western medicine a black box or white box approach?

CHARACTERISTICS OF ECOSYSTEM ECOLOGY – FEEDBACK LOOPS

A system has positive and negative feedback loops. **Positive** loops cause a system to change further in the same direction, like money in a savings account at a bank that earns compound interest. **Negative** loops are corrective. The system changes in the opposite direction like the operation of a thermostat. The temperature gets too high and the thermostat turns off the furnace. The temperature gets too low and the thermostat turns on the furnace.

CHARACTERISTICS OF ECOSYSTEM ECOLOGY – FOOD CHAINS AND TROPHIC LEVELS

A **trophic** (feeding) **level** refers to all the organisms that gain their energy source in the same way. This is the idea of autotrophs, heterotrophs, primary producers, and consumers, ideas developed by Charles Elton, a British animal behavioralist.

CHARACTERISTICS OF ECOSYSTEM ECOLOGY – ADAPTIVE MANAGEMENT

Ecosystem management may include the idea of **adaptive management**, in which uncertainty is acknowledged and the manager changes one's course of action as new information becomes available. It is learning by doing. It does not imply that ecosystem managers have a complete lack of knowledge. It refers to an expectation that ideas, theories, and situations will develop over time, and no one set of ideas is locked in place.

THE AQUATIC ECOLOGY LEGACY

Aquatic ecologists were well ahead of terrestrial ecologists in developing ecosystem ideas (Golley 1993). Almost 50 years before Tansley introduced "ecosystem," limnologists thought of lakes as "microcosms." Like ecosystems, lakes were thought to maintain somewhat of an equilibrium between production and decomposition (Golley 1993).

LEGACY OF BIRGE AND JUDAY

In the 1930s, 1940s, and 1950s, E.A. Birge (1851–1950) and his associate Chancey Juday (1871–1944) completed extensive lake studies in northern Wisconsin, making vast observations about the physical and biological factors in aquatic systems. Birge and Juday described the details of lakes stratified by temperature, central in controlling the distribution of biota (Golley 1993).

Birge coined "epilimnion," "hypolimnion," and "thermocline." Birge and Juday completed what were probably the first whole lake experiments by dumping large quantities of lime in one lake to observe the change in chemistry and organisms relative to a nearby lake.

An understanding of the functioning of a lake as a system came to light when productivity was defined and combined with the idea of a food chain based on Elton's work (Golley 1993). Definitions of niche, food chains, and pyramid of numbers were Elton's contributions, along with thinking of a community in collective terms of predators, mutualists, and symbionts rather than individual species. These ideas were used increasingly by aquatic ecologists.

Juday even made an early attempt at an energy budget over the course of a year for Lake Mendota in Madison, Wisconsin (Golley 1993). He tried to calculate the quantity of solar energy coming in, done by estimating how much was fixed into organisms like **plankton**, the microscopic algae, and protozoans that spend their lives drifting in water currents.

RAYMOND LINDEMAN

Building on the work of other limnologists, ecosystem ecology was applied consciously for the first time in 1942 with the publication of Raymond Lindeman's (1915–1942) monograph, *The Trophic-Dynamic Aspect of Ecology*. Lindeman had done research for the project while a graduate student at the U. of Minnesota working on Cedar Bog Lake with the help of his field technician and wife, Eleanor Hall Lindeman. Because the lake was small (1 m deep and a shoreline of 500 m) a total energy budget could be calculated. This provided the basis for an analysis of succession in lakes. After his doctorate, Raymond worked with limnologist E.G. Hutchinson at Yale (Golley 1993).

R. Lindeman was the first to actually quantify the energy transfer through each trophic level. He devised an energy pyramid, which linked the interactions between the living (plankton and pond vegetation) and the non-living material at the bottom of the lake. A **producer** was an autotroph and could gain its own energy as in photosynthesis. A **consumer** was a heterotroph and obtained its energy by eating autotrophs. **Decomposers**, organisms (usually bacteria and fungi) that obtain energy by breaking down complex organic matter to simple inorganic forms, were known as saprophages. Through his work, R. Lindeman introduced most of the major questions and concepts of modern ecological energetics (Golley 1993). Unfortunately, R. Linderman died of an illness before seeing his monograph published.

THE ODUMS

Eugene (Gene) Odum (1914–2002) and his younger brother Howard were quick to embrace the focus on whole systems. Howard was a doctoral student of G.E. Hutchinson at Yale and shared his class notes with his brother Gene (Golley 1993). With ideas from Hutchinson in mind, Gene made the ecosystem concept the central theme in his 1953 *Fundamentals of Ecology*, the first comprehensive textbook of ecology. Howard wrote the chapter on energy.

The two brothers worked as early as 1954 on the island of Eniwetok, an atoll coral reef in the south Pacific, one of the sites where nuclear testing was done by the U.S. government (Hagen 1992). The goal of the brothers was to study an entire ecological system in the field through the use of radioactive products. Although they knew little at the time about tropical reefs or radiation ecology, they demonstrated that it was not necessary to identify every species. It was possible to produce a balance sheet where energy input equaled energy output based on the trophic level idea. Howard was already investigating a complete ecosystem at a mineral springs in Silver Springs, Florida (Golley 1993). Gene and his colleagues from the U. of Georgia were beginning to study terrestrial ecological succession on the property of the Savannah River nuclear plant located in South Carolina bordering the Savannah River.

THE SAVANNAH RIVER PROJECT IN SOUTH CAROLINA

The Savannah River land was purchased from private citizens who were required to vacate the property in 1951–52 when construction of the nuclear reactors and facilities began. Gene and his students had almost 300 mi.^{2} of abandoned old fields that were undergoing succession at one time. It provided a prime opportunity to learn the principles of succession from an ecosystem perspective.

The overall findings from the Savannah River site were (Golley 1993):

- For the first 7 years, the plant communities changed, gradually becoming more diverse in species and the number of species sharing dominance.

- Primary production was highest in the first years, possibly because of residual fertilizer left in the soil from agriculture.
- Primary production then declined and became relatively constant.

Frank Golley continued the study for four more years after the community had changed from herbaceous to a grass-dominated system. From this and other studies, the theoretical ideas about succession came to center on the following (Golley 1993):

- The ratio of primary production to total biomass of the ecosystem was high in young stages and low in mature stages,
- In mature stages a more complete use of food occurred, as did a greater proportion of animals, and energy cascaded through more levels,
- More biomass was driven into the system over time to reach a more complex state, allowing maintenance of the same biomass with a lower supply of energy, in other words, more energy efficiency,
- This mature steady state was not a Clements-like superorganism. It was a steady state for very specific measures only. The ideas would be further refined by the insights provided by the watershed project at Hubbard Brook, and other big biology projects.

ECOSYSTEM ECOLOGY AFTER 1960 – BIG BIOLOGY

Ecosystem ecology became popular quickly after 1960, taking its place as the younger subdiscipline of ecology after population and community ecology. Before 1960, ecologists were used to working at small scales, completing low-budget research with little collaboration from other ecologists. In the 1960s, ecologists began taking on larger modeling tasks of tracking nutrient and energy flows through whole ecosystems, confident they could do it with the help of the emerging technology of computers. Large projects became known as **big biology** (McIntosh 1986). The government supplied generous funding and provided the feel of ambitious World War II military operations and space program cultures. In what became known as **radiation ecology**, scientists researched radioactive isotopes at government properties to determine how fast isotopes flowed through the environment.

In 1963, a large ecosystem project was launched by F. Herbert Bormann and Gene Likens at Hubbard Brook in New Hampshire. It received a government grant under the auspices of a multi-investigator and multi-disciplinary project. The goal was to model and describe entire watersheds. As a result, the model for terrestrial ecosystem studies became the watershed, interesting because like lakes, water determined the boundaries. Eventually Bormann and Likens manipulated whole watersheds at Hubbard Brook. Nutrient losses in watersheds were examined before and after cutting trees. The study is still in progress and has been extremely productive. In the first 20 years, 500 papers were published from the research area.

THE IBP

This flood of attention toward ecosystems led to a program in the 1960s called **IBP**, the **International Biological Programme** (McIntosh 1985). Initiated by the British, it was meant to be a way biologists in the English-speaking world could direct their combined effort for solving the worst biological problems of humankind such as starvation. Originally, ecology was only planned as a small part of the program. Over time, it morphed toward the main focus emphasizing the source of natural resources. Because it was international, it was hoped that the IBP would be even bigger than the Apollo space program.

The IBP did accomplish a fairly long list of advances (McIntosh 1985). In the U.S., the IBP started in 1967 and ended in 1974 during the height of the environmental movement. This was when all the major environmental legislation was passed (NEPA, ESA, Clean Air Act, Clean Water Act, EPA). Much of the U.S. environmental legislation reflects ecosystem thinking with a focus on pollutants cycling through the environment (air and mainly water), as well as big biology thinking.

For the IBP in the U.S., an effort was made with government funding to set up five different centers of ecosystem study in five different biomes (Golley 1993). The effort was costly and involved approximately 1,200 scientists. The downfall was a failure of scientists to work collaboratively in modeling ecosystems. In conventional science research, grants are awarded on the basis of individual research freedom. The biome projects had central control, relying more on a business or military model directed from the top. The research was powered by massive funding, ambitious goals, and short deadlines without enough time for reflection, appropriate planning, or organization. The director of the largest program, George Van Dyne, epitomized the central control model, but died unexpectedly of a heart attack at age 49.

These mistakes occurred despite a strategic effort at one point to use systems theory to organize the massive threads of information being developed by the IBP scientists. Ironically, one of the problems was using a reductionist approach to model energy transfers, summing its parts rather than considering the whole. An additional shortcoming was an emphasis on energy rather than matter (Golley 1993). Since the IBP, matter has emerged as a conceptually easier entity to study.

In contrast to the IBP, a far better model for studying ecosystems was developed through the Hubbard Brook endeavor, not part of the IBP (Golley 1993). The high productivity and excellence achieved was used as evidence that individual research on ecosystems was more productive than team research. The emphasis at Hubbard Brook was on biogeochemical cycles rather than energy, producing more definitive results. The research at Hubbard Brook developed more slowly, involved fewer people, had a more collegial atmosphere between graduate students and professors at the field setting, and cost less money, which were all strategies of success compared to the IBP. One of the gifts of ecosystem ecology has been a lesson about organization and direction in how to do science at large scales.

THE ENVIRONMENTAL MOVEMENT AND ENVIRONMENTAL STUDIES

The ecosystem concept came at a time when it could influence the environmental movement and be influenced by it (Kline 1997). By 1960, the post-World War II influence had brought unprecedented development, industrialization, and population growth to the U.S. and Europe. Combined with the nuclear threat and pollution issues, the Western culture was poised for societal change.

Inspiring the environmental movement was U.S. scientist Rachel Carson (1907–1964) with publication of "Silent Spring" in 1962, a scientific argument against the increasing dangers of chlorinated hydrocarbon compounds such as DDT and other pesticides. By the mid-1960s ecosystem ecology had a well-established, active scientific community engaged in applying a holistic philosophy that closely fit the social-cultural environment (Golley 1993). The height of the movement came with the first Earth Day in 1970. The history of ecology is intricately intertwined with applied fields that spun off from ecology. Environmental studies and environmental science as well as conservation biology all evolved mainly during the environmental movement.

THINKING QUESTIONS

1. What are the abiotic parts of an ecosystem that are usually studied by ecosystem ecologists?
2. Contrast the ecosystem idea with the superorganism approach of Clements.
3. What characterizes a systems approach?
4. How does one define the bounds of an ecosystem? Why is it difficult to assign boundaries (upper, side, bottom) in ecosystems?
5. What is meant by a steady state in ecosystem ecology?
6. To investigate decomposition of tree leaves in various environments place a handful of dried leaves from one tree species in each of several mesh bags (e.g., that held tangerines from the grocery). Place bags in stream, pond, forest floor, shade, sun, etc. After a few weeks, weigh again after drying the sample in a drying oven. Which decomposed the most based on weight? Compare among different species.

REFERENCES

Bailey, R.G. 2006. *Ecosystem Geography from Ecoregions to Sites*. 2e. Springer, New York.

Freedman, B. 2022. Indicator species. Science. jrank.org/pages/indicator-species.html

Golley, F.B. 1993. *A History of the Ecosystem Concept in Ecology*. Yale University Press, New Haven and London.

Hagen, J. 1992. *An Entangled Bank: The Origins of Ecosystem Ecology*. Rutgers University Press, New Brunswick, NJ.

Kline, B. 1997. *First along the River: A Brief History of the U.S. Environmental Movement*. Rowman and Littlefield Publishers, Washington, DC.

McIntosh. R. 1985. *The Background of Ecology*. Cambridge University Press. Cambridge, UK.

Vellend, M., and J.L. Orrock. 2010. Biological and genetic models of diversity, lessons across disciplines. In: J.B. Losos and R.E. Ricklefs (eds.), *The Theory of Island Biogeography Revisited*. Princeton University Press, Princeton, NJ.

21

Energy

CLASS ACTIVITY: Study a Rainbow

Gooley (2014) observes that rainbows do not appear during the middle of hot summer days. They appear in early mornings or late afternoons. The explanation is that the sun must be at an angle less than 45° from the horizon for a rainbow to form. Furthermore, shadows are shorter than we are tall if the sun is at an angle greater than 45°. Thus, if our shadows are shorter than we are tall, we will not see a rainbow. Begin a rainbow journal and test Gooley's claims. Additionally, Gooley observes that the smaller the raindrops, the less the refraction and the fewer the colors. True?

THE ECOSYSTEM MACHINE

Ecosystem ecology is specifically about the flow of energy (**energy**=capacity to do work) and the cycling of matter (**matter**=anything with mass), framed in terms of nature as a machine. Think first about a car. Where does it get its energy? What work is being done? Where does that energy go? Where does the matter from the fuel go? Nature as a machine looks like Figure 21.1. Apply the same questions.

The machine is powered ultimately by the sun. Beyond that, the machine is divided into grazing and detrital food chains. The **grazing food chain** is herbivore based. The **detrital food chain** consists of microbial decomposers and detritivores. **Detritus** is dead particulate organic material, which includes dead organisms and fecal material. Most energy is funneled through the detrital food chain. Not pictured in Figure 21.1 is the parasite food chain. It is often short, and consists of organisms that parasitize plants, animals, and other organisms.

Terminology used for understanding food chains comes from trophic level pyramids (trophic=food) (Elton 1927). **Trophic level** refers to all the organisms that gain energy in the same way. At the bottom trophic level are **primary producers**, known as **autotrophs** (self-feeders). Autotrophs include **photoautotrophs** (light self-feeders), the plants, algae, and blue greens that photosynthesize. Autotrophs also include **chemoautotrophs** (chemical self-feeders), the bacteria and archaea that may use hydrogen sulfide, elemental sulfur, ferrous iron, molecular hydrogen, or ammonia as their energy source. Think of **photosynthesis** as fixing carbon, converting it from inorganic carbon to a more usable organic form such as sugar.

$$CO_2 + H_2O \overset{light}{\Rightarrow} O_2 + C_6H_{12}O_6$$

In the next higher trophic layer are **primary consumers**, known as **heterotrophs** (different feeders), the herbivores and decomposers that consume plants. In the next higher level are the **secondary consumers**, also heterotrophs, which are carnivores of the herbivores and decomposers. Next higher are the **tertiary** consumers that consume the carnivores.

DOI: 10.1201/9781003271833-21

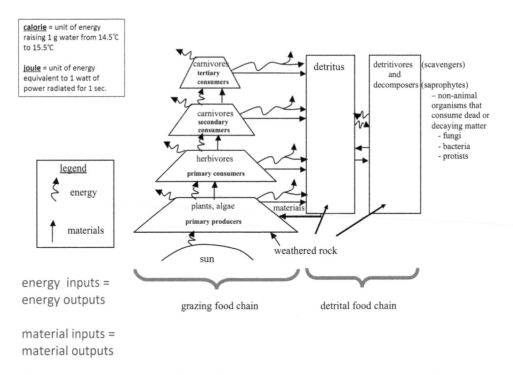

Figure 21.1 An ecosystem machine. Energy inputs match energy outputs and materials in equal materials out.

WHAT IS ENERGY?

Energy can be particles flying out of an atom (think of radioactivity with electrons, protons, or subatomic parts shooting out of an atom as an unstable isotope becomes more stable). Energy can also be wavelengths in the form of **electromagnetic radiation**. Separate names exist for different wavelengths of energy, although all of it is electromagnetic radiation (Figure 21.2). The only thing different about these categories is the length of their waves.

What would you call electromagnetic radiation with a wavelength approximately the height of a child (1 m)? That would be microwave. What about a wavelength as long as two pickups parked together (10 m)? That would be TV waves. Point of a needle (0.00001 m)? That would be heat (infrared). Really? This is fascinating.

Light is the size of small cells. Think about it. Tiny bacteria are not visible with light microscopes, right? Wonder why? Because they are barely visible with light rays. Ultraviolet wavelengths are the size of molecules (10^{-8} m). A confocal microscope that uses ultraviolet is necessary for seeing the smallest parts of cells like microfilaments for instance.

X-ray wavelengths are really short (10^{-10} m), gamma rays even shorter (10^{-13} m). Cosmic rays shorter still (10^{-15} m).

Waves at the shortest end of ultraviolet as well as X-rays, gamma rays, and cosmic rays are known as **ionizing radiation**. These cause damage to living cells because they shoot electrons, knocking other electrons out of molecules including within DNA. This causes mutations in the DNA, which produces out-of-control cell growth, which is known as **cancer**. Use your sunscreen.

WHAT ARE PHOTONS?

Energy is in waves, in packets called photons. **Photons** are chargeless bundles of energy that travel in a vacuum at the speed of light, roughly 300,000 km/s. Why is this important? Because for electromagnetic radiation at the X-ray and shorter scale, wavelengths become meaningless and photons provide a measure of strength.

WAVELENGTHS OR FREQUENCIES?

Along the electromagnetic radiation spectrum, light energy is usually described in wavelengths but some

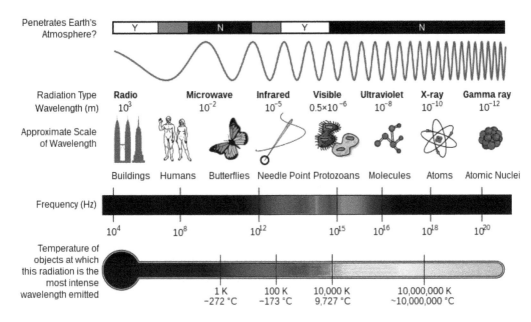

Penetrates Earth's Atmosphere?

| | Y | N | Y | N |

Radiation Type: Radio, Microwave, Infrared, Visible, Ultraviolet, X-ray, Gamma ray

Wavelength (m): 10^3, 10^{-2}, 10^{-5}, $0.5{\times}10^{-6}$, 10^{-8}, 10^{-10}, 10^{-12}

Approximate Scale of Wavelength

Buildings Humans Butterflies Needle Point Protozoans Molecules Atoms Atomic Nuclei

Frequency (Hz): 10^4 10^8 10^{12} 10^{15} 10^{16} 10^{18} 10^{20}

Temperature of objects at which this radiation is the most intense wavelength emitted:

1 K, −272 °C   100 K, −173 °C   10,000 K, 9,727 °C   10,000,000 K, ~10,000,000 °C

Figure 21.2 Electromagnetic radiation spectrum. (*Wiki Commons File:EM Spectrum3-new.jpg, a NASA image.*)

other forms are described in **hertz** (Hz). This is number of waves that pass through a point in a second. Hertz and wavelengths can be converted to one another, but for use in radio and TV the frequency of waves is more important than wavelength.

WHAT IS SOUND, MRI, CT, AND ULTRASOUND?

Sound waves are mechanical waves not electromagnetic because they require matter (gas, liquid, and solid) to vibrate for our ears to detect sound. Wavelength, amplitude, and frequency combine to determine how the sound is heard, but wavelengths of detectable sound for humans are from 17 mm to 17 m. Sound can be converted to electricity by the way, which is why there may be a future in turning the world's waste sound into usable forms of energy (justenergy.com).

Ultrasound is high-frequency sound (20,000 Hz) and mechanical not electromagnetic waves. We cannot hear it. Ultrasound is safe for producing real-time live images, like seeing a developing baby or the inside of a beating heart. Ultrasound use requires little preparation on the part of the patient, and is inexpensive compared to MRI or CT. Ultrasound is being used for surgeries when

tissue needs to be seen in real time. The downside is that gas, liquid, and deep layers of fat disrupt the image. Thus, ultrasound is not useful for all organs (lungs, colon) or all patients.

MRI (**magnetic resonance imaging**) uses radio waves and a magnetic field to detect soft-tissue abnormalities, such as tendon and ligament injuries, brain tumors, or spinal cord injuries. An MRI takes 10 minutes to several hours to produce, but unlike X-rays that take only seconds, MRI does not produce ionizing radiation. Protons in the soft tissues react to the radio waves and give off enough energy to be detected, producing a higher quality image than an X-ray for soft tissues. MRIs are not suitable for patients with implants, like pacemakers or artificial joints because the strong magnet will interfere with the metal.

CT (**computed tomography**) scan uses X-rays at a high level of resolution and has 360° imaging capabilities. Images produce higher resolution than regular X-rays for bones and some soft tissues, but not as high as MRI. CT scans only take a minute and can be used for people with implants, but patients must remain very still. CT is the technique of choice in emergency medicine. Blood clots, organ injuries, and subtle fractures not visible in X-rays can be detected.

BACK TO THE ECOSYSTEM MACHINE – ENERGY CAN CHANGE FORM

Fundamental to understanding the energy part of the machine is the **first law of thermodynamics**, energy cannot be created or destroyed, but can change forms. Forms of energy include light, motion, electricity, sound, chemical, gravity, radioactivity, heat (Figure 21.2), and others. The ultimate source of energy for ecosystems is the sun, but light energy can change its form several times before leaving Earth's atmosphere.

Most of the energy that comes to Earth each day as light is converted to heat and lost at night from the dark side of Earth. Because of greenhouse gases, Earth's atmosphere is currently gaining more energy than what is lost each night. The result is global warming. Additionally, Earth has energy stored in the form of gravity, derived from the sun as the solar system was forming.

In the **second law of thermodynamics**, all energy pathways tend toward **entropy** (disorder). In other words, when energy changes from one form to another it loses some energy, mainly as low-level heat lost to space at night. The useful energy has declined over time to a form that cannot be converted any further.

In both the detrital and grazing food chain, only a fraction of the carbon ingested by the consumers is metabolized into biomass by those organisms. Looking at the ecosystem machine, most energy goes to decomposers, but with each trophic transfer there is some loss of energy dissipated as waste heat. Often ecologists refer to the 10% rule of thumb, meaning that at each trophic transfer only 10% of the energy is stored in the higher trophic level. Ninety percent of the energy is lost as waste heat. In practice, the amount stored varies from 10% to 30% or more.

SUN ENERGY AND ITS EFFECT ON EARTH'S ATMOSPHERE

The sun emits energy through nuclear fusion, in this case, two hydrogen atoms colliding at a time with so much force that it fuses their nuclei and creates an atom of helium (Christensen 2013). The energy given off comes from small amounts of mass converted to energy when the two atoms collide.

The massive kinetic energy (3,000,000°C) required for fusion is provided by the pull of gravity within very large stars including our sun.

Our very hot (6,000 K) sun emits its energy to Earth in several forms (Chapin et al. 2002). Visible light makes up 39% of the total. The rest of the energy is near infrared (53%) and ultraviolet radiation (8%). Near infrared is not heat and cannot be felt. Near infrared wavelengths are used for purposes on Earth for things like remote control.

Of the energy absorbed by Earth's surface, approximately 79% is emitted back into the atmosphere as "far infrared long wave radiation," which is **heat** (Chapin et al. 2002). Water vapor, carbon dioxide, methane, nitrous oxide, and chlorofluorocarbons absorb this heat, hold it for a time, and release it eventually to space. Holding the heat is a lifesaving mechanism because without it, the mean temperature at Earth's surface would be 33°C lower, probably too low to sustain life.

Our current climate change is the result of an increase in these heat absorbing gases. This increase began at the beginning of the Industrial Revolution 150 years ago (Figures 21.3–21.7). The cause of the increase is the burning of fossil fuels at a rate 1 million times faster than they accumulated (Chapin et al. 2002).

Because heat is produced by energy radiating off the Earth's surface, the troposphere is heated from below, not above, and explains why heat is stratified in the atmosphere (Figure 21.8). In contrast, the high temperature of the stratosphere is the result of being heated from above through absorption of ultraviolet radiation via a protective layer of ozone, O_3. Without this protection organisms would be exposed to damaging ultraviolet radiation, permanently damaging DNA.

Chlorofluorocarbons (CFCs) are products now banned but totally synthesized by humans for use as aerosols and refrigerants. They were responsible for causing the "holes" in the stratospheric ozone layer. CFCs have a mean residence time of decades once they reach and accumulate in the stratosphere. Once in the atmosphere, the CFCs destroy ozone, particularly at the poles, accounting for thinning where the transmission of ultraviolet radiation to the Earth's surface is particularly high. Too much UV causes cancer, cataracts, immune system problems, and damage to plants, including agricultural products. The hole at the North Pole is on track to

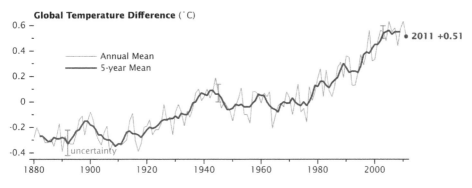

Figure 21.3 Change in annual global temperatures, compared to the average of global annual temperatures from 1880 to 2011. (*Open source, NASA Goddard Institute for Space Studies. Image credit: NASA Earth Observatory, Robert Simmon.*)

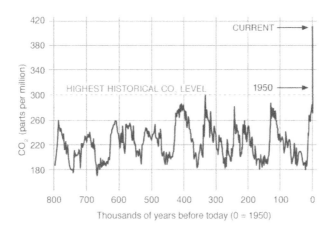

Figure 21.4 Change in global CO_2 concentrations. (*Open source, climate.nasa.gov*)

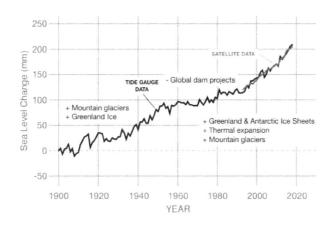

Figure 21.5 Sea level change from 1900 to 2018. (*Open source, NASA's Goddard Space Flight Center.*)

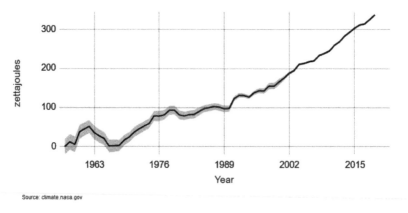

Source: climate.nasa.gov

Figure 21.6 Ocean heat content changes since 1955. (*Open source, National Oceanic and Atmospheric Association.*)

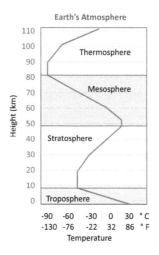

Figure 21.7 Earth's atmospheric temperatures by kilometers from Earth's surface. The troposphere rises to 10 km from the Earth's surface and is the site where most clouds occur and where most airplane travel happens. Occasionally clouds and airplanes enter the stratosphere.

be healed by the 2030s, and at the South Pole by the 2050s (news.un.org).

Above the stratosphere, the mesosphere cools with elevation because the low density of gases has little ability to hold heat (Chapin et al. 2002). However, the topmost layer of atmosphere, the thermosphere, consists mainly of low-density oxygen and nitrogen, absorbing shortwave energy, causing an increase in temperature with elevation.

THE OCEAN ENERGY BUDGET

Whereas the troposphere is heated from the bottom, the ocean is heated mainly from the top. Thus, dense warm water sits on top of cold water maintaining mostly stable layers with little mixing, especially in tropical areas. Even so, if water is salty enough it can sink, as it does in areas near the poles where formation of sea ice excludes salt from ice crystals (Chapin et al. 2002). Examine Figure 21.8 carefully. This increase in the salinity causes cold water to sink forming downwelling areas near Greenland and Antarctica. In contrast, upwelling areas bring nutrients from below, supporting highly productive fisheries. Global currents circulate this water in deep and shallow currents that flow from ocean to ocean known as a conveyor belt.

EL NINO PHENOMENA

El Nino events are large-scale pressure and temperature changes over the equatorial Pacific Ocean.

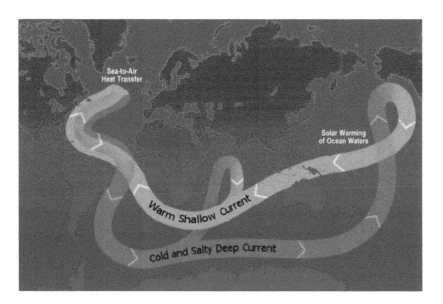

Figure 21.8 Ocean circulation conveyor belt, driven by differences in heat, salinity, and depth. (*Wiki Commons, U.S. Global Change Research Project, Thomas Splettstoesser.*)

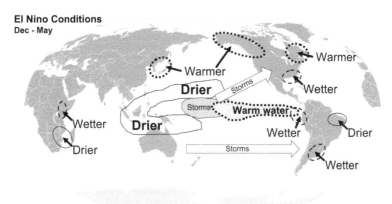

Figure 21.9 El Nino occurs when warm water normally associated with the western Pacific slides eastward, causing worldwide weather changes. (*Weber with Basemap from Wiki Commons.*)

Their appearance is every 3–7 years on average over the past century (Chapin et al. 2002). In most years (the normal years termed **La Nina**), trade winds from the east push warm surface waters of the Pacific to the west. This sets up a deep layer of warm surface water in the west, bringing high rainfall to Indonesia. In the east, upwellings of cold, deep water provide nutrient-rich waters to the coasts of Ecuador and Peru and low precipitation.

During **El Nino**, the trade winds weaken and the warm surface waters move eastward (Chapin et al. 2002). What follows is a shutdown of the ocean upwellings. Ecuador and Peru get more rainfall. Drought comes to Indonesia, Australia, and India. El Nino has been associated with dramatic reductions in anchovy fisheries in Peru, reproductive failure in seabirds and marine mammals, hot dry weather in the Amazon, and storm changes in North America. Warm tropical waters extend into the northern Pacific bringing rain to coastal California and high winter temperatures to Alaska (Figure 21.9).

The North Atlantic oscillation is another large-scale climate pattern. It leads to a warming of Scandinavia and western North America and a cooling of eastern Canada (Chapin et al. 2002). Climatic warming may increase the frequency of El Nino and North Atlantic oscillations, but their underlying causes are poorly understood.

THINKING QUESTIONS

1. List examples of kinetic and potential energy.
2. Does the second law of thermodynamics make ecosystems more and more disordered over time?
3. When looking at a series of hills on the horizon, the farther away the hill, the lighter shade of gray. What accounts for this phenomenon?
4. What causes the appearance of halos around the Moon during winter? What are "sundogs" in the sky and what causes them?
5. What makes the daytime sky appear blue? Why are sunsets orange-red? Why are sunsets more likely to be orange-red than dawns?
6. Wildfires in California have inordinately ravished the Mediterranean coastal region. What has caused the increase in frequency of wildfires in the shrubland of California?
7. Coming out of water on a hot summer day can sometimes produce shivering in humans. What explains this from a physics perspective, and how can one prevent shivering after swimming?
8. Are shadows completely black? Can you see color within a shadow?
9. According to naturalist Tristan Gooley (2014) the smell of smoke or fog on cold mornings is a sign of better than average radio reception that day. Test this hypothesis by measuring the number of radio stations you can receive on FM during a smoky/foggy day versus a clear day. What is it about temperature inversion that produces fog and could possibly allow better reception?

REFERENCES

Chapin III, F.S., P.A. Matson, and H.A. Mooney. 2002. *Principles of Terrestrial Ecosystem Ecology*. Springer, New York.

Christensen, N. 2013. *The Environment and You*. Pearson Press, Boston, MA.

Elton, C.S. 1927. *Animal Ecology*. Sidgwick and Jackson, London, UK.

Gooley, T. 2014. *The Lost Art of Using Nature's Signs*. The Experiment, LLC, New York.

Justenergy.com. Everything you need to know about this electrifying source. https://justenergy.com/blog/sound-energy-everything-you-need-to-know/

News.un.org. 2019. Ozone on track to heal completely in our lifetime, UN environmental agency declares on World Day. UN News Global Perspective Human Stories. https://news.un.org/en/story/2019/09/1046452

22

Matter

CLASS ACTIVITY: Trophic Level Efficiencies

Spanning the divide between energy and matter is the measure of carbon within the ecosystem machine. Carbon accumulates in each tropic level and because carbon burns, this accumulation can be measured in energy (kcal/m²), or biomass (g/m²). Carbon accumulation is called "**production**." For the plant trophic level, **gross primary production** (GPP) refers to the biomass fixed per unit area by photosynthesis (or chemosynthesis). **Net primary production** (NPP) refers to the total biomass stored in plants after respiration is deducted from gross primary production, as in the following equations:

$$GPP = NPP + R$$
$$NPP = GPP - R$$

NPP is what is available for growth, reproduction, fat, or carbohydrate storage in plants. NPP can be measured in kcal/m² or g/m².

    Gross secondary production (GSP) is what is consumed by the next higher trophic level, the herbivores (consumers). Of all that is eaten, only some is assimilated past the gut wall. What is assimilated is known as **net secondary production** (NSP). It is equal to:

$$GSP = NSP + R$$

Step 1: In the box designating foraging efficiencies, write which boxes refer to plants and which to herbivores. Answer this question – of the vegetation that is not consumed, what is its fate? Give three possible scenarios.

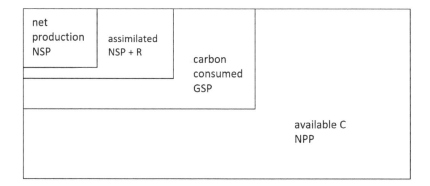

Step 2: Ecologists like to know the efficiencies at which conversions are made between trophic levels. For instance, "consumption efficiency" is the carbon consumed (GSP)/carbon available in the environment (NPP), the amount that makes it to an animal herbivore's mouth. If this were 5.0% in a forest it would mean that of 2,000 g of acorns, 100 g make it to a mouth, the rest goes to decomposers. If the consumption efficiency in a grassland is 23.2% how many grams out of 10,000 g of grass make it to a mouth? What about in a wetland if 52.8% make it to the mouth of a zooplankter eating phytoplankton out of 10,000 g?

Step 3: What about "assimilation efficiency" – the assimilated (NSP+R)/consumed (GSP)? This is what is assimilated across a gut wall. The rest goes to feces. For herbivores the efficiency is on the order of 10%. For detritivores it is in the 20%–50% range. For carnivores it is approximately 80%. For an insect-eating bird population, the production of the prey population is 232.4 g. The total prey captured was 8.5 g. The amount assimilated was 5.6 g. The amount put into net production was 2.72 g. What is the assimilation efficiency?

WHAT IS NUTRIENT RECYCLING

In the previous chapter energy was said to "flow" from the sun through trophic levels, then depart Earth mainly from the dark side. In contrast, matter clearly "cycles" on the global scale, gaining or losing matter only rarely to space (Figure 22.1). **Matter** is anything with mass. **Mass** is anything that takes up space. There are several ways mass can be categorized in nature. The rest of this chapter covers the most prominent schemes.

ORGANIZING CATEGORIES IN BIOGEOCHEMISTRY – COMPARTMENTS

The cycling of matter is studied within the discipline of **biogeochemistry**. The **ecosphere** is the part of the Earth that includes living beings and divided into what are sometimes called **compartments** or reservoirs. The compartments include the:

- **lithosphere**: rocks of Earth's outer crust,
- **hydrosphere**: water of the ground, streams, lakes, and oceans,
- **biosphere**: the detrital and grazing food chains,

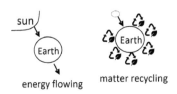

energy flowing   matter recycling

Figure 22.1 Energy flowing from sun to Earth, losing energy each night, and matter cycling on Earth.

- **atmosphere**: divided into the two layers closest to Earth's surface,

  - **troposphere**: spanning the atmosphere from Earth's surface to 10 km high,
  - **stratosphere**: the layer above the troposphere up to 40 km high.

ORGANIZING CATEGORIES IN BIOGEOCHEMISTRY – STATE AND CARBON CONDITION

Other biogeochemistry designations include **gaseous**, **liquid**, and **sedimentary** (rocks and soil) states. Additional dichotomies include **organic** and **inorganic** conditions.

- **Organic** refers to carbon compounds produced by organisms.
- **Inorganic** refers to non-carbon compounds and carbon compounds not produced within living organisms.

  - Non-organic carbon compounds include carbon oxides, cyanides, and crystalline forms of carbon such as diamonds, graphite, and Bucky balls (C_{60} with a shape in a spherical geodesic dome).

NUTRIENT CYCLES

Other categories are cycles for the chemicals and nutrients required by living beings. Examples include the **hydrologic** (water) cycle and cycles for carbon, nitrogen, phosphorus, sulfur, and others. **Nutrients** refer to substances required for growth and activity in an organism. Plants typically acquire nutrients from soil or solution. Carbon and oxygen come to them from the air. Animals typically acquire nutrients from food and liquids. Oxygen comes from the air. **Elements** are building blocks of matter as identified in the periodic table, the chemicals that cannot be broken down in a stable form. Of the approximately 90 naturally occurring elements (Sterner and Elser 2009):

- 11 predominate in living organisms, and of these:
  - 4 make up 99% of living biomass (C, H, O, N),
  - 7 are essential to all living things (Na, K, Ca, Mg, P, S, and Cl),

- 10 or so others are required by most but not all species,
- 8 others are required by some species.

Some elements, like Fe and Mg, may be essential for an organism, but in low quantities. Note that 99.9% of the atmosphere by volume is composed of just three gases, nitrogen, oxygen, and argon (Chapin et al. 2002). After these, carbon dioxide is the next most abundant, but accounts for only 0.0416% (Tiseo 2022). However, this is an increase of 31.4% since 1959, which is astonishing because this change is caused by humans.

The atmosphere is surprisingly homogeneous in its composition of gases around the world and up to 80 km high. This consistency is the result of long **mean residence times** in air. Although the atmosphere's composition is homogeneous, the atmosphere's density is not. Air is most dense at the Earth's surface because of gravity.

HYDROLOGIC CYCLE

Water is involved in the most critical cycle because it carries many nutrients with it, moving them from compartment to compartment (Figure 22.2). **Groundwater** stores water in the ground, or in other words is underground water. Groundwater moves slowly, on the order of a couple meters/year. **Streams** carry surface runoff, moving the great bulk of nutrients. In lakes, most of the nutrients pool. Those lakes with no outflow are considered **endorheic,** in which water evaporates to the atmosphere and nutrients accumulate in sediments, making the water more nutrient rich over time. **Water residence time,** the

average time a molecule of water spends in an ecosystem, can vary from minutes in streams, to years in lakes, to millennia in groundwater (Bade 2009).

For the nutrients carried by water, **wetfall** is the nutrient load within precipitation. Rain is not merely distilled water (Smith and Smith 2001). The nutrient concentration is highest early in a rainstorm and falls as the atmosphere is cleansed. Snow scavenges chemicals from the atmosphere less effectively than rain. Fog has a high ionic concentration. Humid air can carry nutrients to the point that it can be a source of plant nutrition. **Dryfall** can be a source of nutrition, consisting of airborne particles that move by wind, which includes sulfates, nitrates, and potassium.

Within the hydrologic cycle, water may be stored as ice, slowing the cycle. Evaporation and transpiration balance the water in precipitation. **Transpiration** is the water that evaporates from plants through stomata. Water in **aquifers** (water in the spaces between underground rocks) may move as slowly as 3 m/year. The ultimate fate of water is oceans, which hold 97% of the world's water. Only a fraction is available as freshwater, but this is essential for non-marine living beings including humans (Figure 22.3).

> What is the source of energy to move water through the hydrologic cycle?

Residence time as liquid water is 100,000 times longer than in the gaseous state (Smith and Smith 2001). There is 26 times more liquid water than gaseous. Humans are most likely to interfere within the water cycle through dams, irrigation, and cutting trees, all increasing runoff.

CARBON CYCLE

Nothing can live without carbon. It is essential for life, making up cellulose in producers, and the four molecules of life in all living beings (lipids, carbohydrates, nucleic acids, and amino acids). Carbon provides the major conduit of energy for living organisms through ATP and its associates. Carbon also has the massive job of buffering the pH in the oceans and other waterways (Falkowski 2009). The great majority of carbon is inorganically held in carbonate rocks (Table 22.1). It enters the atmosphere and becomes dissolved in water when volcanoes erupt and the sea floor spreads (Falkowski

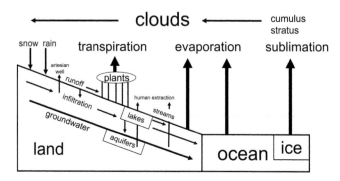

Figure 22.2 Hydrologic cycle.

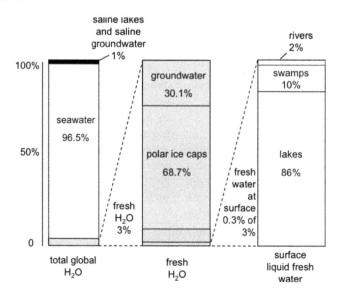

Figure 22.3 Global Water Distribution. Freshwater makes up only 3% of Earth's water, and surface water makes up only 0.3% of that 3%. The "other" category making up 0.9% of freshwater includes atmospheric, biological, and soil moisture.

2009). From the atmosphere, carbon enters the food chain through photosynthesis.

Photoautotrophs, literally light self-feeders, take up carbon as sugar and make it into cellulose. Photoautotrophs require a pigment such as chlorophyll to operate and require biological enzymes to catalyze the reaction. **Chemoautotrophs**, literally chemical self-feeders, are archaea and bacteria that use inorganic C deep in the dark sections of the oceans as their energy source (Falkowski 2009). They must be near vent areas with warm temperatures to survive.

The various autotrophs give back some of this carbon to the water or atmosphere as they respire (Figure 22.4). The carbon within the autotroph

bodies can be eaten by consumers and decomposers. Within them this carbon can be assimilated and put into tissues. Carbon may also exist within the ecosphere as **methane hydrate**, methane bonded with water existing as a crystal and found within pressurized areas of the ocean or in frozen soils of the tundra.

CARBON IN WATER AND ITS EFFECT ON pH

Carbon is 50 times more prevalent in the water of the oceans than in the atmosphere (Falkowski 2009). Cold water holds more dissolved CO_2 than warm water. Thus, as oceans warm through climate change they store less dissolved carbon in

Table 22.1 Carbon in the major reservoirs on Earth (Falkowski 2009)

Pools	Quantity (×1,015 g)
Atmosphere	720
Oceans	38,400
Total inorganic	37,400
Surface layer	670
Deep layer	36,730
Dissolved organic	600
Lithosphere	
Sedimentary carbonates	>60,000,000
Kerogens (shale oils)	15,000,000
Terrestrial biosphere (total)	2,000
Living biomass	600–1,000
Dead biomass	1,200
Aquatic biosphere	1–2
Fossil fuels	4,130
Coal	3,510
Oil	230
Gas	140
Other (peat)	250

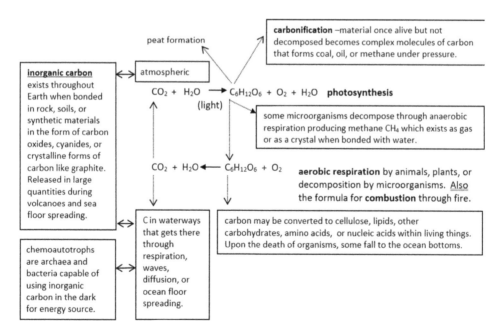

Figure 22.4 The carbon cycle.

the water (Figure 22.4). Wind, turbulence, and time of day (sun energy) can increase the concentration of CO_2 and O_2 in water. Temperature and ice formation can further affect photosynthesis and respiration within these complex cycles. Carbon loading from inflowing streams and groundwater can import large quantities of inorganic carbon.

CARBON DECOMPOSITION IN LAKES

If organic carbon is buried in lake sediments of anaerobic conditions, fermentation and methanogenesis can take place during decomposition (Bade 2009). Fermentation produces acetic acid, CO_2, and hydrogen (H_2). In methanogenesis, hydrogen supplied from fermentation is a source of energy. Acetate is decomposed to CO_2 and methane:

$$CO_3COOH \rightarrow CO_2 + CH_4$$

This is important because methane can then form bubbles as it escapes the sediment and goes to the atmosphere, a process called **ebullition**. This can be a considerable contribution as a greenhouse gas, methane being 20 times more effective than CO_2 at holding heat.

If fixed carbon does not become respired by either animals or through decomposition, it may become fossilized under pressure in a process called **carbonification**. This occurs when flooded areas have dead vegetation that sinks to depths at which it never decomposes, or ocean life dies and sinks to great depths. The carbon becomes part of a complex molecule as coal, oil, or trapped gases under pressure. Millions of years later, this may be burned in combustion as a **fossil fuel**, releasing CO_2 to the atmosphere.

CARBON CYCLE DYNAMICS IN GENERAL

Think about how time of day and season affect the carbon cycle. In terrestrial situations, how does fire affect the system? What about insect infestations? How does leaf fall in general affect the system? Overall, what is the source of energy for the carbon cycle?

At the global level, the carbon cycle has been heavily influenced by humans, in 2021 reaching an increase of over 50% more carbon dioxide in the atmosphere since the beginning of the industrial revolution (Betts 2021). The increase has stimulated the growth of plants in some situations, influencing other cycles because of the plants' increased demands (Vitousek and Matson 2009). Think about how increased temperature affects the overall rates of reactions, especially for decomposition.

NITROGEN CYCLE

Nitrogen is essential for life because it is the basis of amino acids, the backbone of proteins, important for building tissue and for making enzymes that catalyze most of the biochemical reactions within organisms. In particular, nitrogen is a component of the major enzyme for photosynthesis, ribulose bisphosphate carboxylase, one of the most important and prevalent chemicals on Earth. To visualize the nitrogen cycle (Figure 22.5), consider the store of nitrogen in the atmosphere in which air is 78% nitrogen in the form of N_2 or N_3. In the gaseous form, nitrogen cannot be converted to amino acids. Most of it becomes available through the process of **fixation** from **ammonium** (NH_4^+).

Fixation can occur in small amounts within meteorite trails or through lightening, but most of what is fixed naturally on Earth is converted organically by bacteria and cyanobacteria. This happens in the root nodules of legumes (by bacteria in the family Rhizobiaceae) and alder trees (by bacteria in the genus Frankia), then taken up by roots of plants. Other fixers are free-living bacteria, *Azotobacter* or *Clostridium*, as well as the free-living cyanobacteria, *Nostoc*, common in freshwater areas throughout the globe. Diatoms in the ocean, lichens, and nitrogen-fixing bacteria in the guts of termites also fix nitrogen.

In even greater quantities than biological fixation, nitrogen can be fixed by human means (Bade 2009). Anthropogenic sources of fixed N entering the cycle come from fertilizers using the Haber-Bosch process, producing ammonia under high pressure using an iron catalyst. This ammonia is used as fertilizer in agriculture.

Once nitrogen is fixed, much of the rest of the N cycle is aquatic (nutrients in solution of wet soils or waterways) rather than gaseous (Houghton 2009). Ammonium can become bound to soil clays and organic matter or taken up by some plants and microorganisms. It can be oxidized to **nitrates** (NO_3^-) or **nitrites** (NO_2^-) in a process called **nitrification**. *Nitrosomonas* bacteria is the primary source converting ammonium to nitrites in nature (Bade 2009):

$$NH_4^+ + 1.5O_2 \leftrightarrow 2H^+ + NO_2^- + H_2O$$

The bacterium *Nitrobacter* then converts the nitrite to nitrate:

$$NO_2 + 0.5O_2 \leftrightarrow NO_3^-$$

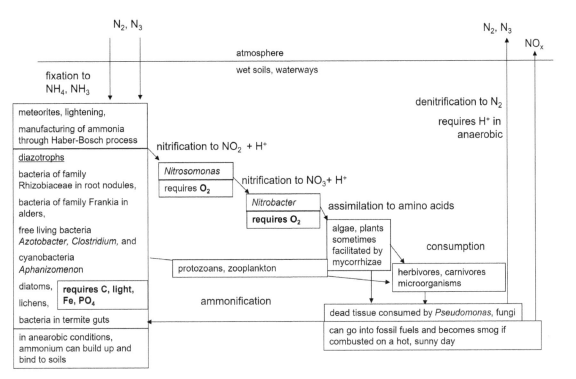

Figure 22.5 The nitrogen cycle.

Nitrification requires oxygen (Bade 2009). In anoxic conditions ammonium may accumulate. Nitrates are highly mobile in soils and can be used by a wide range of plants, algae, and micro-organisms. Some plants are aided by symbiotic relationships with mycorrhizae in taking up the nitrates or nitrites, but not all. Once in the plants and other organisms, conversion to amino acids takes place through a process called **assimilation**. **Amino acids** are the building blocks that make proteins. Herbivores including human vegetarians gain their amino acids by eating plants.

Once plants, animals, and other organisms die, the fate of nitrogen has several pathways. Bacteria (e.g., *Pseudomonas*) and fungi may convert the amino acids to gaseous nitrogen in a process called **denitrification**. The reaction proceeds as:

$$NO_3^- \rightarrow NO_2^- \rightarrow N_2O \rightarrow N_2$$

The reaction requires H^+ and anoxic conditions (Bade 2009). For this reason, wetlands are sometimes constructed specifically for removal of high nitrogen loads near large farms. Bacteria and fungi can also convert the decaying tissues to ammonium in a process called **ammonification**, which goes back to the ammonium pool.

Dead tissues can accumulate in the sediments and become part of fossil fuels. Nitrogen oxides (NO_x) can then enter the atmosphere from high-temperature combustion in automobiles and other engines. On hot days, nitric acid is produced within the soup of hydrocarbons released into the atmosphere through combustion, creating the brown haze known as smog, visible on the horizon near all urban areas. NO_x is a source of acid rain. This nitrogen-rich rain can sometimes increase productivity of ecosystems that are nitrogen limited.

Anthropogenic changes to the global N cycle have been even more dramatic than for the carbon cycle (Vitousek and Matson 2009). Fixed nitrogen fertilizers barely existed in 1950, but humans have more than doubled the quantity of N_2 fixed on land through fertilizer application. Atmospheric nitrogen has increased by 5–10 fold in urban, industrial, and intensive agricultural areas above background levels. Approximately 25% of the anthropogenic fixed nitrogen added to terrestrial ecosystems moves via the hydrologic cycle, driving eutrophication in estuarine and coastal areas. This includes the "dead zone" in the Gulf of Mexico near the mouth of the Mississippi River.

FINAL ANALYSIS

As great of an emphasis as there has been on the carbon cycle in recent years, the cycles of nitrogen, phosphorus, and sulfur have been altered by humans to an even greater extent (Vitousek and Matson 2009). The sulfur cycle has been altered greatly through emissions released from fossil fuels, although this has been mitigated substantially since 1980 through regulation for lower-sulfur fuels (Vitousek and Matson 2009).

THINKING QUESTIONS

1. When someone "burns" calories on a diet, is this an accurate description? Weight is a measure of matter not energy, so what specifically (as an element) is lost when a person loses weight? Where does the weight go? How does it exit the body?
2. When a tree is a tiny sapling it weighs a few grams. Decades later it weighs metric tons. What element does it gain to produce most of this weight? What is the origin of this mass? Hint: If the source was the ground a big hole would appear around every tree. That does not happen, thus it is not the ground.
3. Matter leaves or comes to Earth only rarely during present day. What are ways that mass is permanently lost or gained? Investigate ways that helium is leaving the Earth and why this is a problem.
4. What is the ultimate source of energy for the water cycle?
5. Rain is not distilled water. The water cycle is important not just for water alone but because it carries a rich broth of ions and other nutrients along with it at times, especially in streams. Is the nutrient concentration in rain highest early in a rainstorm or after several hours of heavy rain?
6. What are the ways that humans are most likely to interrupt the water cycle?
7. How would climate, soils, and vegetation influence the pools and pathways of water in an ecosystem?
8. Draw the carbon cycle from memory after reviewing the figure in the book.
9. How does carbon become incorporated into living tissue?
10. How does leaf fall affect the carbon cycle?
11. What does the work of decomposition?
12. How does the increased burning of fossil fuels and forest affect the carbon cycle?
13. How does time of day affect the carbon cycle? How does time of year affect the carbon cycle?
14. If we have iron in our blood, do magnets affect it? Some breakfast cereals have 100% daily recommended levels of iron. Is there enough iron to pick up flakes of Wheaties with a magnet?
15. Think about how humans influence the nitrogen cycle – turning forest into cropland, adding fertilizer to a cropland, using a vehicle. Remember that nitrogen tends to be the limiting nutrient for terrestrial plants and salt marsh plants.
16. How have human activities changed the global phosphorous cycle? What are the primary reasons why the phosphorous cycle is so important for ecologists?
17. Which cycles are under the greatest human influence?

REFERENCES

Bade, D. 2009. Freshwater carbon and biogeo-chemical cycles. In: S.A. Levin (ed.), *The Princeton Guide to Ecology*. Princeton University Press, Princeton, NJ.

Betts, R. 2021. Met office: atmospheric CO_2 now hitting 50% higher than pre-industrial levels. Met Office Hadley Centre and University of Exeter. https://www.carbonbrief.org/met-office-atmospheric-co2-now-hitting-50-higher-than-pre-industrial-levels/

Chapin III, F.S., P.A. Matson, and H.A. Mooney. 2002. *Principles of Terrestrial Ecosystem Ecology*. Springer, New York.

Falkowski, P. 2009. The marine carbon cycle. In: S.A. Levin (ed.), *The Princeton Guide to Ecology*. Princeton University Press, Princeton, NJ.

Houghton, R.A. 2009. Terrestrial carbon and biogeochemical cycles. In: S.A. Levin (ed.), *The Princeton Guide to Ecology*. Princeton University Press, Princeton, NJ.

Smith, R.L. and T.M. Smith. 2001. *Ecology and Field Biology*. Addison Wesley, San Francisco, CA.

Sterner, R.W. and J.J. Elser. 2009. Ecological stoi-chiometry. In: S.A. Levin (ed.), *The Princeton Guide to Ecology*. Princeton University Press, Princeton, NJ.

Tiseo, I. 2022. Global atmospheric concentra-tions of carbon dioxide 1959–2021. https://www.statista.com/statistics/1091926/atmospheric-concentration-of-co2-historic/#statisticContainer

Vitousek, P.M. and P.A. Matson. 2009. Nutrient cycling and biogeochemistry. In: The Princeton Guide to Ecology. S.A. Levin, (ed.) Princeton University Press, Princeton, NJ.

23

Ecosystem regulation

INTRODUCTORY PRINCIPLES

In population ecology the important question was what regulates population size. In community ecology it was what structures communities. In ecosystem ecology it is what regulates the flow of energy and cycles of matter. For starters, think about animal bodies and what regulates their physiology. Body heat, pH, sleep cycles, sugar uptake, these operate within feedback loops. Plants have analogous systems to control photosynthesis. What, if anything, controls these systems? In animals, nervous and endocrine system may be in control, directed within cell nuclei.

What controls feedback loops in ecosystems? Is this even a legitimate question? It should be obvious that ecosystems are not as closed as an individual body. (**Open systems** have input and output with the rest of the world. Closed systems do not.) Of course, an ecosystem is not living as a single entity, nor capable of reproducing.

WHY IS THE WORLD GREEN?

In 1960, Hairston, Smith, and Slobotkin addressed the ecosystem regulation question with another question, "why is the world green." In other words, why are there so many plants on this Earth? Why

DOI: 10.1201/9781003271833-23

Why is the World Green? (HSS)

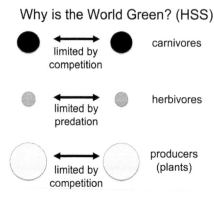

Figure 23.1 The world is green, according to Hairston, Smith, and Slobotkin, because carnivores keep the number of herbivores in check, thus allowing a surplus of plants, a top-down perspective.

do the herbivores not proliferate to the point where they eat all the plants then go extinct? The answer offered by Hairston, Smith, and Slobotkin was that different factors were at work depending on trophic level (Figure 23.1). Predators kept herbivore abundance in check, allowing plants to proliferate, and competition kept green plants, decomposers, and predators in check. This was a top-down hypothesis to the question of what regulates an ecosystem. In other words, higher trophic levels affect the abundance of lower trophic levels. A bottom-up hypothesis would posit that autotrophic production determines the size of trophic levels above it. In other words, nutrient levels and physical conditions regulate all trophic levels above.

EUTROPHICATION IN LAKES

The dichotomy drew attention because one of the great ecology problems of the 1960s was eutrophication and fish kills in lakes. **Eutrophication** literally means true food, which translates to high nutrient content and productivity in a lake. Eutrophication can occur because of pollution leading to excessive production of autotrophs, especially green algae. Because the algae are not consumed by herbivores, they decompose, channeling the nutrients into the detrital food chain instead of the living (grazing) chain. The decomposing bacteria and fungi use up dissolved oxygen in the water during decomposition, leading to fish kills. This prompted research starting in the 1960s on how to control primary production in lakes.

One of the chief questions was whether P (phosphorus) was the limiting nutrient for photosynthesizers. In terrestrial systems the limiting nutrient is often N (nitrogen), but in water, green algae can take up N in several more forms than on land, including NH_4 or NO_2 and NO_3. In other words, algae do not find N to be as limiting as in terrestrial plants.

In contrast, P tends to become unavailable and rare in lakes because waves oxidize it to PO_4^{3-}, which binds with Fe^{3+} and falls to the sediment, where it becomes unusable by plants until resuspension (Figure 23.2). More specifically, when $pH < 7$ the reaction proceeds as $FE^{3+} + PO_4^{3-} \rightarrow FePO_4$, which acts as an iron trap. When $pH > 7$ it proceeds as $FE^{3+} + PO_4^{3-} + OH \rightarrow Fe_x(OH)_y(PO_4)_z$, which is easily resuspended. An anoxic hypolimnion will resuspend P, which occurs during hot summer months when algae growth is at its highest. Overall, if there were a strongly limiting nutrient for producers in lakes, like P, it would support a bottom-up explanation for why lakes become very green in polluted situations.

DAVID SCHINDLER

In the mid-1970s Canadian researcher David Schindler tested the hypothesis that P was the limiting nutrient. His strategy was to place a curtain between two basins of an hourglass-shaped lake in Ontario, called Lake 226. In one basin he dumped N and C. In the other were quantities of C, N, and P. If phosphorus was the problem, the basin with P would bloom. The other basin would not, despite increased nutrient contributions from common pollutants (Schindler 1974).

Results were so dramatic that an aerial photograph (Figure 23.3) was largely responsible for passage of the Clean Water Act of 1972 in the U.S. (McIntosh 1985). The basin with P had a bloom of algae. The other had only minimal growth. The subsequent version of the Clean Water Act focused on controlling P, the largest source of which was phosphate in laundry detergents. Reducing P in detergents greatly reduced P loading in lakes. It was a bottom-up solution. However, as a companion study, Schindler analyzed samples from 66 lakes in Ontario, finding that P loading explained only 48% of the variance in lake primary production. His conclusion was that although the presence of P was a major factor regulating trophic levels, other

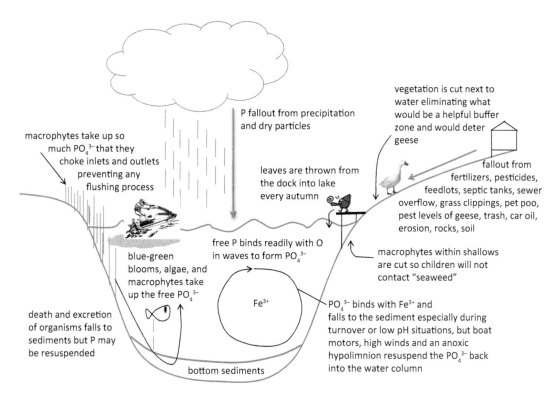

Figure 23.2 The phosphorus cycle in a eutrophied lake.

Figure 23.3 Lake 226 in Ontario, showing the eutrophied half in the background with P added, and no P added in the foreground half. (*biology-forums.com/index.php?action=gallery;sa=view;id=1532.*)

factors must be at work too. Could there be top-down factors as well as bottom-up influencing the extent of the algal bloom?

MORE LAKE STUDIES – LAKE MICHIGAN

Lake Michigan in the 1960s and 1970s provided insight in the top-down versus bottom-up debate with a food chain that included:

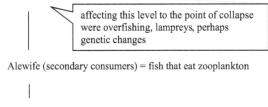

salmon (tertiary consumers) = piscivore, mainly Lake Trout

> affecting this level to the point of collapse were overfishing, lampreys, perhaps genetic changes

Alewife (secondary consumers) = fish that eat zooplankton

zooplankton (primary consumers)

plytoplankton (producers)

> affecting this level was P loading and perhaps effects from the top

The native Lake Trout fishery had collapsed because of a combination of overfishing, lamprey parasitism, pollution, and perhaps genetic changes. With trout gone, prey at the next lowest trophic level proliferated, including abundance of Alewife. This led to a decreasing abundance of zooplankton, which increased the abundance of phytoplankton and algae. Excess P from detergents had exacerbated the phytoplankton and algae. The blooms turned the water brown, murky, and smelly, with mountains of soap bubbles forming at the bottom of dams or anywhere a waterfall occurred. The Alewife were so prevalent, thousands washed up on shore every day. On the public beaches, bulldozers were used to push the dead fish into small mountains of smelly carcasses, while tourists swam nearby (Figure 23.4).

SOLUTIONS FOR LAKE MICHIGAN AND THE GREAT LAKES

Although the trout fishery in Lake Michigan could not be restored directly, replacement species were introduced including Coho Salmon. Simultaneously, new regulations reduced P loads that entered through detergents. Both a top-down and a bottom-up strategy were implemented although ecologists at the time did not use those terms.

Within just a few years, a positive change of tremendous magnitude was evident in Lake Michigan's ecosystem. The secchi disk measurement showed radical improvement. Water went from brown and murky to clear blue. Eventually, the water became too clear after the Zebra Mussel was accidently introduced from the ballast water of ships. These tiny clams filtered nutrients from the water column, further clarifying it.

The reintroduction of salmon created a **trophic cascade**, a ripple effect through the food chain from a predator at the top (Figure 23.5). It is easy to see how the reintroduction of top predators in a lake could be a useful management technique for improving water quality. Note that a trophic cascade for an odd number of trophic levels will produce a green lake and a cascade with an even number will produce a blue lake (Borer and Gruner 2009).

WHOLE LAKE EXPERIMENTS – STEVE CARPENTER

From a theoretical standpoint, the question remained whether it was top-down or bottom-up effects that were the most important. Drawing on observations from Lake Michigan, experiments were initiated in lakes of the northern peninsula of Michigan to observe trophic cascades under controlled conditions. In the first whole lake experiment by Carpenter and his students, the top predator bass were taken from one lake and introduced to a second. Planktivorous minnows were taken from the second lake and introduced to the first. Results were observed for 3 years. A nearby third lake was monitored to record the year to year and season to season changes over the duration as background. The results of the fish swap were expected to follow the simple diagram in Figure 23.6 (Carpenter et al. 1985). The actual results were more complex.

The expected trophic cascade more or less occurred, but several more subtle interactions took place (Carpenter et al. 2010). These included effects of some zooplankton eating other zooplankton, time lags because bass and zooplankton reproduced on separate schedules, dietary changes in bass over the course of their growth from young

Figure 23.4 Dead Alewife in a pile on the shores of Lake Michigan near Milwaukee. (*image.pbs.org*)

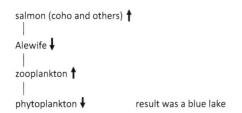

Figure 23.5 The trophic cascade in Lake Michigan in the 1970s. Several salmon species were introduced for an increased fishery. Arrows indicate whether populations increased or decreased after the cascade.

Figure 23.6 The trophic cascade expected in Peter Lake, a whole lake manipulation at the University of Notre Dame Environmental Research Center in the Upper Peninsula of Michigan, U.S. Bass were taken from Peter Lake and put into Tuesday Lake, and minnows from Tuesday Lake were put into Peter Lake. Paul Lake was used as a reference to detect background-level intra- and inter-year changes.

fish to adults, minnows escaping the lake to flee predaceous bass, and unexpectedly high inputs of carbon from terrestrial sources.

WHAT DOES THIS TELL US ABOUT THE REGULATION OF LAKES?

The conclusion was that cascades occur in open water systems of lakes, but within the confines of nutrient levels. Organisms are constrained in their own stoichiometry by the stoichiometry of the lake (Sterner and Elser 2009). **Stoichiometry** refers to the proportion of nutrients making up the content of something. If the lake stoichiometry changes from year to year, the organisms that thrive are changed depending on which is most able to mirror the chemical proportions of the lake.

As for trophic level cascades as a management tool, did the introduction of piscivorous fish produce a blue lake? The answer is yes, even with all the subtle interactions that take place when top predator populations are altered, a cascade can still have a dramatic effect on water quality. Whether a cascade will have this intended effect in every situation is a gamble, but perhaps a gamble worth trying as a relatively easy fix within highly used but degraded lakes (Carpenter et al. 2010).

TROPHIC CASCADES IN BENTHIC ECOSYSTEMS?

While trophic cascades can occur dramatically in open water situations of lakes, can they also occur among **benthic** (bottom, benthos) dwelling creatures? When a crayfish – snail – periphyton food chain was investigated in the littoral zone of a freshwater lake (near-shore submersed plant area), crayfish had a strong behavioral effect on snails, which then considerably affected periphyton abundance (Lodge et al. 1994, Weber and Lodge 1990). **Periphyton** is the community of algae, moss, animals, and protists living on substrates. However, the strong effects of a cascade dissipate once prey reside within bottom soft sediments.

When a benthic food chain consisting of shorebirds – predatory polychaetes (within sediment) – invertebrate detritivores (within sediment) – was studied a cascade did not occur even though the shorebirds caused a significant reduction in benthic prey (Weber and Haig 1997). Subsequent other studies have also found that trophic level interactions within the sediments are not as obvious or dramatic as in the **pelagic zone**, the open water section of a waterway. Within benthic sediments, there are more places of refuge for prey to hide. Additionally, more disturbance occurs interrupting the processes in the food chain. More patchiness overall happens, with more omnivory.

NUTRIENT CYCLES IN TERRESTRIAL ECOSYSTEMS – HUBBARD BROOK

Are their trophic cascades in terrestrial systems? If not, what regulates ecosystem cycles and energy? Consider the eastern deciduous forest at Hubbard Brook in the White Mountains of New Hampshire. Separate watersheds were investigated, some of which had been previously clearcut,

others intentionally cut, others left intact. Results showed that the highest biomass occurred during the **aggradation phase**, the first 140 years when the forest regrows after a disturbance. After aggradation, the forest actually began to lose biomass. Regarding the nitrogen cycle, during the first part of the aggradation phase, nitrogen accumulated rapidly because of the high nitrogen to carbon ratio in pioneer species. After the pioneer stage, the rate of N accumulation slowed. Overall, the nitrogen cycle was essentially closed, meaning that little was lost as runoff.

This makes sense. Although 78% of the atmosphere is nitrogen, it is not available to trees unless fixed. Once in a tree, however, the nitrogen must go somewhere when the tree dies. It mineralizes into the soil, but because it is usually the limiting nutrient in terrestrial systems, it is taken up readily by other plants. Typically, only 5% runs off because it is so quickly taken up by other plants.

A fire or a clearcut, however, can take up to 33% of the nitrogen pool into the air. Because nitrogen is the limiting nutrient, this is a tremendous loss for a forest in a short time. Tree removal increased stream flow considerably. Less uptake by plants and transpiration occurred. Nutrients losses were eight times higher than in intact forests.

Succession in the Hubbard Brook forests included a phase in which there was no net accumulation of biomass. Primary production balanced respiration, and the rate of accumulation of nutrients slowed. Bormann and Likens were hesitant to call this a climax forest, a balance of nature, or even a steady state. Instead, disturbance was so prevalent at large and small scales that Bormann and Likens envisioned the forest as just a cluster of gaps and patches in various successional stages. Water and many of the elements were not in a steady state. The forest was more of an open system than expected in a **shifting mosaic** of patches.

WHAT REGULATES ECOSYSTEMS?

Getting back to the original question of what regulates the cycles in ecosystems, accumulated studies from many types of ecosystems over several decades reveal that it is sometimes:

- the activity of a particular trophic level.
- limiting nutrients.
- disturbance, for example, fire that reduces nitrogen levels.

- one species, for example, a filter feeding mussel that plays a keystone role in recycling nutrients.
- climate and topography as in tropical forests. The tropics have not been glaciated for millions of years. The soil is highly weathered and thus nutrient poor.
- plants changing the ecosystem, for example, alders that change the acidity of the soil as they fix nitrogen. The acidity influences the ability of other plant species to take up nutrients. One nutrient cycle can sometimes influence another.

ALTERNATIVE STABLE STATES

Additionally, ecosystem ecologists have discovered that at times, ecosystems change dramatically with the slightest disturbance. The change occurs at a tipping point that causes the system to enter a phase of self-propagating change until it comes to rest in a contrasting **alternative stable state.** These changes are known as **regime shifts** and are in contrast to gradual trends that are more common in nature.

Examples of alternative stable states include:

- dense mesquite stands in Texas deserts resulting from fire suppression. Ordinarily, mesquite trees are intermixed with grasses or other herbaceous vegetation. Fire is carried by this non-tree vegetation. When fire is suppressed and grazing reduces the green vegetation, the mesquite becomes an impenetrable monoculture. Without vegetation to carry the fire, the fire dies in the bare patches.
- turbid, shallow lakes within one lake district (Scheffer 2009). In some lakes, submersed vegetation may enhance water clarity by shading out competing phytoplankton, taking up the P that would otherwise be available to phytoplankton. In other words, the macrophytes facilitate the growth of more macrophytes in the community through a positive feedback loop that promotes water clarity.

- eutrophication, however, may bring about a tipping point, in which there is so much turbidity that the vegetation no longer has enough light. An abrupt die-off of vegetation takes place, and without the feedback loop, the water gets increasingly cloudy. Thus, similar lakes may have two extremes, one vegetated and clear, the other turbid without plants.
- a brief shock therapy of eliminating the fish may move a turbid lake back to a clear condition (Scheffer 2009). Once planktivorous fish are removed, zooplankton concentrations can increase, reducing the turbidity caused by phytoplankton. The elimination of bottom-dwelling fish would reduce suspended sediments.

ECOSYSTEM SUMMARY

Ecosystems are not closed locally. Some cycles in ecosystems are more open than others. Water, carbon, and nitrogen are more open than phosphorus. Ecosystems are not homogeneous. Ecosystem ecologists have not always appreciated the patchiness of the entity that lies within the black box, a weakness of the ecosystem approach. As fragmentation ensues, ecosystems become less homogeneous. To be neither homogeneous internally nor closed externally is in sharp contrast to the traditional equilibrium view that communities are self-regulatory, self-repairing, and homeostatic (Wu 2009).

By incorporating ideas of landscape ecology, a truer model may be formed. What we need is to draw together elements of all types of ecology. We need to integrate the pattern-based horizontal methods of landscape ecology (Chapter 24), with the process-based vertical methods of ecosystem ecology (Wu 2009). We need to promote the coupling between the organism-centered population perspective and the flux-centered ecosystem perspective. Patch dynamics is the unifying theme, covered in the next chapter.

THINKING QUESTIONS

1. Why is the world green?
2. Give a biochemical explanation for why fish die in highly eutrophied lakes.
3. David Schindler found that phosphorus loading explained 48% of the variance in primary production of lakes. What accounts for the other 52%?

4. How can trophic cascades become a management tool for limnologists? In other words, how can one improve water quality through trophic level manipulation?
5. Predict the trophic cascade effects in an ecosystem if a top predator is introduced or removed in a food web. Mary Power discovered a food web with five trophic levels. What would you expect to find in phytoplankton densities relative to zooplankton densities in Power's aquatic system? In other words, would you expect clear or algae-laden water?
6. What was the effect of clear-cutting on stream flow at Hubbard Brook? Why?
7. According to the results of Bormann and Likens, is there such a thing as a climax forest? Explain.
8. What are examples of alternative stable states other than those found in this chapter? What makes a stable state tip to an alternative state?
9. Do ecosystems work like a gear box? How are ecosystems different from physiological systems and nutrient recycling in a living body?
10. Ecosystem ecology is presented as machine theory applied to nature. This is very different from the approach taken in traditional ecological knowledge. What might be the benefits and drawbacks of thinking about ecosystems in such an analytical way?

REFERENCES

Borer, E.T., and D.S. Gruner. 2009. Top-down and bottom-up regulation of communities. In: S.A. Levin (ed.), *The Princeton Guide to Ecology*. Princeton University Press, Princeton, NJ.

Carpenter, S.R., J.F. Kitchell, and J.R. Hodgson. 1985. Cascading trophic interactions and lake productivity. *Bioscience* 35:634–639.

Carpenter, S.R., J.J. Cole, J.F. Kitchell, and M.L. Pace. 2010. Trophic cascades in lakes: lessons and prospects. In: J. Terborgh and J.A. Estes (eds.), *Trophic Cascades: Predators, Prey, and the Changing Dynamics of Nature*. Island Press, Washington, DC.

Hairston, N.G., F.E. Smith, and L.B. Slobotkin. 1960. Community structure, population control, and competition. *American Naturalist* 94. https://doi.org/10.1086/282146

Lodge, D.M., M.W. Kershner, J.E. Aloi, and A.P. Covich. 1994. Effects of an omnivorous crayfish (*Orconectes rusticus*) on a freshwater littoral food web. *Ecology* 75:1265–1281. https://doi.org/10.2307/1937452

McIntosh, R. 1985. *The Background of Ecology*. Cambridge University Press, Cambridge, UK.

Scheffer, D. 2009. Alternative stable states and regime shifts in ecosystems. In: S.A. Levin (ed.), *The Princeton Guide to Ecology*. Princeton University Press, Princeton, NJ.

Schindler, D.W. 1974. Eutrophication and recovery in experimental lakes: implication for lake management. *Science* 184:897–899.

Sterner, R.W., and J.J. Elser. 2009. Ecological stoichiometry. In: S.A. Levin (ed.), *The Princeton Guide to Ecology*. Princeton University Press, Princeton, NJ.

Weber, L.M., and D.M. Lodge. 1990. Periphytic food and predatory crayfish: relative roles in determining snail distribution. *Oecologia* 82:33–39.

Weber, L.M., and S.M. Haig. 1997. Shorebird-prey interactions in South Carolina coastal soft-sediments. *Canadian Journal of Zoology* 75:245–252.

Wu, J. 2009. Ecological dynamics in fragmented landscapes. In: S.A. Levin (ed.), *The Princeton Guide to Ecology*. Princeton University Press, Princeton, NJ.

Landscape ecology

INTRODUCTION

Landscapes are heterogeneous areas of land, usually square kilometers in extent, composed of interacting ecosystems or patches. The study of spatial heterogeneity in these landscapes is **landscape ecology**. The goal is to examine interactions between spatial patterns and ecological processes. Landscape ecology incorporates metapopulation theory, patch dynamics, and integrates all the subdisciplines of ecology.

Landscape ecology can be used, for example, when considering the 1988 Yellowstone fires in which 33% of the area burned, but it happened in a patchy manner (Figure 24.1). What caused the patchiness? Was reestablishment different in small versus large patches? Did reestablishment occur from the middle of the patch to the edge, or vice versa?

Is a Holstein cow black or white? Of course, the answer depends on spatial scale. At what scale then should we evaluate a fire, hayfield, or watershed if the answer will vary depending on our choice? Landscape ecology studies the geometric shapes of natural elements and incorporates

fractal geometry. What shape and size of reserves will hold the most biodiversity? In asking these questions, landscape ecology considers the conservation value of corridors, fragments, metapopulation dynamics, and edge effects. Other questions that might be asked in landscape ecology: how is climate different within small forest patches versus large? What is the smallest wetland that migrating shorebirds will use? What is the orientation of lakes and the direction of flow by aquatic organisms across a region? Models consider the influence of more light, higher temperatures, and less distance for predators to travel.

MAJOR THEMES

Landscape ecology became a major subdiscipline of ecology after the 1986 book *Landscape Ecology* by Richard Forman and Michael Godron. Its origins were even earlier coming from Europe in the 1930s. Because much of the European landscape had been permanently altered by humans for centuries, only human-dominated landscapes were left to study. The focus was on landscape elements and their arrangements rather than the complex

DOI: 10.1201/9781003271833-24

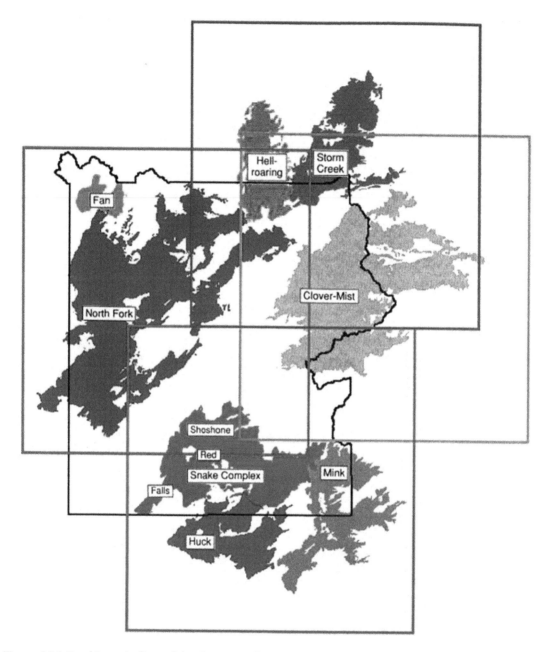

Figure 24.1 Patchiness in fires of the Greater Yellowstone Ecosystem in 1988. (*Rothermel et al. 1994 free access.*)

interactions within communities or ecosystems in a wild state. Worldwide, landscape ecology now finds utility because it incorporates several contemporary themes:

1. It accepts the idea that all landscapes have been influenced by humans and incorporates their presence.

2. Its applications could be at any scale, but it tends to have its greatest value at large scales, larger than watersheds or drainage basins.

3. It is technologically laden, using **remote sensing**, the science of acquiring information about the Earth's surface without being in contact with it. Data is obtained from satellites, aerial photographs, space shuttles, and

Data Source **Data Layers**

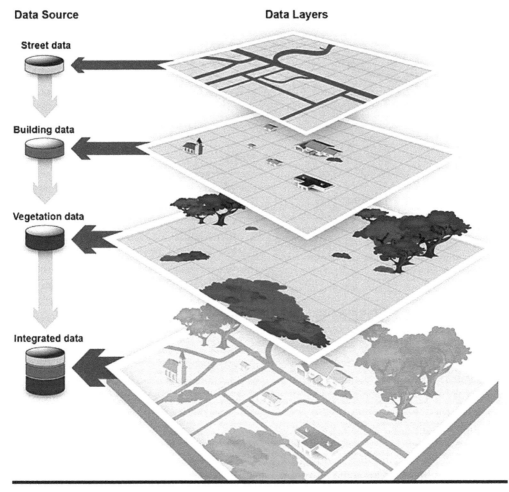

Street data

Building data

Vegetation data

Integrated data

Source: GAO. | GAO-15-193

Figure 24.2 Geographic Information Systems (GIS) uses layered maps, software, and a systems approach for finding data (*Powner 2015 free access.*)

space stations. In other words, remote sensing provides a way to look at ecology as if we are hovering over the ground. Landscape ecology uses technology from **GPS** (global positioning system), Google Earth, and **GIS** (geographic information systems) (Figure 24.2), a powerful set of computer tools used for collecting, storing, retrieving, and displaying spatial information in layered maps. Computer modeling is included. For instance, a GIS can compare scenarios using different forest prescriptions.

4. Landscape ecology works at several scales. It acknowledges that some phenomena at one scale may not apply at another.

5. Landscape ecology uses **hierarchy theory**, a type of logical thinking known since the Roman Empire. Ecologists popularized the idea in the 1960s as a conceptual framework to deal with multiple scales. Top levels can control lower levels, but lower levels cannot control higher levels.

6. Landscape ecology has the ability to integrate ideas from the other major subdisciplines of ecology. By combining theory from metapopulations, island biogeography, patch dynamics, conservation biology, the technology provided by GIS, remote sensing, LiDAR, and hierarchy dynamics, new models producing spatially explicit models are very powerful.

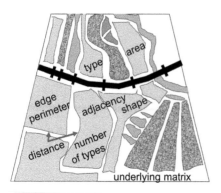

Figure 24.3 The matrix is the surrounding background vegetation most commonly found in the region. Patches are whatever is different. Some of the chief metrics collected by landscape ecologists are shown and include perimeter length, adjacency, and others.

VOCABULARY IN LANDSCAPE ECOLOGY

Several metrics and definitions are used to describe landscapes. **Patches** are the fundamental unit in landscape ecology and are small homogeneous areas that differ from their surroundings (Figure 24.3). They are usually identified by the plant species with the highest proportion of cover (Peters et al. 2009). Beyond patches of interest, the surrounding background is called the **matrix**. This could be considered the most common patch type in which smaller types are distributed. Some of the metrics collected by landscape ecologists include perimeter, adjacency, area of each patch, shape of patches, proximity of patches to other patches, number of patch types in one region, and connectivity between patches.

PROGRESSION OF THEORY IN LANDSCAPE ECOLOGY

Biogeographers of the 1500s first tried to map the world's vegetation blocks. A shortcoming was the impossibility of defining borders at fine scales. Another was that biome-type models were not predictive, only observational. Recognizing these challenges, the goals of landscape ecology now are to:

- Record changes over time in patch boundaries.
- Be ever more accurate and precise in mapping small patches and stands within a region, now with the capability of using infrared cameras.

- Work at several spatial scales at once to understand the evolving story of why things are where they are.
- Take advantage of the best technology and teach ecologists these new technologies.
- Be able to synthesize ideas from all the other major subdisciplines of ecology in its models, finally bringing synthesis to ecology.

1960s

The island biogeography theory of the early 1960s led to an understanding that species richness tends to increase with area and nearness to the mainland or other islands. It taught us that species in small patches are more likely to go extinct. The shortcoming was that it oversimplified the situation on land where a continuum of changes from patch to patch occurred without the defined borders of an oceanic island. Another shortcoming was its focus on equilibrium ideas. In reality, landscapes are ever changing and often increasingly fragmented, not in equilibrium (Wu 2009). Real landscapes have habitat heterogeneity, disturbances, edge effects, multiple interactions among colonizing species, and constraints within the matrix.

1970s

Patch dynamics models introduced in the 1970s offered models built on the idea of terrestrial "islands." Animal behavior could be predicted based on distance to the next patch, resource availability, and crowding. A shortcoming was the neglect of animal behavior. Only one or a few species were considered at a time. The models concentrated on patches, perhaps without enough attention to intervening matrix.

The 1970s idea of metapopulations, a group of subpopulations linked by dispersal, taught that population dynamics could be modeled regionally. From a conservation standpoint, we learned the value of retaining habitat patches even if they were temporarily vacant. A shortcoming was the single species focus, not concerned with neighboring species. Subpopulations were highlighted, but not patch heterogeneity or matrix. Most of all, metapopulation models were not **spatially explicit**. In other words, they were not concerned with the size, shape, and configuration of the patches that held subpopulations, only the size of the population.

Real landscapes have patches not easily defined. Processes are a product of physical, biological, and socioeconomic effects (Wu 2009).

1980s

In the 1980s, one of the more dramatic applications of landscape ecology was the advent of **gap analysis** initiated by the USFWS to inventory biodiversity at the state level in the U.S. Gap analysis uses remote sensing, GIS, and on-ground vegetation mapping to identify hotspots of biodiversity and endemism. It makes recommendations for where to establish a reserve so that all rare species are represented somewhere in a protected area. It was first used to analyze Hawaii, then Idaho, and is now being used in most other states. Even a worldwide analysis of protected reserves has been completed.

The attributes of gap analysis were the spatially explicit contribution, that is, specifically identifying all habitat and nonhabitat elements. Gap analysis demonstrated the power of GIS, not just as fancy layered maps, but as systems to represent realism in modeling and detail over theory such as was used in metapopulations. Its shortcomings were that it required complex GIS computer software and training, large data sets, and ground truthing. As much of an advance as gap analysis was when used in conjunction with GIS, it presented output in just two dimensions, even though nature is inherently three dimensional. It has only been with the benefit of a new technology, LiDAR, that three-dimensional representations are now available through remote sensing. **LiDAR**, Light Detection and Ranging, is modeled after RADAR, which uses sound waves to determine distance to another object. In the same way, LiDAR uses light waves to determine precise distances and produce three-dimensional images.

1990s

More recent models were known as patch-matrix corridor models. They recognized patches, corridors, and the background mosaic roughly equivalent to islands within oceans. They acknowledged that matrix serves an important role. The shortcomings include a representation still based on human-defined patterns, not the scale important

to other organisms. They did not recognize the subtleties of variation that may be important, such as temporary variation.

RECENT AND FUTURE MODELS

More advanced models would simultaneously recognize more than one spatial scale in what are known as hierarchical patch dynamics models (Figure 24.4). Of course, these models add more complexity, both an attribute and a shortcoming. The representations are still based on human-defined patterns. More advanced models, such as variegation models, allow organisms to define their space. They recognize gradients in the environment rather than patches in a matrix. They acknowledge that patch models may not apply in all circumstances. Their shortcomings are complexity and a focus on just one or two species at a time and for only a few variables. They are perhaps too complex for applicable results.

TOP FIVE USES FOR LANDSCAPE ECOLOGY

One of the most common uses for landscape ecology is reserve planning, providing spatially explicit models for conserving corridors, boundaries, and integrating development and humans (Figure 24.5). For instance, in the hierarchical, spatially explicit model devised for conservation of Cougar (*Puma concolor*) in Utah (Stoner et al. 2010), the goal was to quantify land cover types used by Cougar, but additionally where human-Cougar conflict was most likely to occur. The authors were able to produce probabilistic maps as the outcome. Researchers examined multiple behaviors and time periods for the Cougar. The conclusions were that Cougar movements were more extensive (longer distances) and used different cover types at night than when resting in secure cover during daylight hours.

Another use for landscape ecology is recording land use changes over time. For instance, habitat areas in Illinois that could be used by the Greater Prairie Chicken were recorded through progressive maps as in Figure 24.5. Climate modeling is yet a third common use for landscape ecology. Models related to fragmentation are a fourth. Models related to nutrient cycling are a fifth.

Metapopulation Model Patch Dynamics Model Island Biogeography

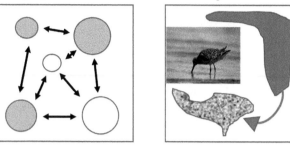
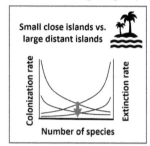

Patch Matrix
Corridor Model

Species Oriented Variegation Model

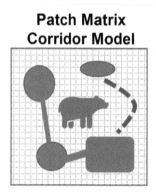
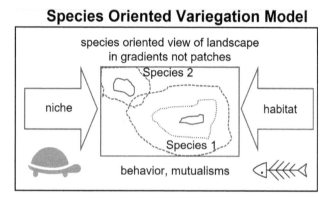

Figure 24.4 The current theory in landscape ecology has progressed from the island biogeography model of the early 1960s to the more recent gradient variegation models.

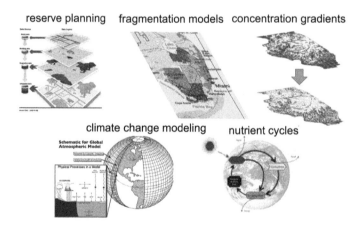

Figure 24.5 Top five uses for landscape ecology.

SYNTHESIS

The equilibrium notion of steady state, stability, and homeostasis has given way to the idea of landscape heterogeneity in both space and time with multiscale linkages of ecological systems (Wu 2009). Landscape ecology very naturally provides a unifying framework for integrating population, community, and ecosystem perspectives in fragmented landscapes. The caveat is teaching the technology to ecologists.

THINKING QUESTIONS

1. What are hierarchical models? For an example read Lodge et al. (1987).
2. Use the internet to find a description of gap analysis and the historic success it found despite early criticism.
3. Give examples of spatial heterogeneity at scales of 1 m, 10 m, 10 km, 100 km, and 1,000 km.
4. What ecological processes are most affected by landscape pattern?
5. Use the internet or library to find examples of landscape ecology studies. How does landscape ecology naturally integrate all the subdisciplines of ecology?

REFERENCES

Forman, R.T.T., and M. Godron. 1986. *Landscape Ecology*. John Wiley & Sons, Inc., Hoboken, NJ.

Lodge, D., K.M. Brown, S. Klosiewski, R. Stein, A. Covich, C. Bronmark and B. Leathers. 1987. Distribution of freshwater snails: spatial scale and the relative importance of physicochemical and biotic factors. *American Malacological Bulletin* 5:73–84.

Peters, D.P.C., J.R. Gosz, and S.L. Collins. 2009. Boundary dynamics in landscapes. In: S.A. Levin (ed.), *The Princeton Guide to Ecology*. Princeton University Press, Princeton, NJ.

Powner, D. 2015. Geospatial data: progress needed on identifying expenditures, building and utilizing a data infrastructure, and reducing duplicative efforts. United States Government Accountability Office.

Rothermel, R.C., C. Richard, R.A. Hartford, and C.H. Chase. 1994. Fire growth maps for the 1988 Greater Yellowstone Area Fires. General Technical Reports INT-304. U.S. Department of Agriculture, Forest Service, Intermountain Research Station, Ogden, UT.

Stoner, D.C., W.R. Rieth, M.L. Wolfe, M.B. Mecham, and A. Neville. 2010. Long-distance dispersal of a female cougar in a basin and range landscape. *The Journal of Wildlife Management* 72:933–939 https://doi.org/10.2193/2007-219

Wu, J. 2009. Ecological dynamics in fragmented landscapes. In: S.A. Levin (ed.), *The Princeton Guide to Ecology*. Princeton University Press, Princeton, NJ.

Wildlife management and habitat ecology

CLASS ACTIVITY: Wildlife Management Plans – farm or field

Goal: Complete an inventory for a farm or field as a first step toward completion of a wildlife management plan.

Wildlife management plans are an extremely valuable resource for landowners who want to enhance their property for conservation, or simply to see more wild, native species. Plans consist of an assessment of a site, a list of goals and objectives, management recommendations, and a timeline with a budget. Plans vary in sophistication from high school 4-H projects to paid consultant reports. Once a few principles are learned, it is surprising how fast wildlife respond to treatments. The right plan brings a site to life whether in a small backyard or on a large farm or forest preserve.

The first step is to complete an inventory for a farm, ranch, field, or forest. A qualitative inventory is best because it can be completed more quickly than a quantitative effort and is sufficient for making recommendations about management. For instance, the quality of tree species diversity can be assessed on a scale of –2 (monoculture) to +2 (highly diverse) just by quick observation based on the evaluator's experience. If the forest is moderately diverse, the surveyor would write +1 in the box. For each of three areas at a site, fill out a worksheet as in Table 25.1, which is written for areas without a tree canopy such as a farm, ranch, or park.

Table 25.1 Worksheet for qualitative assessment of wildlife in farm and fields

Farm/Field Assessment				
1. Current land use:		**Is edge present?**		**Soil issues?**
	domestic animal use		Edge soft and stadium-like and ragged with diversity of canopy layers	Earth moved recently (e.g., reconstructed slopes, terraces)?
	occasional grazing			
	restored prairie			Soils or other physical features noteworthy like sand dunes or soil areas still in pre-settlement form?
	recreation/lawn		Edge is hard and abrupt without a diversity of canopies	
	old field			Any areas where soil moves, slumps, or unstable because of frost heaving or shrinking?
	crop			

(Continued)

Table 25.1 (*Continued*) Worksheet for qualitative assessment of wildlife in farm and fields		
Farm/Field Assessment		
2. **What would plant community be if succession happened without disturbance?**	**Significant rock or brush piles?**	**Waterway/erosion issues?**
	Several fence posts present?	Any marshes, lakes, or ponds? Has the hydrology been altered?
	Vegetated fence rows?	Increased runoff because of disturbance in natural vegetation?
3. **What current management takes place to prevent succession? Mowing? Plowing? Grazing? Fire? Other?**	**Human-created overhangs, open barns or buildings, lofts, outhouses?**	Erosion areas, steep slopes, floodplains, stream banks, shorelines, wetlands
	Pets within close barking range?	Areas where shade from dense vegetation is contributing to the decline of soil-stabilizing vegetation?
4. **Dominant plant species within field itself and field margins.**	**Historic sites, cemeteries, or archeological sites?**	
5. **Rate the plant height diversity within field itself.** -2=even height. . . $+2$= highly diverse height	**Notable animals (invasive or rare):**	Areas where groundwater emerges to the surface as springs, seeps, marsh, or wet ground?
	Runoff from neighboring areas?	Drainage features been modified by straightening, deepening, damaging, draining, drawing, or diversifying?
6. **Are hedgerows present?**	**Notable plants (invasive or rare):**	
length of hedgerow (m)		Stream within 100 m?
average width (m)		Dams, weirs, culverts?
species in hedgerow		Ditching, tiles to dewater poorly drained soils, or deepening and straightening stream channels?
7. **Rate cavity tree and snag abundance nearby:** -2=sparse . . $+2$=abundant		

WHY WILDLIFE MANAGEMENT?

Nature as therapy is a gift. We feel better when spending time in nature, but more so if we use the time to give back to the Earth. Reciprocity is the name for this, but just how do we go about it? One of the most useful applied disciplines in ecology is wildlife management. It identifies the major ways to attract wild animals to a site, and in doing so provides the basics of restoration principles within temperate areas. It is the most useful and relevant topic for relandscaping one's yard and bringing it to life for pollinators and wild animals while eliminating pests.

Wildlife management is so useful that it is a student favorite, yet often not taught within ecology courses, probably because of its early association with game species and hunting. Currently, wildlife management also covers non-game species, which is why wildlife management might now be better defined as "habitat ecology" than anything else.

THE LEOPOLDIAN BASICS OF WILDLIFE MANAGEMENT

Aldo Leopold (1887–1948) founded wildlife management in 1933 with publication of his book, *Game Management*, while he was on the faculty at the University of Wisconsin (Figure 25.1). As such, "**wildlife**" traditionally referred to warm-blooded vertebrate animals, those neither **domesticated** (bred in captivity) nor **feral** (once in captivity, now wild). More specifically, "wildlife" referred originally to game species, especially deer and ducks.

Wildlife management traditionally excluded fish, invertebrates, and organisms outside the animal kingdom. Within game species, four types were recognized: **furbearers** (small game mammals from rabbits to Gray Wolf), **big game** (from deer to Cougar and Grizzly Bear), **upland game birds** (Ruffed Grouse, Ring-necked Pheasant, etc.), and **waterfowl** (ducks, geese, and swans). The major principles of wildlife management are as follows (Leopold 1933, Bolen and Robinson 2003):

1. **An animal's habitat has four components:** *food, water, cover, and usable space.* The beauty lies in the simplicity of this framework. To encourage any species to an area, the four must be present. To discourage a pest from an area, the four must be removed. In this context, **habitat** is what defines the home of an organism, with characteristic architecture and relationships with other organisms. An organism's habitat is species-specific.

2. **To provide for species one must provide habitat –** *if you build it they will come.* The only necessities are an appropriate habitat and a dispersal route from an existing population to that habitat. This was revolutionary when first proposed, contrasting with the European tradition of stocking, as when a fox is intentionally released, then hunted.

3. **The habitat most attractive to vertebrate herbivores (e.g., turkeys, rabbits, quail, deer) is edge. Edge** occurs on the forest/field boundary or forest/waterway boundary (Figure 25.2). It consists of early succession grasses, shrubs, thickets, and herbaceous flowers. This vegetation is knee-high to waist-high and not under a tree canopy. Edge includes **old fields**, areas once farmed now in a grassy, shrubby stage which may be on their way to mature forest if no further disturbances occur. Edge encompasses hedgerows, fence rows, and field margins.

Game species favor edge because it provides cover (hiding places) and amino acids (the building blocks of proteins). Comparatively, the forest interior has low vertebrate diversity unless it takes on

Figure 25.1 Aldo Leopold (1887–1948), founder of wildlife management.

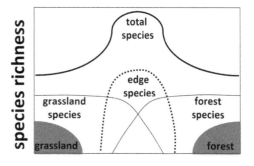

Figure 25.2 Edges often support more species than the grassland core or the forest interior. Individuals from both elements occasionally visit, plus other species are edge specialists. (*Adopted from McComb 2015.*)

the characteristics of **old growth**, with a patchy mosaic of age classes and openings supporting grass, shrubs, thicket, and herbaceous flowers.

FOOD FOUND IN EDGE

Most game species are warm-blooded herbivores (rabbits, quail, deer, and ducks). Herbivores have a problem; they must get the protein they need to build their body mass without eating meat. The problem is solved by eating plants that fix nitrogen, or the growing tips of any plant. Both options have a high **nitrogen to carbon ratio (N:C ratio)**. This nitrogen is either in the form of amino acids or can be converted to amino acids within the herbivore body. Proteins not only make up the muscle mass of an herbivore; enzymes, organ systems, and many of the solutes within animal fluids are proteins.

Nitrogen-fixing plants include clover, alfalfa, beans, lespedezas, and others found in greatest abundance in shrubby/grassy knee-high to waist-high vegetation. Most forest trees are not good food for large herbivores. Mature trees are mainly woody carbon, thus supporting a low N:C ratio. Many of the growing shoots of forest trees are too high for accessibility by ground-dwelling herbivores.

Where forest and field meet, the most valued edge is wide and gradual rather than abrupt. A gradual edge produces a **stadium effect** like the gradual edge of seats in a stadium (Figure 25.3). Edge induced by humans is often sudden and abrupt, caused by roads, tree disease, clear-cutting, development, and row crops. On these edges, low species diversity occurs. Abrupt edges experience sudden winds and greater sunlight as well as

blowdowns, erosion, evaporation, and non-native seeds that come in with wind (McComb 2015).

FOOD TYPES

Specific words for wildlife food include:

- **Forb**: broad-leaved non-woody plant,
- **Browse**: woody vegetation such as twigs and bark,
- **Tuber**: underground stems,
- **Carrion**: dead animals,
- **Nectar**: sugary liquid substance in flowers.

Mast refers to fruits:

- **Hard mast**: nuts and seeds,
- **Soft mast**: fleshy fruits.

Animals that eat specific items include **browsers**, **grazers**, **granivores**, **frugivores**, and **carrion eaters**.

DIGESTION – BIRDS

Within the avian digestive system, the path of food leaving the mouth can be traced sequentially (Figure 25.4).

Food is swallowed into a long esophagus, which includes an expandable portion called the **crop** in some species. The crop's function is storage and, as part of the esophagus, has no ability to digest food chemically. The herbivore digestive system overall is longer than in carnivores because plants, seeds, and grains digest more slowly than meat. This is why herbivores in particular need a crop. The food can be delivered incrementally as the stomach is ready.

The avian stomach consists of two parts, the soft **proventriculus** (glandular stomach), and the hard **ventriculus** (gizzard) supporting a tough corrugated inner surface. Some chemical digestion occurs in the proventriculus. The ventriculus has the same function as teeth, which are absent in adult birds. Birds ingest grit, which rolls the contents of the gizzard, grinding food against the inner surface. The gizzard is muscular and so strong that a turkey can swallow a nut whole and crack a pecan in an hour. In carnivorous birds (owls, hawks) the gizzard forms **pellets**, indigestible rolled-up fur, feathers, and bones from prey. The pellets are regurgitated by the bird, dropped from bird roosts.

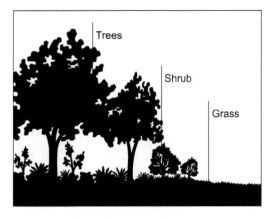

Figure 25.3 The stadium effect produces ideal edge conditions (*G. Giocomin*).

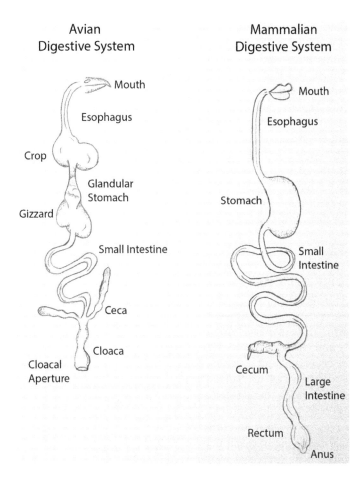

Figure 25.4 Birds have different digestive system features than mammals (*E. Tauber-Steffen*).

The ground up food is moved into the intestine where much of it is absorbed into the blood vessels that engorge these intestines. Blind tubes called **ceca** (singular = cecum) arise from the posterior of the intestine (look at Figure 25.4 again). Bacterial decomposition of fibrous foods and absorption of some water and proteins comprise their function. This aids digestion because cellulose does not break down otherwise. Some grouse and waterfowl species can grow their ceca and intestines longer during the seasons when plants and fiber are most heavily consumed.

Although the liver does not take in ground food directly, the blood vessels around the intestines take what they absorb to the liver, where some carbohydrates are stored, and where blood is filtered for toxins.

Within the digestive tract, material not absorbed by the intestine is passed to the **cloaca**, a common passageway for three products: intestinal waste, either eggs or sperm, and **uric acid**, which is more concentrated than urine and produced by the kidneys. The cloaca deposits these products out of the body through the **cloacal opening**. The droppings of birds are mainly solid because more water is absorbed in birds than in mammals. Droppings are black and white because feces and uric acid become combined in the cloaca.

DIGESTION – MAMMALS

The mammalian food path consists of the mouth, then the pharynx, then the esophagus and stomach. The stomach is multichambered and **ruminant** only in some mammalian herbivores. Ruminant means that microbial fermentation occurs in the stomach (not ceca).

Distal to the stomach in all mammals is the small and large intestines and finally the **anus**, the actual opening to the outside. Unlike birds, mammals have teeth, but not crops, gizzards, or cloaca. Some mammal species have a **cecum** for digesting cellulose. Unlike birds, the size of the intestine does not change seasonally, although intestine length does vary among species. Just as in birds, carnivores have the shortest intestines.

Note that mammals with a multichambered ruminant stomach cannot change diets quickly because their microbial community takes time to form and adjust. Thus, the diet of domesticated cattle and others cannot be changed abruptly. Note that most mammals have gall bladders, but deer do not, making it harder for them to digest fatty foods than other mammals.

ARE SALT LICKS USEFUL? WHAT ABOUT OTHER SUPPLEMENTS?

Salt licks are supplied by wildlife managers to supplement trace elements and calcium, but in most situations, they are not necessary and a waste of money if soil nutrient levels are adequate. Salt licks may help species losing antlers each year if calcium salts are included. However, soil often provides abundant calcium, the reason why deer and other animals can be seen licking the ground. Salt licks have their greatest utility when a manager wants to move or attract mammals to a specific place, like the back pasture for one season. Just realize that salt kills the nearby vegetation wherever it is placed.

Female birds might need calcium during eggshell production, but they can normally just eat small quantities of soil to replenish. Some bird enthusiasts place their discarded eggshells in a container near the bird feeder. Female birds during the nesting period may consume them if offered.

WATER

Vertebrate individual bodies are 80% water, which means they can survive only a few days without replenishing. For this reason, water sources in the eastern U.S. should generally be no farther than a half-mile apart (0.81 km) and generally require 60 inches of rain/year (150 cm) to keep populations healthy (Bolen and Robinson 2003).

Having said this, supplying water is sometimes less important than one might think. Some species (e.g., mice, Northern Bobwhite) acquire all they need from green plants, dew, and metabolic by-products, i.e., breaking down fat within their bodies.

> Write here the chemical formula for cellular respiration:
> Do you see now how water is produced during metabolism?

COVER

Animals need cover for shelter, to afford protection from excess energy loss and weather factors. Some forest types provide more of a roof than others, which prevents heat loss to the sky on a cold clear night. In very snowy areas a **deer yard** can develop, a sheltered place where deer congregate away from drifting snow, usually within a coniferous grove. Cover is necessary as concealment, providing protection from predators or concealment from prey while a predator is stalking. Cover is a relative term. What might be good cover for most songbirds (e.g., dense vegetation waist-high) is not good for doves, which would have a hard time walking through it.

Common sources of cover include:

- **Knee-high to waist-high vegetation** growing during nesting seasons (usually May–July in northern temperate areas). Vegetation at this length is usually in short supply because farmers either cut it for hay or cut it because it is a seed source for weeds.
- **Brush piles**,
- **Snags** (standing dead trees), although they should not be allowed to stand in areas where they fall on people, buildings, or cars,
- **Nest boxes**,
- **Travel corridors** between fragments usually in the form of fencerows, hedgerows, roadside vegetation, and riparian areas,
- **Logs at the sides of ponds**,
- **Snake boards** to provide warm concealment,
- **Snake hibernation mounds**,
- **Shelterbelts** useful in windy and snowy areas. If grown, their placement relative to wind direction is critical to be effective.

SPACE

Animals have three kinds of needs for space, home range, territory, and migratory route. **Home range** is used for living, moving around, escaping predators, locating mates, resting, and sunning. It is where an animal spends most of its time from birth to death, but not for migration. A **territory** is an area defended. Temporary and long-term examples include space on a telephone line, a breeding or feeding territory, or a wintering territory. For migratory route, **migration** is defined as an intentional and directional mass movement by animals, with a round trip return to the place of origin.

Several generalized rules apply to space:

- The larger the animal, the more space required.
- Carnivores need more space than herbivores.
- Home range space must be larger when carrying capacity is low because of poor resources.
- Some individuals share space and some do not according to their evolved social behavior.
- Space needs can change with seasons as sexual conditions (lactating/non-lactating) and feeding requirements change.

WHAT ARE EDGE EFFECTS?

Edge is a controversial topic. While heralded by Leopold, forest interior bird species can be excluded if edge effects penetrate an entire woodlot. **Edge effects** refer to a host of problems created where forest and field meet. Sunlight and wind alter the microclimate such that the forest edge is drier, lighter, and more prone to tree fall. Shade-intolerant vegetation can extend several meters into the forest. Disturbance-adapted plants can occur up to 500 m into forests when their seed is carried through wind, roads, and animals.

Additionally, damage to trees on the edge can occur through the cutting of roots or damage to the bark and cambium, which can create blowdowns and **windthrows** (a domino effect when trees blow down, knocking down more trees). A problem on an abrupt edge is the establishment of **sidewalls** (a tangle of vines and shade-intolerant species that grow in a thick vertical mat) (McComb 2015).

Mid-sized predators are particularly attracted to forest edge. These are generalist feeders such as jays, crows, raccoons, opossums, foxes, squirrels, coyotes, and skunks, which eat eggs and chicks. When areas are too small to support large carnivores, mid-sized animals flourish, creating problems. Bird nests can be raided up to 200–600 m into forest; thus, a small reserve is all edge.

Additionally, the **nest parasite** Brown-headed Cowbird in the U.S. lays eggs in the nests of other bird species. Over the breeding season, females may lay 30–40 eggs, each one in a different nest left for the other species to incubate and raise the chick. Within the nest, cowbird chicks gain an advantage because they hatch in 11–12 days, while most other species need 12–14 days. Cowbird chicks are bigger than the chicks of the other species, starving other birds. Cowbird chicks literally have the biggest mouths, thus getting the most food, and they may push the other chicks out of the nest.

IS EDGE THE VILLAIN?

It depends on the surrounding matrix. Small clearcuts within the forest matrix in Maine create forest interior edge that could add valuable diversity to vast areas of forest (Figure 25.5). Forest interior edges are often immune from weedy vegetation, mid-sized predators, and nest parasites because of their distance from infected areas.

In contrast, small woodlots in Illinois surrounded by row crops could have the same amount of edge as above, but as forest exterior edge. The entire woodlot could become completely saturated with weeds, cowbirds, and generalist predators, making the woodlot a sink for nesting songbirds. Predators find the edge easily because the woodlots are highly visible standing out above the row crops. Whether edge is the villain depends on the species of bird considered. Some species specialize in edge. Many of these are the migrant species showing the steepest declines such as Rufus-sided Towhee, Orchard Oriole, and Chat (Dickson 2008). Whether edge is villain also depends on the type of environmental problem. The problem is not grassy/shrubby/edge vegetation itself, but a combination of:

- Habitat loss (reduction in spatial area),
- Fragmentation (isolation of patches),
- Edge effects (disturbance from surrounding lands).

Fragmentation is the splitting of expansive tracts into isolated patches. It is the Swiss cheese effect

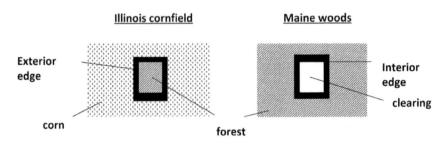

Figure 25.5 Two situations with the same amount of edge have dramatically different characteristics depending on the surrounding matrix. Edge in Illinois surrounded by corn fields will be saturated with mid-sized predator pests. Edge in the Maine woods surrounded by forest will not be saturated with mid-sized predator pests.

when forest remnants become islands within a sea of agriculture or development. It occurs when islands are fragmented by parking lots, roads, and domesticated vegetation. Abrupt edges tend to dominate and are corridors for weedy species rather than wildlife. The process within a forest biome happens in a predictable sequence:

- Roads are created in a forest,
- The forest near the roads becomes farmland,
- Power lines, fences, canals, and vegetation removal by livestock occur on the farmland,
- Farmland tends to become suburb and exurb,
- Exurb become suburb,
- Suburb becomes urban,
- Native vegetation becomes non-native.

So yes, some edge is detrimental to birds, but in the case of Illinois woodlots we would not really want to cut the woodlots down. Forest remnants can be the last bastion for some species of fungi, lichens, bryophytes, vascular plants, and invertebrates in a sea of agriculture or urbanization. Beyond this, when it comes to the reason for the population declines in migrating birds, the main problem is not in the U.S. It is decline of forest during the non-breeding season, most critically in tropical forests. Heavy development pressure, forest destruction, pollution, and human population growth are the major threats.

IS CLEAR-CUTTING DETRIMENTAL TO SONGBIRDS?

Edge includes thickets that can be created by clear-cutting or through the natural creation of small gaps in the forest. Wildlife management in forests now tries to create openings that mimic natural treefalls and support shrubby/grassy vegetation. For bird species dependent on shrubby thickets, clear cuts are a benefit for the first 3–12 years after cutting as hardwoods rootsprout and undergo succession (Askins 2002). Thicket-dependent species are equally abundant in small and large clear cuts.

Overall, the preponderance of peer-reviewed research concludes that small clear cuts are not detrimental to songbird density and diversity. Any type of even-aged management within a forest matrix generally increases diversity of native bird species as long as it is in the temperate region and shrubby/grassy vegetation emerges. Clear-cutting small areas within large forests does not appear to have negative effects on birds, and may increase bird density and abundance, at least for 3–12 years.

After 3–12 years, decades of slow change may ensue before a forest reaches old growth status, which can be another period of high biodiversity. Many forest species do not inhabit forests until at least some of the pioneer trees are near maturity (Apfelbaum and Haney 2010). Epiphytes, shrubs, a host of invertebrates, and some vertebrates may take decades before colonization. Tree species typical of open sites will be different from what establishes under a mature, relatively stable canopy.

So back to the question of whether clear-cutting should be done specifically to increase bird density and diversity. The answer would be clearer if we had a better idea of the optimal ratio of edge to interior forest. We must also weight the benefits for species besides birds. Clearcuts are detrimental for salamanders (Ochs et al. 2022, MacNeil and Williams 2014, Hocking et al. 2013, Petranka 1994), although effect on snakes and turtles may be neutral or positive (MacNeil and Williams 2018). The next two chapters provide more detailed recommendations for managing farm, ranch, open fields, and forests in temperate regions. This is some of the most helpful information for managing yards and landscapes in urban areas for increasing wildlife sitings.

THINKING QUESTIONS

1. Where is the safest place to stand outside in a thunderstorm, in the forest, in the edge, within shrubs, or in an open field?
2. What is the best way for a person to achieve happiness?
3. What places and activities induce you to feel like you are living life to the fullest?

REFERENCES

Apfelbaum, S.I., and A. Haney. 2010. *Restoring Ecological Health to Your Land*. Island Press, Washington, DC.

Askins, R.A. 2002. *Restoring North America's Birds*. 2e. Yale University Press, New Haven, CT.

Bolen, E.G., and W.L. Robinson. 2003. *Wildlife Ecology and Management*. 5e. Prentice Hall, Hoboken, NJ.

Dickson, J.G. 2008. *Wildlife of Southern Forests Habitat and Management*. Hancock House Publishers, Surrey, BC.

Hocking, D.J., K.J. Babbitt, and M. Yamasaki. 2013. Comparison of silvicultural and natural disturbance effects on terrestrial salamanders in northern hardwood forests. *Biological Conservation* 167:194–202. 10.1016/j. biocon.2013.08.006

Leopold, A. 1933. *Game Management*. Scribner's Sons, New York.

MacNeil, J.E., and R.N. Williams. 2014. Effects of timber harvests and silvicultural edges on terrestrial salamanders. *PLoS ONE* 9:e114683. https://doi.org/10.1371/journal. pone.0114683

MacNeil, J.E., and R.N. Williams. 2018. Harvesting our forests – the wildlife debate. Purdue University Forestry and Natural Resources Got Nature? Blog. January 31, 2018. https://www.purdue.edu/fnr/extension/ harvesting-our-forests-the-wildlife-debate/

McComb, B. 2015. *Wildlife Habitat Management: Concepts and Applications in Forestry*. CRC Press, Boca Raton, FL.

Ochs, A.E., M.R. Saunders, and R.K. Swihart. 2022. Response of terrestrial salamanders to the decade following timber harvest in hardwood forests. *Forest Ecology and Management* 511. https://doi.org/10.1016/j. foreco.2022.120159

Petranka, J.W. 1994. Response to impact of timber harvesting on salamanders. *Conservation Biology* 8:302–304.

26

Wildlife management for temperate farms and ranches

ACTIVITY: Wildlife Observations and Research Questions

It is one thing to identify species while birding, botanizing, or tracking. It is another to make generalized observations about animals and plants in context as a wildlife biologist. Research these wildlife biology questions:

1. Which type of brush pile (pine branches, deciduous tree branches, last year's pile) has more tracks around it when observed after a fresh snow?
2. How soon after making a brush pile do tracks appear around it? Sand traps near a brush pile can record tracks.
3. Which has more tracks and scat, edge or forest? A snowy day is an especially good time to observe.
4. Which has more small mammal tracks versus large mammal tracks, a prairie, open forest, or forest with dense shrubs when observed after a fresh snow?
5. Does quantity of downed wood affect number of small mammal tracks? Does it affect number of woodpeckers that inhabit a woodlot?
6. When hiking in the winter look for last summer's bird nests in bare branches. Which height and type of tree has the most nests?
7. Which tree species have the most active cavities?
8. What foods are eaten most commonly at a location based on mammalian scat and bird droppings?
9. In what type of habitat are most birds found in the winter? At what height of vegetation are they found? What is the architecture and other characteristics of the most crowded winter bird areas, equator-facing areas? dense cover? near waterways? near a corridor for flying away? near power lines? near houses?
10. What is the configuration of logs in a waterway that attracts the most basking turtles?
11. What is the phenology of turtle reproduction from mating underwater to egg laying and hatching? If you can watch from late winter through the spring, viewing from a high perch or bluff overlooking a pond, you will get a good start.
12. What is happening below water in a pond as winter approaches? When the first skim of ice appears on a lake, this is a great time to watch activity from a high perch or bluff. The ice prevents small waves and glare, providing a clear window to see plants and animals below the surface. Use binoculars.

DOI: 10.1201/9781003271833-26

WHAT IS THE GOAL OF THIS CHAPTER?

This chapter builds on the principles taught in Chapter 25, providing more concrete instructions for creating wildlife habitat in open canopy areas such as on a farm, ranch, park, or backyard. The chapter stresses the importance of writing a plan to guide one's efforts.

Landowners of even small areas are important wildlife managers whether they realize it or not. Land management can protect and restore ecological integrity no matter how small the property. However, conservation is a process, and a written plan can strategize the abatement of introduced species, polluted water, and harmful disturbances. We would not construct a major building without a blueprint or take a major trip without access to a map. So too, a conservation area will not reach its potential without a plan. There is much to do and learn toward this effort.

FOUR MANAGEMENT PRIORITIES

To provide food, water, cover, and space, a manager should have the following four priorities in mind, in the order written (Bolen and Robinson 2003):

- treat water like gold,
- provide a mosaic,
- protect unique and important habitat features,
- minimize invasive and introduced species.

Whether the area is farm, forest, or something in between these priorities are the same. The management of farms/ranches is described in this chapter. Forest management is covered in Chapter 27.

APPLYING MANAGEMENT PRIORITIES TO FARMS AND RANCHES

Farms and ranches have tremendous potential for fostering wild organism conservation (Bolen and Robinson 2003). They cover vast amounts of land, and at one time the family farm had more wildlife value than today's industrial mega-operations. What changed?

Family farms used to be characterized by a patchwork of crops, barnyards, hayfields, fencerows, orchards, berry patches, conifer stands, woodlots, gardens, recreation areas, wood piles, and outbuildings. Since the 1940s these have given way to monocultures of row crops with few permanent fences. Efficiency and "clean farming" have facilitated the near elimination of field margins with knee-high vegetation, fencerow vegetation, hedgerows, and wetlands. Today, ditches are cleared to serve irrigation systems. Wooden barns with haylofts have been replaced with aluminum warehouses sitting on concrete floors.

Rotational farming, wetland reserves, and fertilization with manure have largely been replaced with reliance on petroleum-based fertilizers, irrigation, and herbicides. With its loss, a culture of local care for the land has been replaced with off-site commercial interests primarily focused on profit. For those landowners who are conservation-minded, what can they do to foster more wild and native species?

TREAT WATER LIKE GOLD

Waterways provide more resources to more individuals of more wild species than any other factor. Thus, protecting areas that still have their original waterways should be the first goal. Restoring damaged waterways and creating new ones should be second and third goals.

Shallow water wetlands are among the most valuable of all resources, yet they have been widely destroyed. Wetlands originally composed 89 million ha of the lower 48 states, but now make up less than 42 million ha (www.emagazine.com/earthtalk.html).

To protect rivers, lakes, and wetlands, a wide, livestock-free vegetated buffer area around a waterway is vital. Livestock in waterways tend to introduce diseases to wildlife and vice versa. Their waste fosters eutrophication. Even a narrow (1 m) buffer of grassy vegetation around the water can take up much of the phosphorus and nitrogen running off a pasture that feeds green algae and blue-greens (cyanobacteria). Further damage occurs because the compaction of soil from livestock near waterways increases runoff.

Even areas only seasonally wet can greatly benefit wildlife if fenced off and kept in tall vegetative cover. For farm ponds, ancillary practices include adding nesting boxes and turtle rafts or logs. Low till practices, as well as planting in valleys and pasturing on slopes are good rules of thumb.

A: Poor Interspersion (1 Covey)

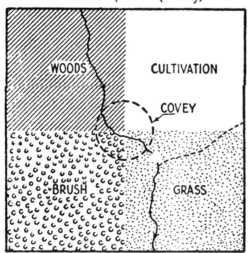

B: Good Interspersion (6 Coveys)

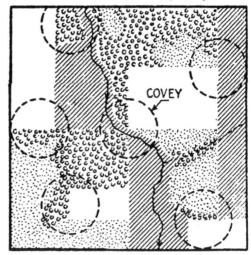

Figure 26.1 Leopold's illustration of interspersion to promote more coveys of quail. (*Game Management by Leopold 1933.*)

PROVIDE A MOSAIC

Aldo Leopold described the mosaic (quilt-like) concept with the aid of an illustration comparing "poor interspersion" to good (Figure 26.1). Both panels of the illustration have the same area of each type of landscape. However, the B panel can support more coveys (groups) of Northern Bobwhite (*Colinus virginianus*) because habitats attractive to the quail are more interspersed rather than in blocks. Likewise, a patchwork of edge habitat, crop types, barnyards, orchards, woodlots, and wetlands will host the most species. Providing vegetative cover throughout the farm or ranch diminishes erosion and keeps water quality high for wildlife. It retains nutrients and water in the topsoil, making fertilizer and water additions less necessary. Knee-high vegetation along grassy roadsides and in field margins provides valuable nesting and feeding habitat and food for pollinators.

PROTECT UNIQUE AND IMPORTANT HABITAT FEATURES

Retaining fencerow vegetation and the shrubby vegetation in hedgerows (Figure 26.2) should be the next goal. Hedgerows and fencerows maintain travel corridors and provide cover for species like rabbits, mice, and wintering birds as well as herptiles and pollinators (Dickson 2008). The

Figure 26.2 The ideal hedgerow profile includes some trees, as well as shrubs and herbaceous plants. Often a mound of soil develops under the hedgerow in long-standing situations.

most effective hedgerows are 3 m wide minimum and have snags for cavity nesters as well as some live trees. Low vegetation within these hedges is essential. If snags are not present nest boxes can be installed for cavity nesters (e.g., bluebirds) along the fencerows, but they should have predator guards if Black Snake is present and other good climbers.

Hedgerows used to be the cheapest form of fencing, acting as live fences. Now in the U.S., barbed wire and electric fencing are the cheapest, if fencing exists at all. In Britain, hedgerows were a cultural tradition for 5,000 years. Different regions of Britain had different hedging characteristics. Contests took place to see who could hoe out or plant the most yards of hedgerows (which explains why rows are not always straight in Britain).

Farmers are sometimes resistant to leaving patches of vegetation because weeds may go to seed and blow into the row crop monocultures. To prevent high mortality rates when mowing, farmers could adjust haying times near nesting season, limit nighttime mowing, or mow around marked nests. Winter cover is vital. Ancillary practices include bluebird and swallow boxes, Kestrel and Barn Owl boxes, brush piles, rock piles, snake hibernation mounds, and snake boards.

A **shelterbelt** is a line of trees or shrubs planted as a windbreak to protect a farmhouse or barnyard. Shelterbelts conserve energy for households and diminish erosion and snow drifts. Shelterbelts are useful in windy and snowy areas but are usually not necessary in areas receiving little snow like the U.S. South. As for the value of shelterbelts to wildlife, the topic needs more research. Opponents claim shelterbelts act as exterior edge, attracting mid-sized predators and nest parasites. If the shelterbelt is planted with non-native tree species, they could potentially spread and foster tree encroachment in prairie regions. This is an especially difficult problem in areas where fires are suppressed. Trees can block the sun in small fields and gardens or provide too much shade for houses in humid conditions. Overall, the placement and necessity of shelterbelts needs careful thought.

Retaining old farm buildings and orchards may serve wildlife, although the soil of many old orchards may be contaminated with pesticides. Nest boxes for bats, Wood Duck, hawks, and bluebirds may be a benefit. Control of domestic dogs and cats is essential. Winter food plots are especially valuable for deer and game birds when they are placed adjacent to a forest. Butterfly gardens, water features, nesting boxes, rain gardens, rain barrels, and bird feeders enhance property values and can be created at nearly any scale in any rural or urban area. Those features that incorporate food, water, cover, and space will be the most productive.

MINIMIZE INVASIVE AND INTRODUCED SPECIES

Each property needs a survey of invasive and introduced species and these species should be categorized into priority groups (Apfelbaum and Haney 2010). The plant species that cannot be tolerated should receive top priority for eradication. Those that are pesky but less prone to monoculture or provide some wildlife value receive lower priority.

The worst species are often waterway plants (e.g., Purple Loosestrife, *Phragmites*) that have the potential to quickly dominate shallow bays and spread their seed to other ponds via boats. Next in line are often upland plant species that are non-native and prickly, brambly, shrubby, or highly poisonous and prone to monoculture like Teasel or honeysuckles. Especially in riparian zones something like Multiflora Rose, Poison Ivy, and Nettle might be tolerated because of its wildlife food value (rosehips are nutritious and the berries on Poison Ivy are well-liked as wildlife food).

Plant eradication efforts should first involve pulling by hand, burning, grazing, or cutting rather than herbicide use. Cutting sometimes makes the situation worse or must be done at the right time. Otherwise, it may spread seed or foster more sprouting. Burning often encourages grasses over forbs. Herbicide application should be the last resort and if possible only dabbed on individual stems after each cut rather than aerial application. In any case, eradication may seem impossible when first encountering a situation, but a good rule of thumb is that if 90% of the individuals are eradicated each year for 3 years, the situation should only need spot control after that.

Morale is lifted when witnessing Mother Nature reward efforts. She will often fill in with native plants on her own once invasive plants are moved out of the way. Pulling by hand may seem time intensive, but if roots and all can be removed, it ultimately is the sustainable solution. Invasive and non-native species of animals must also be considered. Often it is the mid-sized predators such as Raccoon that may be destroying turtle or bird nests.

GOVERNMENT PROGRAMS

Since the New Deal of the 1930s, farm policies and programs in the U.S. have strongly influenced agricultural land use (Warner et al. 2012). Several government programs have encouraged wildlife conservation while increasing crop prices by reducing surplus. Building on the model set by the Agriculture Act of 1956 known as the "Soil Bank" the Food Security Act of 1985 has been known as the *Farm Bill*. One section refers to the **Conservation Reserve Program** (CRP) in which farmers remove highly erodible land from production for 10 or 15 years. Farmers must establish permanent cover in exchange for rental fees (based on local land rental) and 50% of the cost of establishing cover. The fields cannot be harvested except during drought, and permanent vegetation is to be established along water-land interfaces.

The program has clearly been effective (Warner et al. 2012). Populations of pheasant in the Great Plains and nesting ducks in the Prairie Pothole Region of the U.S. have benefited. Nongame species rebounded too, for example, the Henslow Sparrow. Other U.S. farm programs with voluntary easement programs for protecting and enhancing habitat on private lands include:

- Healthy Forest Reserve Program (HFRP),
- Conservation Stewardship Program (CSP),
- Environmental Quality Incentives Program (EQIP),
- Conservation Reserve Enhancement Program (CREP).

Some of the species thought to benefit from these programs in the U.S. include: Northern Bobwhite, Bobolink, Sedge Wren, Red-winged Blackbird, Grasshopper Sparrow, Savanna Sparrow, Field Sparrow, Horned Lark, waterfowl, Meadowlark, Henslow Sparrow, and doves.

Besides government subsidies, Cooperative Extension Services are funded by counties and states. Research advice from land grant universities is used by the Extension Services to promote new methods of farming through field days, demonstration projects, and bus trips to see new innovations. They offer free seedlings and technical advice and provide soil, vegetation maps, 4-H and other youth programs, cooking advice, gardening tips, and in some cases Master Gardener and Master Wildlife programs.

CONSERVATION OF GRASSLAND AT AIRPORTS

Because areas with knee-high and waist-high vegetation on farms are now rare, the best hope for conservation of remaining native species in the eastern grassland may lie in management of airport grassy areas and CRP lands. Ideal management would include sustainable mowing schedules, replacement of non-native grass with native, and prescribed burning. Grass kept somewhat long at airports allows nesting of small songbirds, the species in greatest need of habitat. It discourages gulls, crows, and geese, the large-bodied species that wreak havoc in airplane strikes. Of course, less mowing at airports also uses less fuel and fosters less CO_2 release.

THINKING QUESTIONS

1. When flying over the farmlands of western Kansas the agricultural fields are laid out in perfectly round circles. What accounts for this?
2. Most people find the looks of a farm to be at least somewhat pleasing and usually peaceful, especially compared to the look of factories and warehouses. Why? Are people naturally drawn to nature and farmland? Is there such a thing as biophilia?
3. Why are some farms and farmhouses more pleasing in their appearance than others? Needham (1916) suggests that the homesteads we find attractive are not the ones with the most money lavished on them, but those that fit their environment most perfectly and are planned and planted most perfectly. Is he right? How can you explain this psychologically?
4. Take Needham's (1916) advice and find a back road in agricultural country. From the road, survey the homesteads for their looks. Which "fit their situation, look comfortable, bespeak shelter and privacy, and are arranged with unity and harmony?" What are the features that make them so?
5. In a prairie or grassland, discover the breeding territory of a sparrow by chasing it. Map each place where the sparrow lands. It should look like an approximate circle when finished.
6. Get down on the ground and crawl on your hands and knees like a turtle as it makes its way from a farm pond to a place to nest and back. Determine what obstacles local turtles face and try to eliminate them.

REFERENCES

Apfelbaum, S.I., and A. Haney. 2010. *Restoring Ecological Health to Your Land*. Island Press, Washington, DC.

Bolen, E.G., and W.L. Robinson. 2003. *Wildlife Ecology and Management*. 5e. Prentice Hall, Hoboken, NJ.

Dickson, J.G. 2008. *Wildlife of Southern Forests Habitat and Management*. Hancock House Publishers, Surrey, BC.

Leopold, A. 1933. *Game Management*. Scribner's Sons, New York.

Needham, J.G. 1916. *The Natural History of the Farm*. Comstock Publishing, Ithaca, NY.

Warner, R.E., J.W. Wale, and J.R. Herkert. 2012. Managing farmlands for wildlife. In: N.J. Silvy (ed.), *The Wildlife Techniques Manual*, 7e, Management, Volume 2. The Johns Hopkins University Press, Baltimore, MD.

27

Wildlife management in temperate forests

ACTIVITY: Wildlife Management Plans – Forests

Goal: Using the worksheet provided, complete an inventory for a forest as a first step toward writing a wildlife management plan (Table 27.1).

WHAT TYPE OF FOREST HAS OPTIMAL CONSERVATION VALUE?

Is it old growth? Many people assume that only old growth is worthy of conservation. Old growth forest is often species rich, but optimal forest management is whatever meets the goals of the landowner. The presence of rare birds and salamanders might be what gives local conservation value, or hunted species, or geologic formations. It may be that early succession communities have higher species diversity than old growth.

Where we can find old growth, we should protect it, but first we must recognize it. Old growth forest does not always have massive trees, mainly because old trees are not always large. To be old growth, trees generally need to be at least 100–150 years old without catastrophic disturbance, but no standard age defines it. Furthermore, it is a set of characteristics that describes old growth. The first is the presence of **pits and mounds** (Figure 27.1). These develop from fallen trees, leaving an exposed root with attached soil and a nearby pit on the forest floor. Eventually, a mound forms around the root mass.

Other characteristics of old growth include the presence of trees of all sizes and ages (no missing age classes), with some dead standing trees (snags). Deformities and gaps in the canopy occur when trees fall. Tree trunks may have scars from lightning strikes. The bark of old trees may be heavily plated. Abundant cavities in trees and woody debris (logs) on the forest floor will be prevalent along with native ferns, wildflower species, mosses, lichens, and fungi in abundance. A minimum of invasive plant species would be present. Once we recognize a stand worthy of conversation, the same principles apply as for farms and ranches, with waterway conservation as our highest priority.

HIGHEST PRIORITIES

For conservation of wildlife in temperate forest, and just as in Chapter 26, a manager should have the following four priorities in mind to provide food, water, cover, and space (Bolen and Robinson 2003):

- treat water like gold,
- provide a mosaic,
- protect unique and important habitat features,
- minimize invasive and introduced species.

Table 27.1 Worksheet for qualitative assessment of wildlife in forests

Forest Assessment

Forest type (wood): check one		Ground vegetation:		Area a floodplain?	
	at least 80% hardwood		mainly ferns and native veg. that produces flowers in spring.		several fence posts present, or stone walls?
	mixed (hardwoods 20%–80%)		mainly introduced or pest species that form shrubby waist-high impassable growth or have numerous vines.		significant rock or brush piles?
	at least 80% conifer				Human-created overhangs, open barns or buildings, lofts, outhouses?

Forest type (wetness)		Lumber quality:			
	bottomland		trees are mainly straight and tall.		
	upland		trees are mainly gnarled and crooked.		

Write name of dominant tree species		Average dbh of living trees (cm)?		vegetated fence rows?	
Rate tree species diversity:		Rate tree cavity abundance:			
	−2 = monoculture; +2 = highly diverse		2 = sparse; +2 = abundant		cliffs or caves present?
Rate tree age diversity:		Rate snag abundance:			
	−2 = even age; +2 = highly diverse ages		2 = sparse; +2 = abundant		major noise pollution (vehicles, machinery)?

(Continued)

Table 27.1 (Continued) Worksheet for qualitative assessment of wildlife in forests

Forest Assessment

State of succession. Oldest trees are:	Edge characteristics:	pets within close barking range?
pioneer (1–2 years since cut)	edge not present.	unique or special species or communities? historic sites, cemeteries, or archeological sites?
old field (3–8 years since cut)	edge within a forest matrix.	
old field (3–8 years since cut)	edge on the outside of the forest.	
	edge soft and stadium-like, ragged with diversity of canopy layers.	
mature (>20 years since cut)	edge hard and abrupt without diversity of canopies.	

Number of layers in the stand:	Rate downed wood abundance: look especially for large logs.	recreational areas?
at least three layers (ground veg., tree understory, upper canopy). only 1 or 2 layers	−2 = sparse; +2 = abundant	free-roaming domestic animals or pets?

Rate masting (soft and hard) quality of area:	Considerable problem and pollution areas (erosion, degraded habitat, chemical discharge, garbage, waste)? Describe.	air quality problems (car exhaust, road dust, factory discharge, manure, dust from plowing)?
2 = sparse; +2 = abundant		

(circle those present)
oak, hickory, beech, walnut, birch, maple, dogwood, cherry, apple, juniper, sumac, roses, yarrow, mayapple, blackberry, poison ivy, virginia creeper, greenbrier

Other notable plants	Notable animals

Figure 27.1 A pit and mound forms in the following sequence. Tall trees in an area of high water table or shallow soil become vulnerable to windthrow. If the tree falls, a ball of soil is pulled up, leaving a pit. As the tree decays, the soil and rotting root system become a mound. Pits and mounds make conditions poor for growing timber but are exceedingly rich for wildlife.

TREAT WATER LIKE GOLD

There is no mystery where the most animals live in forests. Food, cover, water, and space proliferate in **riparian zones**, the vegetated zone around wet areas (Figure 27.2). Riparian zones are home to plant species that require readily available water with their roots in aquatic soils. Riparian zones also serve as a corridor for wildlife movement.

Having a well-vegetated riparian zone prevents erosion, important because sedimentation is the single greatest form of water pollution. Riparian zones should not be harvested for trees, especially in steep areas, nor should mowing take place next to the water. Logging roads and skid trails should be kept away from water and the number of stream crossings for roads should be minimized (Yahner et al. 2012).

Livestock should be kept out of riparian zones and waterways for several reasons. Grazing causes erosion, reducing vegetative cover and compacting the soil. Once compacted, the soil is susceptible to pest vegetation. Livestock grazing disturbs burrowing animals and produces animal waste that washes into the waterway. In some cases, light grazing under a forest with a natural grass understory can be beneficial, for example, in Longleaf Pine forests with a Wiregrass ground cover. Even in this situation careful monitoring and a rotation system are strict requirements.

As for wetlands with a tree canopy and standing water (**swamps**), their value can be enhanced by keeping domestic animals out. Placement of basking logs, platforms for waterfowl and turtles, and Wood Duck boxes with predator control devices enhance wildlife value.

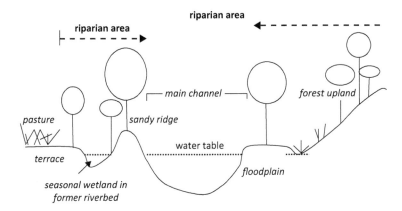

Figure 27.2 The cross section of a stream bed and riparian area is similar in forests the world over. (*Adapted from McComb 2015.*)

WHAT IS THE IDEAL WIDTH OF A RIPARIAN ZONE?

The ideal depends on the circumstance. Certainly 30 m (approximately 100 ft.) is a minimum for width of a riparian zone, but 100 m or more is optimal (Broadmeadow and Nisbet 2004). Salamanders have been known to use uplands as far away as 200 m, and the wider the riparian zone, the more salamanders use the area (Petranka and Smith 2005). The same is true for birds. The best rule is for the riparian zone to be wide enough from one side to the other so one cannot see through it in winter (Bolen and Robinson 2003). Anything narrower may prohibit animal species with wide diameter territories.

A consistent riparian zone width is difficult to achieve beside rivers (McComb 2015). There may be multiple landowners abutting the shore, each with a different width under conservation. In mountain valleys an optimal riparian zone width is difficult to prescribe because of varying steepness and flash floods within the valleys.

Within the waterway itself, species biodiversity increases if the substrate and water velocity vary. In other words, diversity is fostered by boulders and **coarse woody debris** (fallen branches and logs) left in the stream to make pools, riffles, undercut banks, and cascades. Each microhabitat has a different community of organisms. Microhabitats and periodic floods increase diversity as long as the water is clean (McComb 2015). Forests are natural water filters if the riparian area is wide around the waterway. In fact, the nutrient levels in forested streams sometimes become too low to support species. In other words, the water becomes ultra-filtered by the forest.

Light plays an important role in supporting life within a waterway. If a stream has a closed canopy in a deciduous forest, the most productive time may be the winter when light can get in and promote algae and moss growth. Algae is the base trophic level for the rest of the food web. During the summer, gaps in the canopy along the river may enhance species diversity in an otherwise forested area, but they may also increase water temperature (McComb 2015).

PROVIDE A MOSAIC

The second principle wildlife managers must keep in mind is to provide a mosaic of habitat types and patchiness throughout the forest (Bolen and Robinson 2003). The more types provided, the more species the forest will support. There are four subtypes of diversity that lead to this mosaic: vertical, horizontal, life-form, and physical diversity. **Vertical diversity** refers to layers in a forest: ground, shrub, understory, and overstory stratification within a stand (Figure 27.3). Species may specialize on one layer during one season, and another later, like the Cerulean Warbler in N. America, using the canopy for breeding, and the shrub layer for postbreeding (Yahner et al. 2012).

It is especially important to retain some overstory trees in harvested (logged) areas for use in perching, nesting, and foraging. Woody debris should be present on the forest floor, but not so much that it prevents germination, makes the forest impenetrable, or enhances fire risk. In general,

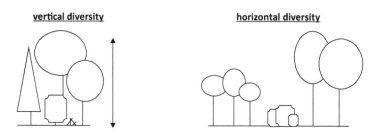

Figure 27.3 Vertical diversity is layering within a forest stand. Horizontal diversity is a variety of heights across a landscape.

Figure 27.4 Vertical diversity under the canopy in this Indiana forest and horizontal diversity across the landscape.

logs are beneficial. They mitigate erosion and retain invertebrates. Large logs are better than small in this regard.

Horizontal diversity refers to variability across the horizon and across stands in the form of different patches and successional stages (Figures 27.3 and 27.4). Horizontal diversity is created by patches of varying ages, species composition, successional stages, and **aspect** (direction slope is facing such as south-facing and north-facing stands). The opposite of horizontal diversity is tree farms with homogeneous stands of evenly spaced trees (Figure 27.5). Certain tree or shrub species are beneficial because they form small patches of monocultures within the forest. For instance, small rhododendron patches in N. America provide good cover for bears and other wildlife when found within a larger matrix of closed-canopy tree species.

Life-form diversity refers to tree forms – different ages, sizes, conditions, fruit types, and species, all in one stand (Figure 27.6). Tree species diversity is important because it provides a variety of tree architectures and a diversity of foods for wildlife. Fissures or scaliness in the bark of old trees is life-form diversity and can provide for species like the Brown Creeper that forages within cracks. Rough bark supports more lichens and bat habitats than smooth bark.

Figure 27.5 White Pine monoculture in North Carolina.

Life-form diversity includes a variety of foods consisting of hard mast, soft mast, browse, buds, and grasses, all within one patch. Ideally, a variety of soft mast species would provide fruits throughout the seasons rather than a profusion that ripens all at once. The best hard mast is in the form of naked nuts and seeds rather than those with thick coverings such as walnut and hickory. American Chestnut was probably the best hard mast in the eastern U.S. because of its equal masting each year (Clark and Pelton 1999). Oak species have unequal masting in their acorn production because they bloom early in spring when frosts nip the flowers.

To ensure more equal masting, trees from both red and white oak species should be provided in forests where they are native. Red oak acorns take

Figure 27.6 Tree burls provide physical life-form diversity. Infections cause the tree to build a growth around it and can become so large and heavy that they cause the tree to break, Washington County, Indiana.

2 years to develop and white oak take 1. Thus, a combination of red and white oaks is ideal because it protects against total oak failure in any one year. Middle-aged oaks do the heaviest masting; very old and young are not as productive.

Snags can be categorized as hard or soft. Optimally, 1–6 hard snags and as many soft snags as possible should be included per acre (Bolan and Robinson 2003). **Hard snags** are standing dead trees with slow decomposition. **Soft snags** are species with crumbly centers that decompose within 2–3 years. Hard snag species in the eastern U.S. include Sourwood, Black Locust, Red Mulberry, White Oak, Cherry, Black Walnut, and Sycamore. Soft snag species include pines, willows, cottonwoods, red oaks, and beeches.

Snags provide nesting, feeding, and perching sites for at least 85 species of cavity nesting birds in N. America (Bolen and Robinson 2003). Many foresters remove snags in the belief that it reduces fire risk, reduces the breeding of pest insects and fungi, and is safer for hikers. Many concerns may or may not apply, but snags left next to a parking lot or building can fall and cause damage.

Snags may burn, but the risk is minimal in riparian zones where they are needed the most. Overall, artificial nest boxes may be installed in a managed forest to provide cavities, but on balance, it is less expensive to leave snags. Snags can

be created by girdling deformed and dying trees. **Girdling** can be accomplished by cutting a ring above the base of a tree into the bark all the way through the cambium, but not through the xylem in the center (Figure 27.6). This stops phloem from reaching the roots, resulting in tree death. The process can be hastened by injecting herbicides into the ring.

Life-form diversity includes pits and mounds that develop when trees fall. The exposed pits on the forest floor are amazingly attractive for small mammals like mice. Additionally, the trunk and roots act as a nursery of fertility for seedlings. Decaying stumps provide animal homes and organic matter that eventually forms a fungal garden. As the stump decomposes, fungi feed on it creating a ring on the forest floor, sometimes called a **fairy ring**, providing a rich food source for wildlife.

Physical diversity is the fourth subtype of diversity in the forest mosaic. Hardscapes such as rock piles, cliffs, caves, old building sites, bridges, cut banks, snake hibernation mounds, and brush piles create valuable habitat (Figure 27.7). Artificially created brush piles can be good habitat, but only when alternative habitat is lacking. A brush pile on a golf course or lawn can be extremely helpful, but where dense thickets are present, brush piles are unnecessary (Bolen and Robinson 2003). Having

Figure 27.7 Physical diversity even includes abandoned human habitations, now occupied by wildlife.

some live material included with dead material in brush piles makes them more usable. Older piles attract rats and can be burned once the live material is brown. Optimally, 2–3 piles/ac. Could be created in rows as large as 25–50 ft. long, 15 ft. wide, 5 ft. high in habitats such as golf courses (=5–8/ha, 8–16 m long, 5 m wide, 1.5 m high).

PROTECT UNIQUE AND IMPORTANT HABITAT FEATURES

Unique areas like rocky outcrops or temporary pools used by salamanders should be protected from harvesting or use by cattle (Yahner et al. 2012). Old forest stands need preferential conservation for the forest interior they protect.

SHRUBBY THICKETS IN EASTERN N. AMERICA

Gaps do not always become grassy after a disturbance in a forest (Askins 2002). Thickets of woody shrubs (like blackberry) and small trees may eventually form instead (Figure 27.8). Like a grass stage, the thicket stage is short-lived. The only permanent shrubby habitats that form are in bogs, pocosins (shrub bogs), and other areas where soils are too waterlogged to support trees. The lesson learned is that shrubby thickets have a different community of animal species compared to grasslands or forests. When thickets are lost, the dependent animals are lost too. Thus, thickets need special conservation attention.

Thicket formation is achieved through the action of beavers, wildfires, localized insect outbreaks, and flooding along wooded streams (Askins 2002). Because natural disturbance regimes have diminished, the thicket-dependent animals have become rare. Thicket creation under power lines is helpful (Askins 2002). Since the 1970s some electric companies have developed a technique that creates permanent woody thickets under their lines to prevent invasion of tall trees (Figure 27.9). Initially, this requires girdling young tree or cutting them, then spraying each stump with herbicide. More research is needed to determine whether these corridors promote the transfer of mid-sized predators and cowbirds to the forest interior.

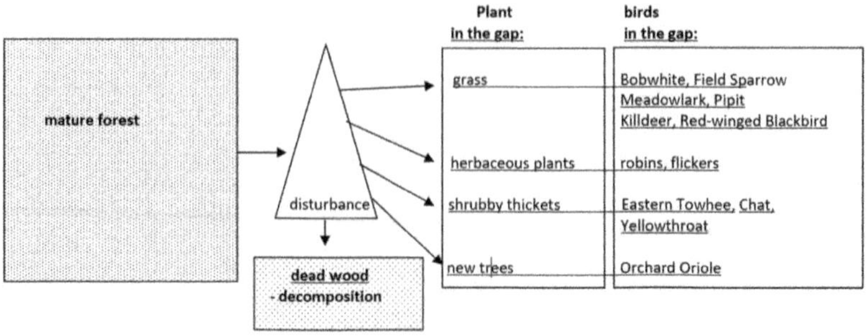

Figure 27.8 During succession, vegetation converts to one of several possibilities. Which one depends on the seed bank and other conditions in a forest after a major disturbance.

Figure 27.9 These trees were intentionally girdled under powerlines at Gorges State Park in North Carolina to promote the establishment of permanent thickets.

MINIMIZE INVASIVE INTRODUCED SPECIES

A controversial subject within wildlife management is whether introduced plant species that become invasive are good for wildlife. The animals often eat these plants readily because they are tasty, delicious, and sometimes nutritious (rose hips). However, this spreads their seed farther and wider. What is not debatable is non-native introduced **animals**, which are almost always a problem. Especially difficult are mammals like armadillos and wild boar because of their rooting behavior, causing dislodged vegetation, erosion, and competition with native burrowers.

MINIMIZE HUMAN INTERFERENCE

As much as possible human interference should be minimized (Bolen and Robinson 2003). Lights, barking dogs, loud music, and off-road vehicles must be minimized.

READY TO WRITE A FOREST MANAGEMENT PLAN?

Table 27.1 is a check-off form to be used in assessing a forest as a first step in writing a conservation management plan. The best plans include landowner participation from the start. Any plan will fail if it is at odds with the realities of the personnel involved and the complex world in which adequate cash flow and profitability are essential (Warner et al. 2012). Unanticipated problems will always occur when working on the land. They provide an opportunity for reevaluation and updating as conditions change.

THINKING QUESTIONS

1. Worldwide in virgin flood plains the contour of the riparian zone next to a river is the same as in Figure 27.2. In particular, one can often find a mound right next to the river, and a seasonal wetland on the upland side of this mound. What accounts for this phenomenon? Which soil is older, the soil at the bottom of the seasonal wetland, or the soil in the mound?

2. What are some hypotheses to explain why some species of deciduous trees keep their leaves for the winter? What is the name for this phenomenon?

3. Walk across a log while wearing a heavy backpack. This is in preparation for wilderness excursions, medical mission trips, walkabouts, or military service in which you might have to cross a stream on a suspended log or walk over a makeshift bridge through very polluted water. Set up a narrow board or log in the backyard and practice crossing while wearing a heavy backpack or carrying bulky equipment.

4. Become aware of forest indicators that identify adverse weather events from over the years. Fallen branches and trees in a forest (including pits and mounds) tell the history of storms. Walk a forest looking for the following indicators, then examine the weather history for your region and see if you can pinpoint the date of corresponding storms. The indicators are: a majority of trees have fallen in the same direction. This would indicate a single storm event, usually coming from the direction of prevailing winds. Many branches are down, but they have not fallen in one particular direction. This might indicate an ice storm, which brought down many branches in one event. Several trees are dramatically twisted and broken. This can indicate a small tornado.

5. Are leaves less prickly at the top of a tree than the bottom? Tristan Gooley (2015) has observed that leaves of Holly trees are less prickly near the top, and ivy leaves are less pointy. Do you observe these phenomena for prickly leaved trees or vines in your region? Poison Ivy is purported to have the same phenomenon. True?

6. Do woodpeckers retain pair bonds in winter? Pileated Woodpecker is unusual in that males and females retain pair bonds all winter and the male and female share a territory. They call to each other frequently during the day and stay in separate, but nearby cavities at night. This phenomenon was discovered by an undergraduate college student (Kellam 2003). If you have Pileated Woodpecker in your region, see if you can find these paired mates during winter. If it is summer, see if you can determine the size of the territory and compare it to territory size in Missouri and Oregon (Renken and Wiggers 1989, Bull and Holthausen 1987). Do red-bellied woodpecker or other woodpecker species have the same winter pair bond behavior as pileated woodpecker?

7. Which supports more vines, rough or smooth barked trees? What explains your results? How does this affect the ecology of a forest?

8. For cavity openings in trees, do they all tend to face the same direction? For those cavities excavated by animals, in which direction (N, S, E, W) do you find the openings for the cavities? Is the direction random? If the trees are near a waterway, does the direction of the waterway affect the direction of the opening? Do they face the Sun at any particular time?

9. Set live traps for pit and mound sampling. Determine which has a higher population of small mammals, pits, mounds, or unaltered forest floor? Sample the two areas using small mammal traps (live traps) and make a comparison.

10. What is the purpose of squirrel leaf packs found in deciduous trees during winter and built by Fox Squirrel and Gray Squirrel in N. America? Do squirrels nest in these leaf packs? Are the squirrel leaf packs more prevalent in some types of trees compared to others (species, size)? Do these squirrel species nest in cavities? During what time of year do these squirrel species mate and bear young? What does squirrel dung look like? Why do we not see it more commonly? Where is it prevalent?

11. Investigate the making of humus on the forest floor. Do the leaves on top of the detritus decompose as much as the leaves closest to the ground? Oak leaves contain tannin. Do oak leaves decompose at the same rate as other species? What is the difference between humus, hummus, and Hamas?

REFERENCES

Askins, R.A. 2002. *Restoring North America's Birds*. Yale University Press, New Haven, CT.

Bolen, E.G., and W.L. Robinson. 2003. *Wildlife Ecology and Management*. 5e. Prentice Hall, Hoboken, NJ.

Broadmeadow, S., and T.R. Nisbet. 2004. The effects of riparian forest management on the freshwater environment: a literature review of best management practice. *Hydrology and Earth Systems Science* 8:286–305.

Bull, E.L., and R.S. Holthausen. 1987. Habitat use and management of Pileated Woodpeckers in northeastern Oregon. *Journal of Wildlife Management* 57:335–345.

Clark, J.D., and M.R. Pelton. 1999. Management of a large carnivore: black bear. *Ecosystem Management for Sustainability: Principles and Practices Illustrated by a Regional Biosphere Reserve Cooperative*. Lewis Publishers, New York.

Gooley, T. 2015. *The Lost Art of Using Nature's Signs*. The Experiment, LLC, New York.

Kellam, J.S. 2003. Pair bond maintenance in Pileated Woodpeckers at roost sites during autumn. *The Wilson Bulletin* 115:1860192. https://doi.org/10.1676/02-098

McComb, B. 2015. *Wildlife Habitat Management: Concepts and Applications in Forestry*. CRC Press, Boca Raton, FL.

Petranka, J.W., and C.K. Smith. 2005. A functional analysis of streamside habitat use by southern Appalachian salamanders: implications for riparian forest management. *Forest Ecology Management* 210:443–454.

Renken, R.B., and E.P. Wiggers. 1989. Forest characteristics related to Pileated Woodpecker territory size in Missouri. *Condor* 91:642–652.

Warner, R.E., J.W. Wale, and J.R. Herkert. 2012. Managing farmlands for wildlife. In: N.J. Silvy (ed.), *The Wildlife Techniques Manual*, 7e, Management, Volume 2. The Johns Hopkins University Press, Baltimore, MD.

Yahner, R.H., C.G. Mahan, A.D. Rodewald. 2012. Managing forests for wildlife. In: N.J. Silvy (ed.), *The Wildlife Techniques Manual*, 7e, Management, Volume 2. The Johns Hopkins University Press, Baltimore, MD.

28

Conservation biology

CLASS ACTIVITY: Investigate the Biophilia Hypothesis

In efforts to conserve biodiversity we can tap the immense love humans have for nature and individual species. If we can connect people emotionally to individual lives, plights, and locations, we are likely to find some success in conservation efforts. What explains this psychological attraction to nature, pets, and a fascination with diverse life? Investigate the concept of biophilia by asking the following:

- Do humans prefer certain nature images more than others? Show pictures to a group of human subjects.
- Is the view from a hillside looking down on a lake more satisfying than the same view looking down on a landscape without the lake?
- Is an image of the sea with a sailboat in the distance more satisfying than the same sea without a boat?
- Is a boat with an engine as satisfying as a sailboat?
- Is an image of the woods as satisfying as the same woods with a trail leading into the distance?
- Is the image of a beautiful building with green architecture as satisfying as a field of flowers?
- Is an image of any other structure as satisfying as a cabin nestled within a natural scene?
- If preferences are strong, what explains this? Why do images of puppies and kittens get so much attention in general? What accounts for the popularity of images that show two species of animals cooperating with each other? Likewise investigate biophobia. Do humans react with the same fear when shown images of a gun as spiders, snakes, or rats? Which kills more people?

WHAT IS CONSERVATION BIOLOGY?

The formation of conservation biology was guided in the late 1970s by Michael Soulé (1936–2020) from the University of Santa Cruz in California. He led this movement because neither wildlife management nor ecology were directly addressing the loss of species and biodiversity. Like wildlife management, conservation biology takes a **conservation**, not **preservation** approach. Conservation uses active management of the land when necessary, unlike preservation that takes a hands-off, lock-it-up-and-throw-away-the-key philosophy.

Conservation biology focuses on the protection of **biodiversity**, that is, **biological diversity,** the diversity of species, alleles, gene complexes, populations, communities, habitats, ecosystems, landscapes, endangered phenomena, biomes, and biological cultures. **Endangered phenomena** include events like bird migration, autumn leaf color, and simultaneous blooming of

DOI: 10.1201/9781003271833-28

flowers in Madagascar at a wide geographic and sometimes multi-species scale.

The primary goal of conservation biology is to keep species **extant**, that is, living somewhere on Earth. The opposite is **extinct**, to cease to be alive anywhere on Earth. **Extirpated** refers to the local extermination of a species or taxonomic group, but individuals still exist in some part of their geographic range. Beyond simply conserving species richness that might include invasive species, the organisms of special interest are **endemic**, native species restricted to a particular geographic region.

CHARACTERISTICS OF CONSERVATION BIOLOGY

Conservation biology identifies itself as a **crisis discipline** (Soulé 1985). Like emergency room medicine, action must sometimes be taken without complete knowledge during a crisis. Waiting could mean inaction, which could lead to extinction. Embedded within conservation biology is the expectation of political involvement among its proponents as activists or in some applied way. Conservation biologists should give advice from an objective and scientific standpoint, but not remain silent.

A further characteristic is **multidisciplinarity** with the inclusion of:

- **natural science**: genetics, ecology, evolution, geology, chemistry;
- **social sciences**: psychology, sociology, anthropology, economics, policy, environmental law;
- **humanities**: ethics, religion.

Conservation biology includes whatever discipline is necessary to combat complex biodiversity problems. Evolution, genetics, and population ecology are included because the genetic composition of a population is constantly evolving. Genetic variation allows a population to survive disease, climate change, and other environmental problems. It provides the potential for species to adapt to future change through natural selection.

Conservation biology is **value-laden**; unlike most sciences that are supposedly value-free and objective, conservation biology professes its values as proposed by Soulé:

a. **diversity of organisms is good**: hundreds of millions of people each year visit zoos, plant shows, dog shows, cat shows, state fairs, or have an obsession with life lists for birds. It demonstrates a fascination with biodiversity.
b. **ecological complexity is good**: complex coevolutionary relationships and rare ecological phenomena are a work of millions of years of evolution. This deserves our respect as a legacy to the future.
c. **evolution is good**: it is the driving force of all present and future diversity.
d. **diversity has intrinsic value**: this is value in and of itself, beyond any utilitarian value.

CONSERVING SPECIES DIVERSITY

The species level of diversity has received the most attention for several reasons. The species concept is easier to comprehend than something like genetic variation or endangered phenomena. Even a child recognizes distinctions between cow, horse, and dog. Additionally, it was the U.S. Endangered *Species* Act of 1973 that preceded the advent of conservation biology. From the start, the law provided an inherent accounting system for the science, and focused on two early questions: how many species exist, and what are the current extinction rates?

Obtaining answers has been difficult because the majority of the world's species have not yet been named. Among prokaryotes, 10 billion bacterial species may exist, if it is legitimate to even classify bacteria this way. Most do not have names or even a well-known classification system (but see *Bergey's Manual of Systematic of Archaea and Bacteria* (BMSAB) for the most widely accepted current authority).

Of the eukaryotes, 2,161,755 have formal names as recorded by the International Union for the Conservation of Nature (IUCN Table 1, December, 2022). The **IUCN** is a non-profit group chartered by the United Nations to count existing threatened species and help protect them. It is the global authority on the status of the natural world. Of the named species 70% are invertebrates, with just over a million insects, half of them beetles. Flowering plants include 369,000 species.

The number of named species is just a portion of total species. Nematodes, fungi, ferns, protists, and segmented worms are known to be undercounted.

Anyone who takes the time to walk a forest, and brush the bark of a few trees with a dry paint brush into a cup of alcohol would likely be in possession of unnamed species.

THE GAP BETWEEN THE NUMBER NAMED AND THE NUMBER EXISTING

The deficiency between the number of named species and the number existing is called the **Linnaean shortfall.** During the interim before all species are named, the danger is that more species will fall into **Centinelan extinctions**, species going extinct before they are named. This phrase was coined by E.O. Wilson in reference to Centinela, Ecuador, in the Andes. Scientists found 90 endemic plant species in a preliminary survey, but the land was cleared for agriculture before the group could finish their documentation.

EXTINCTION RATES

Without knowing the total number of species, extinction rates can only be estimated. Evidence indicates we are in a period of massive species decline. Researchers have put forth several estimates of extinction rates, all criticized heavily no matter what strategy was taken to calculate the estimate (May 1988, 1992, Erwin 1991, Stork 1993, and several others). As a summary, the extinction rate was estimated to be 10% per decade calculated by these researchers (reviewed in Stork 2010). This estimate is now known to be too high. Half the world's species would have gone extinct in the last 40 years since these estimates were proposed. It has not happened. For instance, El Salvador has lost 90% of its forest, but only three of its 508 forest bird species. Species may be on borrowed time and their extinctions may be imminent, but many species seemed to have gone to other habitats or are sustaining themselves at minimal levels in what remains.

POPULATION DECLINE

Because extinction rate estimates were criticized so heavily, researchers now take different approaches to monitor biodiversity decline. One successful strategy has been to monitor thousands of populations for hundreds of species worldwide and measure their growth or decline over several years (WWF 2020). In the Global Living Planet Index 2020, population size of all species plummeted an average of 68% between 1970 and 2016. For mammals, birds, amphibians, reptiles, and fishes, 3,741 monitored populations representing 944 species have declined by an average of 84%, equivalent to 4%/year since 1970. Most of the declines are seen in freshwater amphibians, reptiles, and fishes. They are recorded across all regions, particularly Latin America and the Caribbean. A 94% decline for the tropical subregions of the Americas is the most striking result observed in any region. The conversion of grasslands, savannahs, forests, and wetlands; the overexploitation of species; climate change; and the introduction of alien species were key drivers. The report further concluded that freshwater biodiversity declined faster than biodiversity in oceans or forests (WWF 2020). Almost 90% of global wetlands have been lost since 1700. Humans have altered millions of kilometers of rivers.

IUCN RED LIST

Another tactic to monitor decline is IUCN's Red List in which the most vulnerable species are placed in the "threatened" category (Table 28.1). To be threatened, the species must sustain a 90% reduction in abundance over the last 10 years or three generations, whichever is longer, and its threats must be high risk.

The conclusion of the IUCN is that 27% of all mammal species, 13% of all bird species, 21% of reptiles, and 41% of amphibian species are on the brink of extinction. For all assessed species approximately 29% are on the Red List. This is shocking because it is the percentage of species in imminent threat of extinction.

Furthermore, in the spring of 2019 the UN published a major report from the Intergovernmental Science-Policy Platform on Biodiversity and Ecosystem Service (IPBES 2019). It concluded that a million species are at risk of extinction. The report further concluded that:

- 75% of the land-based environment and 66% of the marine environment has been significantly altered by humans.
- more than 33% of the world's land surface and nearly 75% of freshwater resources are now devoted to crop or livestock production.

Table 28.1 The percentages of species on the December 2022 IUCN Red List are in the right-hand column

Taxa	Number of described species	Red list
Mammals	6,598	27%
Birds	11,188	13%
Reptiles	11,733	21%
Amphibians	8,536	41%
Fish	36,367	Insufficient data
Gymnosperms	1,113	42%
All assessed species		29%?

Use Table 1 from the IUCN website to see updates.

- approximately 60 billion tons of renewable and nonrenewable resources are now extracted globally every year, having doubled since 1980.
- up to $577 billion in annual global crops are at risk from pollinator loss.
- 100–300 million people are at increased risk of floods and hurricanes because coastal habitats have been lost.
- 33% of marine fish stocks are being harvested at unsustainable levels.
- urban areas have more than doubled since 1992.
- plastic pollution has increased tenfold since 1980.
- more than 400 ocean dead zones exist, a combined area greater than that of the U.K.
- since 1970 the global human population has more than doubled from 3.7 to 7.6 billion.
- the number of invasive non-native species per country has risen by 70% since 1970.
- since 1980 greenhouse gas emissions have doubled, raising average global temperatures by at least 0.7°C and are expected to increase in their impacts over the coming decades.

THE SIXTH GREAT EXTINCTION EVENT

From data by the WWF, IUCN, and IPBES we can conclude that we are in the sixth great extinction event of the last 600 million years (Figure 12.6). The fossil record shows that:

- one mass extinction event occurred shortly after the evolution of the first land-based plants, about 450 million years ago.
- another occurred approximately 350 million years ago, which led to the formation of the coal forests.

- two major extinctions occurred during the Triassic, between 250 and 200 million years ago.
- the best studied extinction event occurred at the end of the Cretaceous, 65 million years ago, caused by an asteroid impact, which led to the extinction of most of the dinosaurs.

With 25% or more of measured species at imminent risk of extinction, this is the same order of magnitude as the other extinction periods.

WHAT ARE THE CAUSES OF THE CURRENT EXTINCTION CRISIS?

The current high extinction rates are human caused. According to the 2019 report from the Intergovernmental Panel, the five greatest reasons for extinction in order starting with the worst are:

1. changes in land and sea (i.e., habitat loss, degradation, and fragmentation),
2. direct exploitation of organisms (i.e., overhunting, overfishing),
3. climate change,
4. pollution,
5. invasive non-native species.

HABITAT LOSS

The loss of tropical forests is probably the greatest contributor. In species-area curves from island biogeography, the rule of thumb is that 50% of the species go extinct when 90% of the habitat is destroyed. In rain forests worldwide, 80% of all species live in tropical rain forests, even though this comprises only 6% of the land surface (WWF. panda.org). Over the last 100 years, development has reduced land in a natural state within these areas by 50%. Over the next 50 years, rain forest destruction will continue to be the single greatest cause of species extinctions. If deforestation levels proceed, the world's rainforests may be extinct in 100 years (Bakar 2019). Indonesia is currently the most deforested nation, with other severe losses in Thailand, Brazil, the Democratic Republic of Congo, and other parts of Africa. In the most developed countries including the U.S., coastal flooding and forest fires have offset conversion of former farmland to forest.

For coral reefs, 75% are now threatened (https://www.wri.org/), which is a dual tragedy for vertebrates as well as invertebrates – 30% of the world's marine fish inhabit coral reefs, yet reefs cover a mere 0.2% of the ocean surface. Seemingly small-scale events can be cataclysmic in very small areas. One of the worst culprits for reefs is sedimentation from tropical forests as they are logged, particularly in southeast Asia, east Africa, the Caribbean, and eastern Pacific.

Destruction is occurring on the ocean floor caused by trawlers, which does the greatest physical damage to ocean ecosystems overall (amnh.org). Trawling perhaps does even greater damage than what is happening in tropical deforestation because it involves even more frequent disturbance. The problem is that almost all the world's fisheries are concentrated on the continental shelves and a few upwellings. On average, these areas are trawled once every 2 years with some areas trawled 5–50 times per year.

Other areas suffering biodiversity loss are islands, hotspots, tropical forests other than rain forests, grasslands, mangroves ecosystems, areas undergoing desertification, and areas dammed by rivers. The loss comes from outright destruction, and fragmentation with associated edge effects. At first, animals may be displaced from the modified habitat. They crowd into fragments, which leads to a short-term increase in abundance for small areas. This phenomenon is known as **crowding on the ark** or the displacement effect.

OVEREXPLOITATION – HUNTING AND OVERHARVESTING

One-third of all avian extinctions and one-fourth of all mammalian introduction since 1600 have been the result of overexploitation. While hunting and overfishing have now been controlled within the borders of most developed nations, in developing nations the **wild-meat crisis** continues to be mired in complicated socioeconomic issues. By-catch in oceanic fisheries is another problem without a worldwide solution.

GLOBAL CLIMATE CHANGE

Average temperature by the year 2100 is expected to increase by 2.9°C–6°C (5°F–10°F) compared to 2022 unless emissions increase more slowly and then begin to decline (climate.gov, 2022). As of 2022 the temperature has increased by 1°C (2°F) since the late 1800s (Lindsey and Dahlman 2022).

Consider this for comparison: since the last ice age 10,000 years ago, global temperatures have increased by 4.5°C taking place over a 5,000-year period. This level of change was enough to melt the great continental glaciers from Indianapolis to Alberta and raise sea levels by 100 m.

INTRODUCTION OF NON-NATIVE SPECIES

Introduced invasive species often replace a community that has high species richness. One of the problems is hybridizing a species out of existence. For example, the greatest threat to recovery for the Red Wolf is the potential for hybridization with domestic dogs and coyotes. Likewise, the introduction of Mallard to New Zealand and Hawaii has led to hybridization and the consequent decline of endemic duck species. Not only can hybridization destroy a species, it often leads to dead ends in breeding because the hybrids are not always fertile.

Introduced species bring their parasites. The rinderpest virus of 1889 was introduced via livestock in Africa. In 10 years it was responsible for extirpating 90% of Kenya's buffalo population, which caused secondary effects on predators and starvation among humans. Introduced species have brought canine distemper and rabies to endangered African wild dogs. Mountain Gorilla are at risk from measles and other human pathogens. The monetary cost is high from introduced species, costing over a billion dollars in the U.S. per year.

How did introduced species become so prevalent? Many were introduced through **naturalization programs**, deliberate attempts mainly by Europeans to surround themselves with familiar species. For instance, 70 European Starling were introduced to Central Park beginning in 1800.

This species is now one of the ten most common birds in the U.S. and it takes over the cavities of native species.

The most common animal introductions have been rats, mice, cats, dogs, rabbits, cows, goats, pigs, deer, waterfowl, pheasants, pigeons, and other game birds (Lomolino et al. 2016). Islands and Australia have received the most introductions to the point where 50% of all extant species on islands are introduced.

Beyond deliberate introductions, the second greatest problem is failure to eradicate the species when the problem is first detected. The Monk Parakeet in Florida was introduced as a pet in the 1960s. It could have been eradicated early in the process by shooting, but public sentiment was against it. By 2000 it occupied 15 states and numbered 100,000 in Florida alone. In contrast, the Giant African Snail was brought to Miami in 1990, infesting 42 city blocks. An intensive campaign was mounted with poison, publicity, and hand picking, resulting in 100% eradication.

A massive campaign to remove rats from islands in New Zealand gained success by the mid-1980s. It required a large-scale poisoning campaign including aerial spread of baits by helicopter. Ultimately, rats were removed from more than 90 islands covering 19,000 ha. Native species made a dramatic recovery. Success was attributed to a public education campaign that publicized the non-lethal effects on non-target species.

UNDERSTANDING EXTINCTION – THE MOST EXTINCTION-PRONE SPECIES?

The species most vulnerable to extinction have a narrow geographic range and low population numbers (Figure 28.1). Examples include the Leon Springs Pupfish, which at one time lived in only one spring that measured 3×15 m. The Checkerspot Butterfly at one time lived only near the end of one runway at the Los Angeles airport.

Species with chronically small populations, but large geographic ranges are also vulnerable. For instance, large predators at high trophic levels that require extensive home ranges are at high risk of extinction. Organisms with large body size, whether predators or not, are vulnerable if they have a long generation time. For instance, the Blue Whale has one offspring every 2–3 years. Within a species group like whale, often the species with the largest body size is the most vulnerable – the largest whale, largest lemur, or largest moa. Extinct examples include the Great Auk, Steller's Sea Cow, Tasmanian Wolf, and birds of prey.

Species with low dispersal rates or narrow habitats are vulnerable, such as species with particular nesting habitats that lead to low dispersal like cavity nesters (e.g., Red-cockaded Woodpecker) or colonial nesting animals (seals and sea lions, sea turtles, sea birds). Seasonal migrants are vulnerable because they need several habitats during a yearly life cycle (e.g., Whooping Crane). Species with narrow diets are vulnerable because they depend on unreliable resources (e.g., frugivores and nectarivores). Species with little evolutionary experience with disturbances, disease, or competitors are vulnerable. The classic example is the Stephen Island Wren, which was endemic to one island in New Zealand. A single cat belonging to the lighthouse keeper killed every individual.

WHAT ARE THE MOST EXTINCTION-PRONE HABITATS?

Islands are at risk. Since 1600, as many as 50%–75% of all species going extinct have been on islands. Animals on islands have little evolutionary experience with predators. For instance, on an island off the coast of Chile, Darwin discovered a species of native fox (later to be named Darwin's Fox) that was so tame Darwin could have clubbed it over the head with his geologic hammer. He could also touch hawks with the butt of his rifle. Many bird species endemic to islands are flightless because of the lack of predators. Thus, the introduction of predatory species is a primary cause of extinction. Moas, the Great Auk, and elephant birds have all been victims.

Freshwater ecosystems are at great risk. The lakes in the Victoria region of Africa have lost 200 endemic fish species. In N. America 40% of fish species are extinct or in danger of extinction. In California 7% are already extinct and 56% are in danger. In the southeast U.S., the Tennessee River drainage has the highest mollusk (clams and snails) diversity in world. Because of impoundment construction and channelization of rivers, 40%–50% of these mollusk species are now

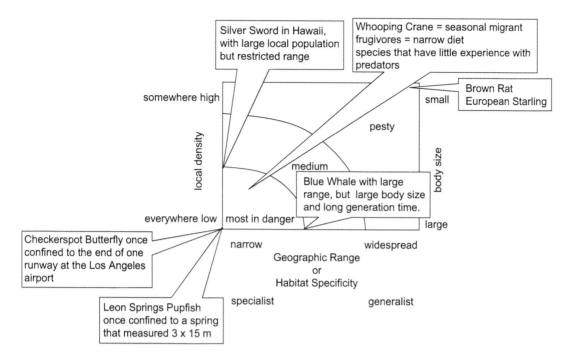

Figure 28.1 Different ways a species could be considered extinction prone. Some species have a widespread geographic range but are not generalists. For example, the Osprey has a wide geographic distribution, but narrow habitat specificity.

extinct or endangered. The Coosa-Tallapoosa drainage in Alabama has one of the highest snail diversities, but this is now in danger. Rivers are vulnerable because of energy production, sewage disposal, irrigation, transportation, bathing, and washing.

MANAGING LAND FOR PROTECTION OF BIODIVERSITY

With perhaps a million species at risk, what can be done? A multi-dimensional strategy will be needed rather than one focus (Figure 28.2). The first priority is to protect nature in reserves, but different strategies exist when choosing which sites to protect.

IRREPLACEABILITY APPROACH TO MANAGING LAND – HOTSPOTS

In the language of land conservation, **irreplaceability** refers to areas with unique circumstances. An area inhabited by the only population of a species would be an example of an area with high endemism. Thus, conservation of hotspots is one of the primary solutions.

COMPREHENSIVENESS APPROACH TO MANAGING LAND – GAP ANALYSIS

The goal of the **comprehensive approach** is to determine where gaps in ecological protection occur. **Gap analysis** was developed in Hawaii in the 1980s using GIS to identify areas underrepresented in the existing reserve system. It compared the distribution of protected areas with the distribution of native species, vegetation types, and landownership, then identified areas to acquire in the future.

A global gap analysis was completed after the *Fourth Congress on National Parks and Protected Areas* in 1992. Participants concluded that tropical forests and marine areas were not well represented and called for protection of at least 10% of each major biome by 2000. Since then, congresses on national parks and protected areas have continued to take place. The IUCN has initiated a Green List program that maintains a global database of protected and conserved areas. Over 250,000 recognized sites protect 15% of the Earth's land surface and 7.4% of the oceans (IUCN.org). Many more conserved areas exist that are not yet on the list as of this date.

Solutions

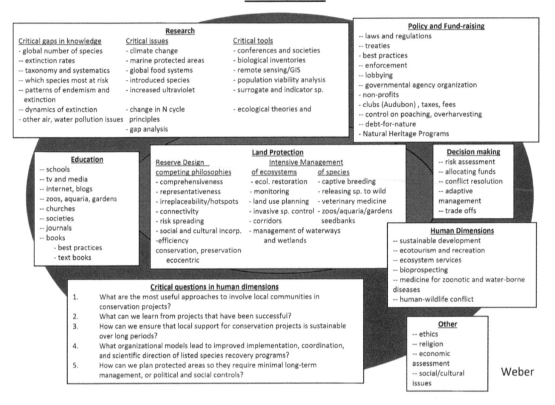

Figure 28.2 Solutions for the biodiversity crises require a multi-dimensional strategy as outlined in this illustration. Central to the list of solutions is land protection.

REPRESENTATIVE APPROACH TO MANAGING LAND – BIOSPHERE RESERVES

The **representative approach** ensures that every unique biogeographic type is represented in a protected area. The idea of **biosphere reserves** came from the *Biosphere Conference* in 1968 held by representatives of UNESCO (United Nations Education, Science, and Cultural Organization). The conference grew out of an initiative in the 1960s called the *International Biological Programme* convened among the English-speaking countries to work together on protecting the environment, eradicating disease, and ending hunger. The participants envisioned a worldwide network of protected ecological areas for research and conservation. As a result, UNESCO began designating biosphere reserves in a program called *Man the Biosphere* (MAB) beginning in 1974. The group identified 193 biogeographic types on Earth. Their goal was to establish five reserves for each biogeographic type.

Approximately 738 reserves have been established in 134 countries with 47 in N. America and 7 in the U.K. (UNESCO.org/MAB).

The controversial philosophy of the program includes the goal of protecting species diversity, ecosystem health, and the cultural values of the local people. What is controversial is including sustainable economic development for the local people who are allowed to make their living off the land. These economic enterprises must be in ways that are compatible with conservation. For MAB sustainable development is meant to serve as a model for rural development in the surrounding area. UNESCO realized it was better to integrate people in the reserve so they could manage the habitat and improve conditions for species. They assumed that people who lived on the land for a long time had a stake in the area and would protect it. This is entirely different from the philosophy of the National Park Service in the U.S., which tries to keep people from living on the land to prevent the tragedy of the commons.

The structure of the reserves fits the sustainable development philosophy. Three zones resemble an archery target when mapped. In the center is a **core**, in which there is to be little human disruption. Ideally, only ecological monitoring is allowed. Outside the core is a **buffer zone** where more human activities are allowed with sustainable development. Activities such as birdwatching, research, education, or harvesting of medicinal plants are allowed. Government officials would ideally work with the local people to develop models of sustainable rural development, for instance, in building ecotourism. Outside the buffer area is a **transition zone** where human settlement is allowed.

The biggest criticism of the program within the U.S. and worldwide is that biosphere reserve designation is a gratuitous honor and rarely creates new reserves. It places the designation over existing protected areas, most of them national parks. Worldwide, 76 are Ramsar sites (wetlands) and 70 are World Heritage sites. Everglades National Park has all three designations.

CONNECTIVITY APPROACH TO MANAGING LAND – WILDLANDS NETWORK (ORIGINALLY KNOWN AS WILDLANDS PROJECT)

Initiated by Dave Forman, one of the founders of Earth First, with Michael Soulé and Douglas Tompkins, the Wildlands Network is an expansion of the biosphere reserve concept except larger in scope. The goal is to establish areas large enough to support large-bodied carnivores. This is done by connecting discreet areas with corridors. For instance, northern areas of Canada and Alaska are considered giant reserves. Large areas of the Rockies and Appalachians are being restored to serve as other core areas. Sustainable development is incorporated to make buffer and transitional zones around the cores. Unlike biosphere reserves, an emphasis is placed on corridors between reserves. Cities are to be outside of the wildlands reserves, phasing out existing cities in the core areas. It seems like a grandiose idea, but the plan is to acquire land slowly by attaining easements as people die, taking a long-term approach (100 years). The program has had some surprising success in the U.S. including the Yellowstone to Yukon (Y to Y) project, and the Eastern Wildway which would connect the Adirondacks, Appalachians, Great Smoky

Mountains National Park, and Everglades. Further descriptions are online.

BEYOND MANAGING LAND – OTHER SOLUTIONS FOR CONSERVING BIODIVERSITY

Laws prohibiting take, habitat conservation plans, CITIES, check stations at borders, treaties among countries, and periodic UN meetings are essential to conserving biodiversity. A WWF (2020) report (Bending the Curve) devised by the WWF and more than 40 universities as well as conservation and intergovernmental organizations concludes that we can halt and reverse terrestrial biodiversity loss from land-use change, but it will require an unprecedented and immediate focus on both conservation and a transformation of our modern food system. The report provides a roadmap on how to achieve this.

CONCLUSIONS

In the end there is much that all of us can do to lessen the biodiversity crisis. Unlike climate change, individuals can play a large part in the solutions to this crisis. Conservation biology is not just a science for professionals. It is an opportunity for all to experience a sense of immense achievement in small actions. Transforming a yard into a native grassland, creating a pollinator garden, volunteering at a nature preserve or zoo with a captive breeding program for listed species, educating others via social media or environmental education in churches and schools, or becoming active politically – this brings success. Not only do other species benefit, applied work in the outdoors has immense personal rewards. The reciprocity of healing self while healing the Earth is contagious and addictive. To directly work with individuals in need and to form a personal relationship with the land gives meaning to life in our time on Earth. While the global reports on the biodiversity crisis provide statistics and guidelines for what it will take to bend the curve, they do not address the deep emotional consequences and opportunities that each of us bears in dealing with this topic. For suggestions on how to stay emotionally healthy and connected to the land read Weber (2020) and other articles in the journal *Ecopsychology*. To learn the steps involved in restoring the Earth at small and large scales, Chapter 29 lays out the important principles of restoration ecology.

THINKING QUESTIONS

1. What is the difference between a preservation and a conservation approach? Provide an example of a government entity with a preservation approach rather than a conservation approach.
2. What does it mean for a species to have intrinsic value? What is instrumental value in reference to a species?
3. Research the difference among what conservation biologists label as uniqueness value, cuteness value, cultural value, spiritual value, and existence value regarding species. Give an example of each.
4. What is an umbrella species, flagship species, and indicator species? Give an example of each.
5. In a country of your choice, what percentage of the land is protected by a government entity or other type of nature area?
6. In the U.S., besides national parks, what are national wildlife refuges, and how are national parks and refuges different?
7. What is the difference among the areas designated by the Ramsar Convention on Biodiversity, World Heritage Sites, and Biosphere Reserves?
8. Investigate the major means of protecting marine sites. How far off the coast can one country declare protections as a marine protected area?
9. What is the primary way that most people get their education about biodiversity? Is it from books? TV? Social media? What is the best way to effect change regarding biodiversity issues?
10. How many people per year visit zoos compared to major sporting events?
11. How can we plan for protected areas so they require minimal long-term management?
12. What are the most useful approaches to involve local communities in conservation projects?
13. Who owns the wild species in your country, state, province, and county? What is the history of legislation regarding legal protection for wild species? What government entity has the responsibility for protecting species? Is it legal, for instance, to kill a migratory bird in your region? What entity regulates the taking of hunted species?
14. What is the history of the Endangered Species Act in the U.S.? Do other countries have a similar law? What penalties exist for violating the law? What is the difference between an endangered and threatened species under the law? How is this different from the threatened category for the IUCN?
15. CITIES is the Convention on International Trade of Endangered Species of Wild Fauna and Flora. It is an international agreement (treaty) signed by 183 countries. What is the goal of this treaty? How are treaties supported by law? What happens if an individual violates the agreement? Has the treaty been effective regarding the biodiversity crisis?
16. What species owe their existence to captive breeding? What circumstances make captive breeding for a listed species a last resort option? What are the advantages and disadvantages of captive breeding? What programs are in place within zoos to coordinate captive breeding for listed species such as Orangutan, Cheetah, and Tiger? Where does captive breeding for rare plants occur if it is not in zoos? What are seed banks?
17. What species have been released to the wild after captive breeding? What are the best practices for a successful release and a sustainable existence for a species back in the wild?
18. Take a dry paintbrush and cup of ethanol to the woods. Brush the bark of trees with debris dropping into the cup of alcohol. Scan the debris for organisms using a microscope. Count the number and types of organisms observed.

19. Research a species listed as endangered or threatened by the U.S. Endangered Species Act that does not have a recovery plan. There are several even though the law requires that a recovery team leader be assigned and a plan written. Become the advocate for this species. Write a recovery plan including the use of a population viability analysis as described in Chapter 15 and using the software named Vortex. Use existing Recovery Plans as examples with the very particular format that is used. Alternatively, choose a candidate species that is familiar within your region or a species not yet on the list but should be.

REFERENCES

Bakar, A.N. 2020. Introductory chapter: today's national parks (NPs) and protected areas (PAs) for a sustainable future. In: A.N. Bakar and M.N. Suratman (eds.), Protected Areas, National Parks and Protected Areas for a Sustainable Future. IntechOpen. London. https://doi.org/10.5772/intechopen.77900

Bergey's manual of systematics of archaea and bacteria. 2015. W.B. Whitman (ed.). *Multi-Volume Online Source That Is Continuously Updated.* doi:10.1002/9781118960608. ISBN 9781118960608.

Erwin, T.L. 1991. How many species are there? revisited. *Conservation Biology* 5:330–333.

IPBES. 2019. Summary for policymakers of the global assessment report on biodiversity and ecosystem services of the Intergovernmental Science-Policy Platform on Biodiversity and Ecosystem Services. IPBES Plenary at its seventh session (IPBES 7, Paris, 2019). Zenodo. https://doi.org/10.5281/zenodo.3553579

IUCN, International Union for the Conservation of Nature, https://www.iucnredlist.org/resources/summary-statistics#Summary%20Tables

Lindsey, R., and L. Dahlman. 2022. Climate change: global temperature. https://www.climate.gov/news-features/understanding-climate/climate-change-global-temperature

Lomolino, M.V., B.R. Riddle, and R.J. Whittaker. 2016. *Biogeography.* 4e. Sinauer, Sunderland, MA.

May, R.M. 1988. How many species are there on Earth? *Science* 241:1441–1449.

May, R.M. 1992. Bottoms up for the oceans. *Nature* 357:278–279.

Soulé, M.E. 1985. What is conservation biology? *BioScience* 35:727–734.

Stork, N.E. 1993 How many species are there? *Biodiversity and Conservation* 2:215–232. https://doi.org/10.1007/BF00056669

Stork, N.E. 2010. Re-assessing current extinction rates. *Biodiversity and Conservation* 19:357–371. doi.org/10.1007/s10531-009-9761-9.

Weber, L. 2020. An ecologist's guide to nature activity for healing. *Ecopsychology* 12:231–235. http://doi.org/10.1089/eco.2019.0077

WWF. 2020. Living planet report 2020- bending the curve of biodiversity loss. In: R.E.A. Almond, M. Grooten, and T. Petersen (eds.). WWF, Gland, Switzerland. https://www.zsl.org/sites/default/files/LPR%202020%20Full%20report.pdf

29

Restoration ecology

CLASS ACTIVITY: Comparative Techniques Used in Restoration Ecology

Practice these monitoring techniques for measuring the effectiveness of restoration (Apfelbaum and Haney 2010):

- A timed meander search can be used to evaluate species diversity in treated versus untreated areas. An observer wanders through an area identifying plant species in 1-minute intervals. The cumulative number of species (vertical axis) is plotted against time (horizontal axis). The place on the graph where the curve reaches an asymptote provides a measure of species richness.
- Macroinvertebrate sampling can be used to monitor the restoration of water quality in streams.
- Spring breeding surveys may be completed for bird and amphibian vocalizations.
- Pitfall traps can be used to record presence of salamanders and large terrestrial invertebrates.
- Moth surveys can be done at night by hanging a vertical white sheet in front of a light source and observing the flying insects that come to the sheet.
- A sweep net can be used to collect and record insect diversity in herbaceous growth of grasslands or shrublands.

RESTORATION APPROACHES

Ecological restoration is the process of assisting the recovery of an ecosystem that has been degraded, damaged, or destroyed (Gann et al. 2019). It allows land managers to repair ecological damage and rebuild a healthier relationship between people and nature. Restoration activities can be costly, in time and money; consequently, large-scale restoration must be considered a method of last resort. It is more cost effective to prevent damage in the first place. Thus, restoration should be considered part of a continuum of management options from prevention and conservation of areas still in good condition, to true restoration of other areas.

A spectrum exists from simply repairing damaged contours to recreating natural habitat with abundant native species (Holl 2020, Harris and van Diggelen 2006). Within this spectrum are the following options (Figure 29.1):

- **reclamation**: improving the structure and function enough to make the land fit for cultivation, but native species are not restored nor brought to a place of sustainability. At very large scales, reclamation may be the best that can be done.
- **rehabilitation**: the restoration of certain ecosystem functions for the provision of particular ecosystem goods and services that stakeholders deem necessary. An example is the reduction of flood risks through the creation of water retention areas. This could improve processes of an ecosystem that are vital for a particular

DOI: 10.1201/9781003271833-29

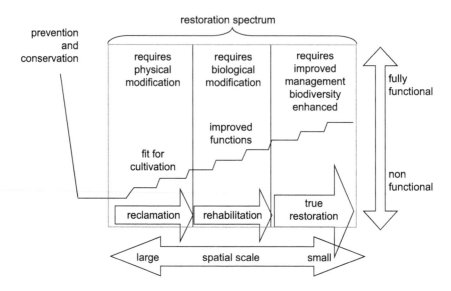

Figure 29.1 Degree of reclamation, rehabilitation, or restoration by spatial scale. (*Adapted by Weber from Harris and van Diggelen 2006.*)

land area, but not increase native biodiversity. Rehabilitation can be accomplished up to intermediate spatial scales.

- true **restoration** consists of a reconstruction of a previous time for that land area (or a close resemblance), with many of the same species as in healthier times. It is often only at small scales that this is possible.

WHAT IS NEXT AFTER IDENTIFYING SPATIAL SCALE AND PLACE ON THE SPECTRUM?

Abiotic factors are most likely to limit how far an area can be taken from degraded to restored. This is why restoration ecologists look at ecosystems in a particular way with particular components (Figure 29.2). The scheme tends to combine facets of ecosystem and landscape ecology, putting both subdisciplines into focus for applied work. For instance, if topography or water table is not sufficient it may not be possible to provide a particular suite of species. Thus, structural and functional quality need to be restored first (Apfelbaum and Haney 2010, Harris and van Diggelen 2006).

Several key steps must be undertaken when attempting a restoration project. These mirror the major principles set out for restoration ecology in the *International Principles and Standards for the Practice of Ecological Restoration* (Gann et al. 2019)

and the practices of other experts (Hobbs 2009, Apfelbaum and Haney 2010):

- Identify the values of the stakeholders and set clear goals.
- Be open to many types of knowledge beyond just scientific, including TEK, local ecological knowledge (LEK), and practitioner knowledge of those doing the work.
- Research the local reference ecosystems for comparison purposes (accomplished through inventory, surveys, historical research, interviews, mapping, observation of photos, and comparison to nearby areas that are intact). Take into account environmental change compared to reference areas.
- Identify the factors limiting system recovery and make a plan taking advantage of natural recovery potential, residual species, and knowledge of the functional traits of species likely to inhabit the restored site. Do small-scale introduction experiments if necessary.
- Develop a plan with associated success criteria and seek the highest level of recovery attainable.
- Initiate restoration activities staying within the context of larger regional constraints beyond the property in question.
- Monitor to evaluate success with measurable indicators. Use adaptive management as things commence.

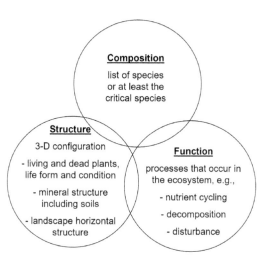

Figure 29.2 Species restoration is dependent on the quality of structural and functional restoration for a site. If topography or water table is not sufficient, complete restoration of a native species community will not be possible. (*Adapted by Weber from Apfelbaum and Haney 2010.*)

VALUES OF THE STAKEHOLDERS

Setting goals for restoration must take place in a societal context (Gann et al. 2019, van Andel and Grootjans 2006). After all, if major objections are filed by neighbors, funding sources, politicians, and the general public, what is achieved? Inevitably, different stakeholders have different values for different projects, but it is the job of the restorationist to identify a group's commonalities for a particular project and lead the group toward specific goals. Restorationists are likely to find three archetypical attitudes toward nature present in the stakeholders (Harris and van Diggelen 2006), and if divergent groups can at least find a common ethic, this is the first step in common ground:

- the **wilderness concept** in which humans should not play a large role in manipulating the land except in legislation and protection. This view is especially supported in large, low-density areas with wide expanses between cities such as in the western U.S., Russia, and parts of Africa.
- the **Arcadian concept** characterized by a preference for semi-natural ecosystems or a pastoral attitude based on the tradition of human interaction with nature, especially dominant in Europe. It harkens back to Gilbert White. Human interaction is required, but in ways that encourage managed hedgerows, slight drainage, and mowing. Simple agricultural systems encourage small patches of habitat in various stages of succession.
- the **functional concept** referring to intensively managed systems specifically for use by humans to provide resources. This is the dominant view of nature and includes agriculture, forestry, fisheries, and ecological engineering.

SETTING GOALS FOR THE RESTORATION EFFORT

Reverting an ecosystem to its previous state before the disturbance or before human intervention may seem like the ideal. However, ecosystems are naturally dynamic (Hobbs 2009). The land may have had human intervention for centuries (Holl 2020). Past composition may be unknown, or current reference systems may not be available. Often, money or societal goals are limiting. Restorationists and their clients may just have to accept limitations, focusing goals on a particular functionality rather than complete restoration. Setting goals inevitably requires decisions within the following context (Harris and van Diggelen 2006):

- **extent**: is an entire ecosystem to be restored or just certain species and components within an existing system?
- **species goals**: is the target an indicator species, keystone species, listed species, community group, or native species?

- **ecosystem functions**: is the desired goal rees-tablishment of certain materials, energy flows, specific standing crops? Are services for people the target such as groundwater for drinking, fish stocks, fixation of CO_2, recreation, or intrinsic value of species and ecosystems?
- **ecosystem structure**: is the desired goal spe-cific ecosystem architecture, or certain trophic food webs?

THE DIAGNOSIS AND TREATMENT PLAN?

As in health care, restoration requires a diag-nosis and treatment plan for execution, and the diagnosed problem may lie in either the biologi-cal or the physical and chemical characteristics. In some cases, the cause of degradation may be obvious such as in mining areas. In others, subtle problems may emerge such as nutrient runoff or persistent flooding. In this context **thresholds of irreversibility** may have occurred, in other words, irreversible damage. For example, there may be excessive acidification, iron depletion, or perma-nent wetland drainage. It may not be possible to go back to the initial species composition.

The existing system might be locked in an **alternative stable state**, a relatively stable eco-system structure or composition that is different from what was present before disturbance. During destructive periods it might have been easy to degrade an ecosystem past a threshold, moving it into the degraded state, but difficult to restore it to a less degraded state, as in an acidified or eutro-phied lake. The source of the excess nutrients may be identified and ameliorated, but existing nutri-ents may recycle in perpetuity.

REFERENCE ECOSYSTEMS FOR THE PROJECT

Reference ecosystems provide a model and target. Historical photographs, written records, or nearby sites may serve as a reference. Note that all records of the past suffer from **moving target syndrome**. Succession and other changes may have taken place over time, leaving us to ask, should prescriptions be set for where the area was or where it would be now? Have key species gone extinct? Should resto-rationists try to produce a shifting mosaic or are

even-aged stands sufficient for now? Has climate change rendered a past phase obsolete through changing water regimes? Is there resistance by the neighbors to reintroducing certain species?

THE RESTORATION EXECUTION

Once a plan is in place, the actual work often involves:

- restoration of hydrology,
- reduction of pollution, toxins, or waste,
- restoration of vegetation structure, that is, establishing a ground cover where previously there was none, eradicating invasive weedy species and replanting natives, or providing buffer strips and corridors between fragments.

HYDROLOGY (APFELBAUM AND HANEY 2010)

Hydrology is the way water enters and leaves the landscape. Much of what happens in restoration involves disabling tiles, removing levees (earthen dams), and restoring ditches to a more natural and meandering state. Permits are generally required for manipulations of stream channels and other waterways. The problem with ditches is increased flow, which tends to down-cut the channel. This can be corrected through use of "**grade controls**," logs, rock structures, fallen trees, strong rhizomatous grasses, shrubs, or even concrete. Besides adding meanders, they help create pools of relatively still water and rif-fles of faster flowing water.

Vegetative cover around the stream must be present to intercept moisture (even fog) and retain it within the soil that otherwise would suffer from erosion and runoff. Small depressions, pits, and mounds could be added to the riparian zone for water retention. In farmland, contour tillage, ter-races, cover crops, and maintenance of sod-water-ways may be necessary.

Tiles consist of clay or plastic conduit placed under the surface of farm fields to drain the soils. Crops and many invasive plant species grow bet-ter in these dewatered areas because the soil becomes more oxygenated and more available to roots. Destruction of these tiles brings back the natural hydrology. Wetland plants are adapted to less oxygenated, more saturated soils. Without the tiles, wetland plants and water-retaining soils release water more slowly to the streams like a living sponge. Some water percolates to the groundwater, eventually replenishing wetlands. The pH of the soil becomes more reduced with less oxygen.

Backhoes and subsoilers (heavy vertical blades pulled behind a large tractor) can be used to break up the tiles. Alternatively, filling the auxiliary ditches into which the tiles drain may disable them. Once the tiles no longer function, flood events and meandering become more frequent in the main stream. Heavy grazing and tilling may have compacted the soils, which can persist for decades, but sometimes come back to reasonable health in 2 or 3 years with aggressive management. Equator-facing and steep slopes have less moisture and may require earlier planting and additions of mulch or additional cover crops to retain rain or snowmelt.

In urban areas, catchments that retain runoff from parking lots and roofs can be designed to grow native vegetation, cleansing the water and assisting in infiltration or evaporation. The catchment itself may be a site that attracts birds, butterflies, and amphibians.

POLLUTANTS

The most common chemical problems involve an alteration of pH, or the management of pollutants coming from acid deposition around coal mining, fertilizer spills, or runoff from feedlots (Apfelbaum and Haney 2010). Adjustments to soil chemistry can sometimes be managed through rehydration once tiles have been disabled. Prescribed burning can increase pH by releasing calcium and magnesium through ash.

Contamination from off-property sites, like road salt or runoff from parking lots, highways, or agricultural operations, may require the construction of buffer areas. Alternatively, **biofilters** can be built, shallow wetlands with small contoured dams on drainageways, vegetated with wetland species. If nutrient loading is excessive, for example, when managing excessive manure from farms, sediments from the contaminated site may periodically be spread on agricultural fields during dry periods.

Bioremediation is the use of rapidly growing plants to absorb contaminants from soil or water. If the contaminant is fertilizer, the plants can be harvested and used for other purposes like mulch or compost. If the plants take up toxic substances, they need to be treated as toxic waste. Even just stabilizing the soil with plants may increase the organic matter, tying up the contaminants.

Contaminated streams may flush themselves and the problem solved through dilution if the source of the contaminants is removed. Sometimes, a contaminated terrestrial area can be covered with soil and planted with native species as long as it does not contaminate groundwater. Burning the debris may contaminate the air.

BIOLOGICAL TECHNIQUES

One way to rebuild the soil is to plant cover crops and till them back into the soil as **green manure**. This provides a quick increase in soil carbon as long as these crops are short-lived, fast-growing species such as buckwheat. The cover crops can also nurse native, longer-lived species back to the land. Once the native plants begin to grow, they build soil by retaining mulch and fostering the establishment of invertebrates and microbes.

From here the assumption of restorationists is that by restoring vegetative structure, animals will recolonize on their own. However, some animal species are ecosystem engineers, playing important roles that can have dramatic ecosystem effects (Hobbs 2009). These are the keystone species such as beavers, or large herbivores in Africa, or digging marsupials in Australia. Elephants in Africa, for example, knock over and feed on particular types of trees, converting savanna or forest to a more grassland-like state. Digging marsupials affect water infiltration, enhancing plant establishment.

Cattle may be introduced to maintain an oak savanna in the U.S. Midwest. Cattle may be fenced out to encourage greater recruitment of cottonwood trees in old stands along rivers in the western U.S. Cattle would ordinarily halt the growth of tree seedlings, but once cattle are eliminated, a chisel implement could be pulled behind a tractor, cutting some cottonwood roots, stimulating the growth of new shoots.

WHAT ARE THE INDICATORS OF SUCCESS?

Monitoring progress is essential, but often overlooked or done poorly (Apfelbaum and Haney

2010). It is important because stakeholders would not know if a restoration project were successful unless monitoring were done through time. A common monitoring technique includes the use of **exclosures**, areas fenced off from treatments to be used as a reference. Transects or plots can record percent cover, vegetation density, and light quantity. Stream flow can be measured with a gauge or by watching something like an orange peel flow from one point to another, recording time per distance. If the velocity and the cross-sectional area are known, a volume per unit time could be calculated.

Soil moisture can be measured by recording sample weight before and after drying (100°C), left in the oven until the soil reaches a constant weight, usually taking 2 or 3 days. The percentage lost in weight is equivalent to the percent weight of moisture in the soil. Aerial or ground photographs from permanent stations are useful for recording changes. Mapping of specific problem areas or insect/disease areas is helpful when good field marks are included for reference.

EXAMPLE: A LANDSCAPE ECOLOGY UNDERSTANDING OF RESTORATION

Landscape function can be represented as a continuum from the least damaged landscape still highly functional to the most damaged, highly dysfunctional. Rehabilitation, say of a mine site, starts with the physical component, typically reshaping the landforms, smoothing spoil heaps into smoother surfaces along contours (Tongway and Ludwig 2009). Once the physical contours are restored, biological restoration may involve microbial introductions and establishment of groundcover vegetation. Eventually, the rip lines of the physical contours flatten as the vegetation becomes developed. At some point a critical threshold is crossed when the landscape becomes self-sustaining. As landscape function increases, the system has more capacity to withstand stress and disturbance. This is the **buffering capacity** or **resilience** of the system. It could be measured in how the system responds to climate and weather. Watching land come back to life like this is extremely rewarding.

THINKING QUESTIONS

1. Why may it not be possible to restore a complete species list to an area compared to the past?
2. Why is it important to understand the values of stakeholders?
3. Use the internet to research the concept of LEK. What does it involve?
4. Why is it particularly important to add soil organic matter in restoration projects?
5. What are three common monitoring techniques used in restoration projects?
6. Investigate the concept of adaptive management. What is meant by this concept and give an example of how it might take place.
7. How does the resilience of an ecosystem influence its sustainability in the face of environmental change?

REFERENCES

Apfelbaum, S.I., and A. Haney. 2010. *Restoring Ecological Health to Your Land*. Island Press, Washington, DC.

Gann, G.D., T. McDonald, B. Walder, J. Aronson, C. Nelson, and eleven other authors. 2019. International principles and standards for the practice of ecological restoration. (2e) *Restoration Ecology* 27:S1–S46. https://doi.org/10.1111/rec.13035

Harris, J.A., and R. van Diggelen. 2006. Ecological restoration as a project for global society. In: J. van Andel and J. Aronson (eds.), *Restoration Ecology*. Blackwell Publishing, Malden, MA.

Hobbs, R.J. 2009. Restoration ecology. In: S.A. Levin (ed.), *The Princeton Guide to Ecology*. Princeton University Press, Princeton, NJ.

Holl, K. 2020. *A Primer of Ecological Restoration*. Island Press, Washington, DC.

Tongway, D.J., and J.A. Ludwig. 2009. Landscape dynamics. In: S.A. Levin (ed.), *The Princeton Guide to Ecology*. Princeton University Press, Princeton, NJ.

van Andel, J., and A.P. Grootjans. 2006. Concepts in restoration ecology. In: J. van Andel and J. Aronson (eds.), *Restoration Ecology*. Blackwell Publishing, Malden, MA.

30

Aquatic ecology

CLASS ACTIVITY: Protection of Water Resources

Waterway protection is usually thought of as the responsibility of government entities, but it does not take much for ordinary residents to educate their neighbors about conservation practices near lakes, streams, and shores. Steve Carpenter suggests an idea in which he traveled to different lake communities in Wisconsin, showed them slides of lakes with no conservation practices, medium-level practices, and superior practices (Carpenter 2008). He asked the community which of the three choices they preferred for their area 20 years into the future. He then provided information and education on how to attain better conservation. Consider using his same approach, but first educate yourself about the following ideas.

1. Blue-greens, also known as cyanobacteria, are prokaryotic organisms like bacteria. Blue-greens have wreaked havoc in lake systems for billions of years (first life on Earth 2.5 billion years ago) (McComas 1993). The specific blue-greens responsible for most blooms are Anny, Fanny, and Mike, also known as *Anabaena*, *Aphanizomenon* spp., and *Microcystis* spp., which cause blue-green blooms. What are they and how can water be managed to minimize them?
2. Seventy years ago, lake management meant fish management or control of water levels alone. Since the 1960s, our understanding of lake chemistry has greatly improved. The result has been more emphasis on control of nutrient levels, especially phosphorus. Investigate the phosphorus cycle in lakes and how it leads to eutrophication. How do outboard motors and carp resuspend P through stirring of sediments?
3. In contrast to blue-greens, green algae is eukaryotic, classified in the Kingdom Protista. Green algae includes the microscopic phytoplankton, giving the water column a cloudy brown or green appearance. Green algae can also grow in filamentous form underwater on substrates. While living, green algae provides O_2 through photosynthesis, but it dies on hot days forming yellow-colored mats of dead tissue floating on the surface with the appearance of green dreadlocks. Its decomposition is done by microorganisms, mostly bacteria, that take up dissolved oxygen from the water column as they respire, using so much dissolved oxygen that it kills fish. Can this biomass be collected while on the water surface and used as fertilizer or something else useful?
4. In what is known as the eutrophication chain of events, lake home residents exacerbate eutrophication in several ways. They may cut their lawn short, right up to the waterfront, which increases erosion, with P clinging to soil particles that go into the lake. Residents then fertilize their lawns for lusher growth, with much of the P running off into the lake. This increases the growth of macrophytes, algae, and blue-greens. Many people then use

DOI: 10.1201/9781003271833-30

herbicides to kill the macrophytes or remove them by pulling and raking. Fish have less cover to hide and their food chain is simplified. The quality of game fish declines. Lake residents then demand that the state release more fish in their lake. Many people release pet goldfish into the lake. These are introduced carp that feed on the bottom, stirring up bottom sediments and destroying fish beds. Grass clippings or raked leaves may be dumped into the lake or nearby wetlands. Can education kiosks help enlighten the community?

5. The macrophyte zone is often destroyed by cottage residents surrounding the lake in efforts to enhance swimming, boating, or just viewing the lake. Not realizing the important ecosystem function of the plants, each landowner may only destroy a small stretch, but if everyone does the same, this littoral zone is gone along with the fish habitat and other important functions. Can social media be used to educate?

6. If you collect a water sample from a lake or river and put it in a glass jar then let it sit on a window, there are several things you will learn about the waterway. A greenish tint indicates high algal content. A brownish tint indicates sediment. If most of the sediment falls to the bottom by the end of the week, most of the waterway's turbidity is caused by stirring from fish, wind, waves, water flow, or outboard motors. If the water is still cloudy after a week, the cause is the chemistry constitution of the water. Settleable turbidity can be controlled by regulating outboard motors, establishing weed beds, and controlling bottom-feeding fish, which stir sediments and destabilize shorelines. Algae could be controlled by limiting phosphorus input and by enhancing the populations of predatory fish.

INTRODUCTION

Aquatic ecology was not covered in the biome section because Shelford and Clements did not recognize aquatic systems as biomes in their scheme. Biome classification is based on plant types that occur in the late succession state, and because most of the world's water is too deep and dark for plants, it is outside the bounds of biome designation. Shallow areas of lakes and oceans have photosynthesizers (think marsh and kelp forests), and surface waters have floating microscopic organisms, but these areas make up only a tiny percentage of all waterways. Instead of biomes, a better classification scheme for waterways is based on soil or sediment type, exactly the idea of ecoregions.

The ecology of waterways is different from land and water ecosystems in other ways besides plant life. Most aquatic organisms spend their lives suspended above the sediment surface, unlike the situation on land in which very few organisms do this. In water, even the largest organisms have little contact with the sediment, and no large trees are totally submersed. Coral

reefs provide treelike structures, but corals are animals, not plants.

WITHOUT BIOMES, HOW ARE WATERWAYS CLASSIFIED?

A variety of abiotic and biotic classification systems are in use. The overarching divisions are freshwater versus marine, and within freshwater, stream versus lake and wetland. Beyond this, more specific classifications are described below.

WATERWAY CLASSIFICATIONS – SALT CONTENT

Saline refers to the concentrations of various salts, which is approximately 35 for oceans (see Table 30.1 regarding unitless measures). Compared to oceans, freshwater has less than 0.5 of salts. **Brackish** refers to intermediate salt concentration between freshwater and saline. **Brine** has a value >50.

Ocean salts are not just NaCl, although this is dominant. Calcium carbonate ($CaCO_3$) is present as are gypsum (calcium sulfate), magnesium (Mg), potassium (K), and others. Because oceans have salts,

Table 30.1 Categories based on concentration of salts

Freshwater	Brackish	Saline	Brine
0–0.5	0.5–30	30–50	50+
Lake Michigan	Ocean estuaries such	Atlantic Ocean	Great Salt Lake (which
Lake Malawi	as Chesapeake Bay	Pacific Ocean	can reach 270)

Salinity is measured by the practical salinity scale of 1978, in unitless numbers measured by conductivity. Sometimes "units" are assigned as practical salinity units (psu). A value of 35 psu is equal to approximately 35 parts per thousand (ppt or ‰) in the old, more familiar way of measuring salinity, often measured in the field with a refractometer. A value of 35 ppt (‰) is equal to 35 g of salt per 1,000 kg water.

the list of animal phyla in the sea is quite different from freshwater. Some phyla overlap, but oceans have unique phyla. This is one reason why **limnologists** (limn=lake in Greek) and **oceanographers** have traditionally been trained separately, at least in biology.

CLASSIFICATION OF FRESHWATERS

Freshwater ecology is mainly divided into standing or running water. **Lotic** refers to flowing water called streams by aquatic ecologists, even for rivers as large as the Amazon. **Lentic** refers to standing water called lakes by aquatic ecologists, even for lakes as small as a tree hole, or within a pitcher plant, or in a mud puddle. "Pond" is a lay term. The science of standing freshwater is **limnology**, including not just the biological aspects, but physical and chemical factors too.

Other definitions:

- **swamps** are lakes under a tree canopy.
- **marshes** are aquatic areas shallow enough to have herbaceous (nonwoody) vegetation such as cattails and sedges that extend above the water surface.
- **peatlands** are accumulations of soil and partly decayed vegetation. It takes cool temperate freshwater areas of the northern hemisphere to establish peatlands, and slow decomposition. Peatlands are further divided into bogs and fens. Other divisions and synonymous terms include mires, pocosins, moors, muskegs, and peat swamp forests.

CLASSIFICATION OF WETLANDS

A **wetland** is shallow and has intimate contact with the terrestrial ecosystem around it (known as the land-water interface). Many wetlands are **vernal**, meaning temporary, usually in the spring rainy season. Note that "wetland" and "waterway" are

not the same. Three criteria must be met to qualify as a wetland: water standing for at least a certain number of days, vegetative species characteristic of aquatic conditions, and soils characteristic of aquatic conditions (hydric soils) (Environmental Laboratory 1987). According to the Army Corps (Environmental Laboratory 1987), wetlands do not include artificial systems as in irrigation and drainage ditches, grass-lined swales, canals, detention facilities, wastewater treatment facilities, farm ponds, landscaped ponds, or ponds made in the construction of a road. Wetlands may include artificial wetlands intentionally created from non-wetland areas to mitigate the conversion of wetlands.

Wetlands are often small, but highly productive, with more biomass and species richness per unit volume than other aquatic ecosystems. In fact, the reason wetlands are often plowed over, drained, or filled is either because the rich soil makes good farmland, or the water is rich with animal life (mosquitoes and other insects), creating a nuisance for humans. Federal laws in the U.S., along with additional regulations in some states exist to protect wetlands. Still, multiple exemptions exist for farming, ranching, and forestry activities.

ECOLOGY OF STREAMS AND RIVERS

From here this chapter summarizes the major ecological factors in the flow of water from streams to lakes to the ocean. For streams most of the nutrients coming into a lotic system come from the land. Thus, a stream cannot be considered separate from the land. A **basin** is the land area contributing to the flow for only one stream. A **watershed** is a body of land that includes the basin for all order streams that flow into one river or lake. A stream is probably best characterized by following it from its source to the mouth where it dumps into the sea.

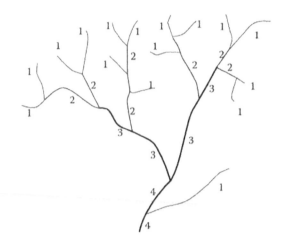

Figure 30.1 A stream order of 1 is assigned to headwater streams with no other branches entering. Two headwater streams must come together to form a second order. Two second-order streams must come together to form a third order. (*Reprinted with permission from McComb 2015.*)

Ecologists use the stream order system as in the following configuration (Figure 30.1):

- a **first-order stream** is a headwater stream without any tributary,
- a **second-order stream** is formed when two first-order streams converge,
- a **third-order stream** is formed when two second-order streams converge.

Orders cannot be increased with the entry of a lower order stream. In other words, if a third order and second order come together, it does not become a fourth-order stream. It remains third order. The largest rivers in the world are classified as 6 or 7.

CHARACTERISTICS OF STREAMS FROM SOURCE TO MOUTH

In what is known as the **river continuum concept**, a stream changes in its characteristics from upstream to downstream as follows:

Steepness: A stream is often steepest in its headwaters in first- or second-order streams. Once a stream reaches flat areas downstream it begins to meander and take on qualities of a lake. As the stream gets closer to the sea it spreads out into a fan shaped with several channels forming a **delta**, each channel taking on the characteristics of smaller streams. Streams in a delta near the ocean are often influenced by tides and salinity and take on some characteristics of oceans.

Oxygen: More dissolved oxygen occurs upstream if the water flows over rocky, whitewater stretches. Downstream, the water becomes lentic-like (lake-like), gaining oxygen at the surface if there are waves and photosynthesizing organisms. Deoxygenated water occurs dark and deep below the surface.

Temperature: Temperatures are typically coldest upstream where the water has little opportunity to stand in the sun, and a closed forest canopy occurs overhead. This is especially true in the summer if the headwater stream runs through a deciduous forest. Compared to lakes, water temperature in streams varies little with season because of the close association streams have with the ground. Stream water is derived from ground and surface waters staying relatively constant throughout the seasons, often in the 50°F–60°F (16°C) zone in temperate regions. It explains why children who walked barefoot to school used to stop at streams during winter to warm their feet. The constancy of the water temperature is the reason why streams are less likely to freeze than lakes. It makes streams dangerous to walk upon in the winter even if nearby lakes have thick ice. Even pools where ice tends to form first are a dangerous hazard for walking.

Substrate size: In upstream areas, boulders, rocks, and sand are often more prevalent and produce turbulent flow. Fine sediments like clay and silt only accumulate in areas of low flow because slow water provides an opportunity for heavier particles to settle to the bottom. Ecologists refer to **coarse particulate organic matter** as CPOM, more prevalent in fast-flow areas. **Fine particulate organic matter** (FPOM) is more prevalent in areas of slow flow. **Dissolved organic matter** (DOM) forms from decaying leaves and is found throughout streams. **Coarse woody debris** consists of branches, sticks, and fallen trees, found throughout streams, providing microhabitats for biota.

Microhabitats: More heterogeneous microhabitats occur in streams than lakes. Even within a small stretch, physical heterogeneity in streams provides a wide variety of places for organisms to live. **Riffles** (not ripples) are shallow stony fast-flowing sections, which tend to have periphyton covering rocks. **Pools** are deeper areas carved out by fast-flowing water.

Curves in the river create **undercut banks** on the fast-flowing side, characterized by unique biotic communities compared to the rest of the stream. Organisms living in the undercut areas take advantage of dark shade interfacing with the bank. On the slow-flowing side of the curve, **areas of deposition** form (Figure 30.2). Sediment taken away from the undercut bank in fast-flowing parts of curves now settles on slow-moving parts. The slow-moving water allows heavier particles to drop out on the slow-moving side, opposite the undercut bank and just downstream.

Because of this continual cycle of undercutting and deposition, curves move downstream with time and become exaggerated unless the flow hits bedrock. Rocky areas in mountains can maintain consistent shapes because they resist this

curving, but streams in flatlands tend to meander. The continual cycle explains why islands tend to move downstream too, with the oldest areas on the upstream side of the island, and the youngest sediments on the downstream side.

Disturbance events and biota: Flow varies tremendously in streams, from occasional catastrophic levels during storms and floods, to tranquil periods during droughts. Floods bring a flush of nutrients to the flooded terrestrial areas, carrying nutrients from stream bottoms. Velocity is usually highest downstream rather than upstream even though turbulence and sheer stress may be higher upstream where the stream is narrow and steep. Within any given stretch of stream, velocity is highest at the upper surface, far from the banks, and lowest near the banks and bottom.

Insects can flow downstream to their advantage. **Drift** is the behavioral strategy insects use to allow the current to take them from an unsuitable location to a more suitable patch. Entomologists have studied this by placing drift nets in streams and measuring what flows into the nets. On a normal day 0.5% may move downstream in drift. In special circumstances, such as when food is scarce or a disturbance occurs, up to 43% of fauna may move downstream in 1 day. When not drifting, stream animals have adaptations for holding fast. Fish, insects, and salamanders tend to be streamlined. Others are dorsoventrally flattened to fit into small cracks and crevices, or cling to surfaces where friction slows the flow. This is the strategy of mayfly and stonefly larvae. Blackfly larvae attach by a sticky substance, then form a web that strains food. Insects can use ballast against displacement, for instance, as caddisflies build a case made of stones or sticks. Other insects have suckers. Many produce sticky silk that adheres to rocks, also used to capture food.

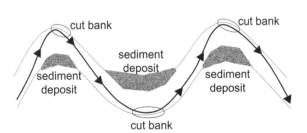

Figure 30.2 Undercutting and deposition within a stream on successive curves. These give and take sediment, eventually exaggerating curves and moving them downstream.

WHAT ARE THE FOOD SOURCES FOR ANIMALS IN STREAMS?

Autochthonous (internal to stream) sources of nutrients include terrestrial plants hanging into the water, macrophytes, and periphyton. Many streams in forested areas have a tree canopy that shade the light, especially in summer. In deciduous forests, streams actually get the most light in the winter and animal life may have its most growth in the winter when larval stages take advantage of abundant periphyton growth.

    Allochthonous (external to stream) sources of nutrients include leaf packs and coarse woody debris. This is how most lotic ecosystems get most of their energy. In deciduous forests, high levels of input come in the autumn from fallen leaves. Bud scales and flowers fall in other seasons. Wind and snow break off branches at all times of year, and rain in summer brings nutrients as water drips off overhanging vegetation.

HABITAT CLASSIFICATION WITHIN LAKES

Habitat classification is based on where a species lives in a lake (Figure 30.3).

- The near-shore **littoral zone** stretches from where the water begins on shore to where it is too deep for plants to have access to sunlight.
- The **limnetic zone** (also called pelagic zone) is the open water volume in the middle of the lake where organisms swim or float rather than touch bottom. The limnetic zone is as deep as

light can penetrate. The **compensation level of light** is the depth at which oxygen use in respiration balances oxygen production in photosynthesis.

- The **profundal zone** is the volume beneath the limnetic zone. It has little light and few photosynthesizing organisms.
- The **benthic zone** is the area where the lake and the sediment interface. The **benthos** is the community of organisms that live in and on the bottom.

STRATIFICATION WITHIN LAKES

Within large northern lakes, the water becomes **stratified** (layered) as a result of density and temperature, although these layers may not be present in small lakes. For temperate lakes during the summer, the surface water is warmest forming the **epilimnion**, with cold water toward the bottom forming the **hypolimnion** (Figure 30.4). A zone of rapid transition between them creates the **thermocline**. The reason the epilimnion and hypolimnion do not mix is because the warm water on the top "floats" on more dense cold water below. In deep lakes the water at the bottom may be 4°C even during the summer. Of course, hypolimnion water at 4°C is a theoretical concept. Measure the water temperature at the bottom of a real lake or of any lake water under the ice and see for yourself whether it is at 4°C.

    During daytime in summer, the epilimnion is characterized by high light, high temperature, and high photosynthesis. Plants include **macrophytes** (large aquatic vascular plants) near the shore.

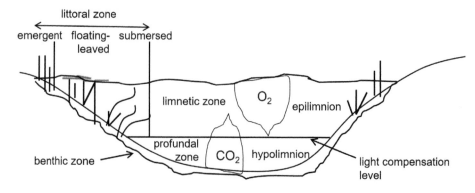

Figure 30.3 Littoral, limnetic, benthic, and profundal zones in lakes. The compensation level is the water depth at which photosynthesis no longer occurs because of darkness, which may or may not coincide with the thermocline dividing the epi- and hypolimnion.

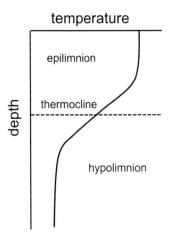

Figure 30.4 Typical temperature profile during summer in temperate lakes that are deep. Water is at 4°C sinking to the hypolimnion.

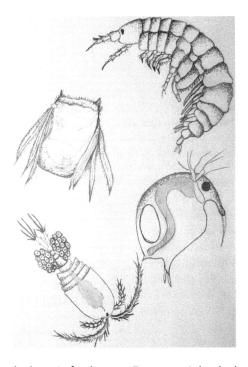

Figure 30.5 Examples of zooplankton in freshwater. From top right clockwise, an amphipod, cladoceran, copepod with eggs, and rotifer (*E. Tauber-Steffen*).

The three types of macrophytes include: **emergents**, **floating leaved plants,** and **submersed macrophytes**. **Periphyton** (literally plants around) also occurs.

Periphyton is the community of organisms (algae, moss, and animals) living on **substrates** (surfaces) such as rocks, plants, and even **sediment**. Small animals, bacteria, algae, fungi, and protozoans can all be part of this community. Periphyton is what makes surfaces feel slippery under water. Often periphyton appears green because of the algae and moss, but it is sometimes colorless, invisible, and slippery.

In the epilimnion **zooplankton** are the microscopic animal organisms in the water column that feed on phytoplankton (Figure 30.5). They include protozoa and crustaceans. Zooplanktors are primarily at the mercy of water currents for their movement

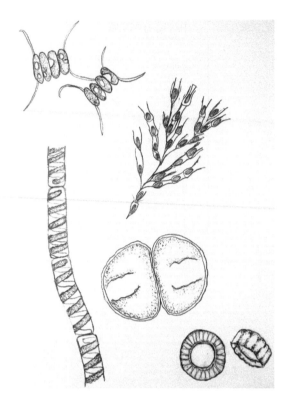

Figure 30.6 Green algae, in freshwater. From top left clockwise *Scenedesmus* (green algae), *Dinobryon* (green algae that photosynthesizes and eats bacteria, also makes water smell like cat urine or fish), *Cosmarium* (in group that shared an ancestor with plants that evolved to live on land), *Cyclotella* (a diatom of marine and freshwater, oligotrophic waters), *Spirogyra* (unbranched chains that form dense mats of floating algae in freshwater with trapped bubbles and slippery touch) (*E. Tauber-Steffen*).

within the epilimnion. **Phytoplankton** are microscopic single-celled algae, often forming tiny colonies that move under the control of water currents (Figure 30.6). They include single-celled and multicelled algae and blue-greens. Because they are not all plants their proposed name change is photoplankton (Margulis and Chapman 2010).

IS THE EPILIMNION WELL OXYGENATED?

Yes and no. During the day in the non-frozen months the area above the compensation point is well oxygenated because of photosynthesis. At night the oxygen dissipates because of respiration, but exceptions exist. Anytime there is wave action oxygen is added to water.

$$CO_2 + H_2O \xrightarrow{\text{light}} C_6H_{12}O_6 + O_2 \ \left(\text{photosynthesis}\right)$$

$$O_2 + C_6H_{12}O_6 \rightarrow CO_2 + H_2O \ \left(\text{respiration}\right)$$

IS THE HYPOLIMNION WELL OXYGENATED?

No. With little light penetration, little photosynthesis, and abundant decomposition the oxygen is depleted. Microbes sometimes produce methane (CH_4) rather than CO_2.

$$O_2 + C_6H_{12}O_6 \rightarrow CO_2 + H_2O \ \left(\text{decomposition}\right)$$

THE WORLD EXPERIENCED BY PLANKTON – REYNOLDS NUMBER

Note another interesting physico-chemical property of water and its effect on small organisms. The smaller the organism, the more it experiences the stickiness of water. This effect is measured by something called the **Reynolds number**, a dimensionless ratio between inertial and viscous forces. Larger objects have a higher Reynolds number as the intended information is on the next page for "large swimming whale " etc (Lampert and Sommer 2007):

Because small organisms have decreasing Reynolds values, very tiny beings experience water as humans would if we were bathing in thick molasses. We can and would swim slightly to change our positioning and access to patches of food. Mainly we would stay put and be forced to rely on diffusion to meet our needs for food and gases. We would be at the mercy of overall currents that move us in flows much larger than ourselves, but our power to swim to other currents would be hopeless.

large swimming whale	10^9
trout in a stream	10^5
escaping zooplankter	10^2
ciliate swimming	10^{-1}
filtering appendages on a zooplankter	10^{-3}

NUTRIENT CONTENT CLASSIFICATION OF LAKES

Oligotrophic lakes are contrasted with eutrophic lakes in Figure 30.7. **Eutrophic** literally means true food, that is, high nutrient content, usually referring to phosphorous as the limiting nutrient. **Eutrophication** means to gain more nutrients, a process that can happen naturally in a lake as it ages. Humans can advance this process through introduction of pollution, grass clippings, fertilizer, raked leaves, and erosion. Note that lakes can become more oligotrophic over time with constant flushing of a lake, as when rivers run through them.

Hypereutrophic lakes occur when phosphorus levels exceed 100 µg/l and high nitrogen levels are present (Brönmark and Hansson 2018). This situation is found in polluted situations from human discharge or lakes with large bird colonies (Figure 30.7). Note that in water, primary producers are usually limited by phosphorus, not nitrogen, and just because phosphorus is high it does not mean nitrogen necessarily is and vice versa.

WHAT CHARACTERIZES SALTWATER ECOSYSTEMS?

Unlike a lake, most of the volume of oceans is unlighted, limiting primary production. Oceans can be 10 km (6.2 miles) deep in places and mostly dark. The salinity in oceans is approximately 35 ppt, but the concentration is not homogeneous. Patchiness is caused by evaporation, precipitation, wind, mixing of different currents, coastal influences, and sinking to the ocean floor. Because the water is saline it makes the living organisms more buoyant than in freshwater.

The sodium in NaCl comes mainly from the Earth's crust in terrestrial sources. The chlorine comes mainly from HCl in underwater volcanoes. Overall, **cations** exceed **anions** in the ocean, making seawater weakly alkaline at pH 8.0–8.3 and strongly buffered, a factor that is biologically important. Other ions have varying concentrations. Nitrogen and phosphorous tend to be limiting nutrients just as they are in lakes and become

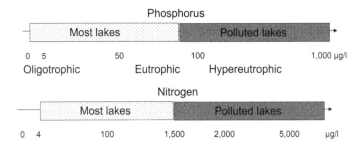

Figure 30.7 Total phosphorus concentrations categorized by lake type (above) and total nitrogen concentrations (below). "Most lakes" refers to lakes not affected by human activities. Numerous exceptions exist. (*Modified from Brönmark and Hansson 2018.*)

depleted at the surface where most organisms occur.

In oceans, freshwater molecules tend to freeze, leaving the salt behind in concentrated liquid water. Saltwater reaches its greatest density at 2°C–3°C and therefore does not drive as much spring and fall turnover as lakes. A thermocline develops in tropical oceans, but not temperate, and its presence prevents exchange of nutrients between top warm water and cold bottom water. As a result, less production occurs in tropical water than in temperate areas despite warm tropical surface waters.

The water pressure in the depths of oceans is many times greater than in terrestrial or lake situations, affecting the distribution of living organisms. Either organisms live only at the surface or are adapted to pressure at great depths. Only a few species can span both. Sperm whales and some seal species can dive to great depths and return to the surface without difficulty.

TIDES

Tides occur in lakes as well as oceans, but their effect in lakes is so small it usually cannot be detected. The gravitational pull of the sun and the moon causes tides. As the Earth turns eastward, tides move westward. Thus, anywhere on Earth, objects experience two high tides per day, one by the gravitational pull of the moon, and one because of the lack of a gravitational pull by the moon on the side of the Earth not facing the moon where water bulges.

Because the moon revolves in a 29.5 day orbit around the Earth, the average period between high tides is 12 hours, 25 minutes. The sun also has gravitational pull, but it is a smaller force than that of the moon. Solar tides are masked by lunar tides except for two times during the month. During a new moon and full moon extra high tides occur, known as a **spring tide** (from sprungen=brimming fullness). At this time, the Earth, moon, and sun are nearly in line and the combined pull of the sun and the moon is additive. A quarter moon pulls at right angles to that of the sun, canceling each other's effect, creating little difference between high and low tides. This is a **neap tide** from the Scandinavian word meaning barely enough.

Tides from around the world are variable, partially canceling each other in some locations, for instance, in the Pacific and Indian Oceans. In the Gulf of Mexico only one daily tide occurs instead of two. On the coast of Maine in the Atlantic, two huge high tides occur per day. Variations are a result of differences in the gravitational pull of the moon and sun, and the angle of the moon in relation to the axis of the Earth, winds, depth of water, and the contour of the shore.

MAJOR ECOLOGICAL COMMUNITIES IN OCEANS

Unlike land, the open ocean does not have clearly defined communities. Exceptions are the subtidal kelp forests found in cold temperate regions and the Sargasso Sea dominated by floating **sargassum weeds**, a form of brown algae. The most abundant ocean organisms overall are phytoplankton, zooplankton, benthic organisms, and nekton. **Nekton** are zooplankton feeders that can move at will in the water column. Small fish, predatory sharks, whales, and penguins are members.

Hydrothermal vents are rich in the diversity of unique deep sea life forms confined to within a few meters of the vent system. Polychaete worms that lack a digestive system produce their own energy from parasitic bacteria powered by sulfur compounds. Deep sea springs along volcanic ridges in the ocean floor of the Pacific near the Galapagos Islands were discovered in 1977. The springs heat water to 8°C–16°C, considerably higher than 2°C in surrounding water on the ocean bottom.

Cold seawater vents flow down through cracks in volcanic areas. They react chemically with lava giving up some minerals but enrich others. Heat comes from lava, with water reemerging through mineralized chimneys high above the sea floor. The result is **white smokers** (milky rich in zinc sulfides under 300°C) and **black smokers** (clear water from 300°C to 450°C) as sulfur compounds precipitate making the water appear black.

The most productive communities are **upwelling areas** mainly found on the western and southern sides of continents. As water is pushed northward toward the equator by winds from the south, water is deflected from the coast by the Coriolis effect. Cold surface waters move in and bring nutrient enrichment to sunlit areas. Thus, anywhere there is upwelling, a commercial fishery can be found.

Tidal marshes of the U.S. east coast appear as a waving grassland for hundreds of kilometers along the sea. They begin to form first as tidal sands or mudflats colonized by algae then eventually **cordgrass** (*Spartina alterniflora*). The organisms are adapted to salt and can rid themselves of it by having salt-secreting cells in their leaves. This leaves crystals on the leaves washed by high tides. Hollow tubes leading from the leaf to the root allow oxygen diffusion. At higher elevations, soil is poorly drained and anoxic. Below ground, material accumulates as peat and *Spartina* above ground is shorter and not as green than on the ocean front. Growing among this *Spartina* are salt-loving plants: Glasswort (*Salicornia*), Sea Lavender, and Spearscale. In the **high marsh** areas, *Spartina patens*, Salt Meadow Cordgrass, and Spikegrass (*Distichlis*) thrive. This community forms a tight mat with dead growth lying beneath the live growth, keeping other plants from invading. The highest part of marsh is dominated by Black Needlerush, flooded only in the highest tides. It is the most fresh because of rains. Draining the high marshes are tidal creeks, heavily populated by alligators in the U.S. South.

THINKING QUESTIONS

1. For animal phyla that you recognize, name three that are confined to the sea.
2. Which do you think has more nutrients, upstream or downstream waters? Explain.
3. How would ecosystem function change if a stream were blocked by a dam?
4. Think about how a beaver forms a lake. Draw a picture and use words to show whether the lake forms upstream or downstream from the dam.
5. Draw a cross section of a lake and indicate where each habitat is found (littoral, limnetic, etc.) and temperature stratification (epilimnion, etc.).
6. In your own words, carefully and completely explain the phenomenon of how animals can walk on the surface of water.
7. Explain the spring and fall turnover phenomenon in northern deep lakes.
8. What is usually the limiting nutrient in lakes?
9. Which gases dissolve in water? Why is turbulent water white? Explain how pressure affects gases and the potential consequences of gas trapped in the bottom of a lake.
10. What is the difference between alkalinity and conductivity?
11. Why is carbon dioxide so important in water?
12. What is a greentree reservoir and how can it provide important habitat for ducks? What precautions need to be taken?
13. What are best practices to enhance species of walleye, pike, trout, perch, crappie, bass, and bluegill in temperate lakes of the U.S.?

REFERENCES

Brönmark, C., and L. Hansson. 2018. *The Biology of Lakes and Ponds*. 3e. Oxford University Press, Oxford, UK.

Carpenter, S. 2008. Seeking adaptive change in Wisconsin's ecosystems. In: D.M. Waller and T.P. Rooney (eds.), *The Vanishing Present: Wisconsin's Changing Lands, Waters, and Wildlife*. The University of Chicago Press, Chicago, IL.

Environmental Laboratory. 1987. Corps of Engineers wetlands delineation manual. Wetlands Research Program Technical Report Y-87-1. US Army Corps of Engineers. Waterways Experimental Station.**

Lampert, W., and U. Sommer 2007. *Limnoecology: The Ecology of Lakes and Ponds*. 2e. Oxford University Press, Oxford, UK.

Margulis, L., and M.J. Chapman. 2010. *Kingdoms and Domains: An Illustrated Guide to the Phyla of Life on Earth*. Elsevier, Amsterdam, The Netherlands.

McComas, S. 1993. *Lake Smarts, the First Lake Maintenance Handbook*. Terrene Institute, Washington, DC.

McComb, B. 2015. Wildlife Habitat Management: Concepts and Applications in Forestry. CRC Press. Boca Raton, FL.

31

New perspectives in biogeography

CLASS ACTIVITY: Testing the Validity of Rapoport's Rule, Geographic Range Size Tends to Increase with Latitude

Activity: Choose a species group (e.g., birds, orchids, ants, reptiles, whales) and assess the size of the geographic ranges for species in relation to latitude. Use simple field guides, such as Peterson Guides, to determine the east-west diameter of range sizes (in km) for species near the poles versus near the equator, and for distances in between. Does Rapoport's rule hold up in your analysis? What caveats and exceptions would you put in your analysis? What did you learn in your assessment of geographic ranges?

THE RELEVANCE OF BIOGEOGRAPHY

Geographic range, the range of a species, is considered the fundamental unit of study in biogeography. An understanding of geographic range not only allows an evolutionary understanding. It helps form predictive models and conservation strategies. It provides insight when we want to stop the growing range of a newly introduced species, or at least steer it away from a biodiverse hotspot. It could help slow the shrinking of the range for rare endemic species, or at least direct the range toward climate safety. Biogeography is a natural partner with landscape ecology to integrate the subdisciplines of ecology. It may bring insight to theoretical concepts. In particular it may provide more organization to community ecology, a subdiscipline that has struggled to find a unified theory, if it would use geographic range as its unit of observation.

SUBSECTIONS OF BIOGEOGRAPHY

We begin by recognizing that contemporary biogeography focuses on several aspects including ecogeography, the study of "name rules." For example, contemporary research includes the study of Bergmann's rule in humans, linguistic diversity of Pacific Island humans, and body size of primates.

The study of species hotspots is a biogeographic topic too, with principles already identified for the phenomenon of hotspot location as described in Chapter 9. For instance, areas with the greatest diversity tend to be in the tropics and mountains. Areas with the highest endemicity tend to be on large, isolated islands in the tropics and subtropics. Recent research extends to the converse of high density, the gaps. **Porosity** refers to low-density areas in the distribution of a population. Gaps tend to become more prevalent as one moves away from the center of a geographic range. This nonrandom pattern presumably occurs because of decreasing habitat suitability toward the periphery.

Contemporary biogeography includes **areography**, the study of variation in the size, shape, and patterns of overlap among geographic ranges. As a discipline, areography was conceived by Argentine ecologist and biogeographer Eduardo Rapoport in a little book published in Spanish in 1975 and

DOI: 10.1201/9781003271833-31

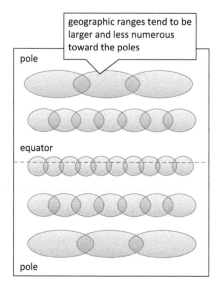

Figure 31.1 Rapoport's rule, an example of an aerographic pattern.

English beginning in 1982 (Rapoport 2013). Rapoport was able to reveal fascinating patterns just by studying geographic range maps in published books. It allowed him to suggest hypotheses about the mechanisms that lead to species' distributions.

In Rapoport's rule, geographic range size tends to increase with latitude (Figure 31.1). The concept has many exceptions, but the pattern is exhibited to some extent by mammals, birds, mollusks, amphibians, reptiles, trees, fish, crayfish, amphipods, and beetles. As an explanation for the pattern, it may be that only organisms with a broad range of tolerances for variable climates can live in high-latitude environments. Whatever the case, the pattern can be striking for groups in which it occurs.

THE CONSTRAINT HIERARCHY IN CHAPTER 9

Before introducing new advances, we circle back to the constraint hierarchy in Chapter 9. Recall that the hierarchy explains which factors provide limits for plants or animals at the global, regional, and local levels. At the global level, Liebig's Law of the Minimum and Shelford's Law of Tolerance were presented. At the regional level dispersal, vicariance, and speciation were highlighted as important processes. It was only at the local level that phenomena like competition, predation, and

symbiosis came to prominence in explaining why species were where they were.

THE NICHE CONCEPT

With expanded vocabulary introduced in Chapters 10–30, geographic range is considered a spatial reflection of a species' niche. If a species has no limits imposed by geographic barriers or colonization abilities, it could occur wherever habitat conditions suited the needs of a species' niche. This theoretical unlimited space is the **fundamental geographic range**, equivalent to the fundamental niche. The **realized geographic range** is what is constrained by geographic barriers, colonization abilities, and other factors.

LARGE SPATIAL SCALES

With increasing latitude, many species are pushed to their most physiological limits, suggesting temperature limitation. At the least, populations may be found only at low elevations and equator-facing mountainsides at high latitudes. In these cold places, it might be competition with another species that limits a population rather than cold itself if the competitor is superior at tolerating cold. Perhaps species tend to be limited on the poleward side of their range by abiotic influences, and on the equator-facing side by a range of biotic interactions. However, we need to understand that at large spatial scales several overarching principles affect the limits for geographic range, rather than a single limiting factor as in Liebig's law. A combination of abiotic, physiological, behavioral, and ecological capacities is at work.

REGIONAL SPATIAL SCALES

At regional spatial scales niche variables alone do not explain all patterns of distribution and abundance. As landscape ecologists have shown, habitat conditions are not equal in all patches across the range. A gradient in habitat quality may be present. Some areas may be **source habitats**, producing surplus individuals that disperse to other areas. Other areas may be **sink habitats**, making it necessary to receive immigrants to maintain population abundance. What may be a source habitat during one year may be a sink habitat during a drought.

Consistent with island biogeography theory, suitability for colonization may be dependent on distance to the "mainland," size of the area, but also age of the island as we learned in Chapter 9. Adequate habitat may exist without residents at all, or with intermittent occupation as predicted by metapopulation theory. Sink areas may absorb excess individuals during irruptions or serve as a link between two subpopulations in a corridor. Very small subpopulations, even if in favorable patches, may go extinct by chance if caught in an extinction vortex. The smallest subpopulations are likely to occur on the periphery of the range and contribute to dynamic shifts in range boundaries.

Furthermore, individuals are often clumped (aggregated) in patches where niche requirements best match the habitat offered within the geographic range (Lomolino et al. 2016). Common species are typically several orders of magnitude more abundant at some sites than others. Results from the N. American Breeding Bird Survey reveal that most species occur in low numbers at several sites, and high numbers at a few sites. Instead of being evenly spaced over their range, individuals reside in a small proportion of their range. For the majority of common songbird species in N. America, more than half the individuals occur at fewer than 20% of the sites within their range.

The abundance of a population may vary substantially over time. Populations at least for vertebrates tend to be boom and busty. The most remarkable are species that show irruptions, such as locusts, mice, grasshoppers, voles, lemmings, and other plague-prone species. When the irruption occurs during favorable periods, some species break out of their source habitats, aggregate in swarms to forage over areas as great as 1,000 times their source patch such as in the Red Locust (Lomolino et al. 2016). Other species increase by orders of magnitude but stay in one place as in lemmings, setting themselves up for repeated bust and boom cycles. Others show wide fluctuations in abundance, but expand the scale of their migration as individuals, as in the Snowy Owl.

The reasons for spatial variation are twofold. Abundance is typically high near the center of the geographic range, and near zero at the boundaries. This occurs for nearly all species, probably reflecting optimal conditions near the center and gradual spreading as abundance grows. Sometimes coastlines and other hard geographic barriers provide exceptions to the rule. Certain species are coastline specialists.

Many habitats are dependent on disturbance, but a large area is required to experience it. For example, grasslands are generally dependent on either stampeding herbivore herds (obviously requiring large spaces) or fire to fend off encroachment by woody shrubs and trees. Fire occurs if the space is large enough to host indigenous cultures who intentionally set fire, or seldom-occurring dry strikes of lightning, which set off large-scale fire. Lightning is usually accompanied by rain which douses fire, or dry strike fires are now extinguished by humans. Thus, most remnant tallgrass prairie reserves are now dependent on prescribed burns and lawn mowers. In other cases, disturbance removes dominant plants or sessile animals, allowing fast-growing but competitively inferior species to colonize. This results in a patchwork of micro-successional stages. In forests, single tree falls may create this effect.

LOCAL SPECIAL SCALES – INTERSPECIFIC INTERACTIONS

Botanical gardens and zoos are evidence that individuals can survive, grow, and reproduce under wider ranges than they encounter in their natural geographic ranges. However, organisms can survive only if they are protected from competing plants, herbivores, and pathogens – a result of biotic limitations. For many species in the wild, the limiting factor may be **diffuse competition**, the combined effect of numerous competitors for a shared resource.

In most cases, species distributions are influenced by a combination of interactions, such as herbivory, parasitism, disease, mutualism, predation, and abiotic influences. Mutualists and facilitators can have just as much influence on a species' distribution as competition, but keep in mind overall for interspecific interactions that a species cannot adapt to each passing species. Organisms are most likely to respond to just a few strong competitive, predator-prey, mutualistic, or pathogenic species. In rare instances, a single keystone species, ecosystem engineer, obligate mutualist, or obligate parasite may influence community structure.

HOW BIOGEOGRAPHY CAN FUNDAMENTALLY CHANGE OUR IDEAS ABOUT COMMUNITIES AND ECOSYSTEMS

The definition for an ecological community is highly arbitrary consisting of a community of species that live together in the same place, defined taxonomically or by trophic level (Lomolino et al. 2016). Boundaries are assigned arbitrarily by the researcher. This vagueness influences the species composition to be found, and the particular interactions that arise. Likewise for ecosystems, researchers try to study naturally confined areas like lakes, but few ecosystems are actually closed. For instance, the main mineral source fertilizing the Amazon basin is dust from the Sahara. Half this dust comes from a dry lakebed northeast of Lake Chad, an area 0.5% the size of the Amazon basin and nearly on the other side of the globe.

Compared to communities and ecosystems, geographic range theoretically has less subjectivity and could bring order to the study of interactions. If we assume that organisms with overlapping geographic ranges have adapted to one another in similar physical environments, we can designate communities as regions and attach geography to our delineation. This might provide a more uniform spatial scale for communities and ecosystems.

Ecoregions are already designated as ecological divisions of the Earth based on responses of organisms to climatic conditions (climate, soil, topography, stream volume, and life form). Although ecoregions were not set up to reflect the evolutionary history of regions and their biotas, they may be a close-enough approximation. With advances in satellite imagery, remote sensing, and further study of climates, ocean currents, and species ranges, ecoregions could become a more common unit of study in ecology. A nested, hierarchical scheme consisting of divisions and provinces at different spatial scales is already built into the idea of ecoregions, which lends itself readily to the constraint hierarchy at different special scales.

THE FUTURE OF BIOGEOGRAPHY

Biogeography is a visionary aspect of ecology. Reintegration of the ideas of biogeography with evolution and conservation biology may make up the ecology of the future. A challenge still left for biogeographers and landscape ecologists is to develop **anthropogenic biomes** now dominating many regions of the earth. These are major vegetative types maintained by human civilizations, including dense settlements, villages, croplands, rangelands, and managed forests. Together they cover 75% of the Earth's ice-free land.

THINKING QUESTIONS

1. Are there plants or animals that are essentially immortal, that is, they live forever?
2. Are there species whose geographic range is the entire globe?

REFERENCES

Lomolino, M.V., B.R. Riddle, and R.J. Whittaker. 2016. *Biogeography*. 5e. Sinauer, Sunderland, MA.

Rapoport, E.H. 2013. *Areography: Geographical Strategies of Species*. Pergamon Press, Oxford, UK.

32

Wicked problems

CLASS ACTIVITY: Adapting to Climate Change

Chapter 32 is the most important section of the book, but one of the most difficult to read. To help put yourself in a hopeful and positive frame of mind, the following activities are suggested.

1. For adapting to the realities of climate change individually and psychologically, review the principles of reciprocity (healing self while healing Earth) in Chapter 1. Find further peer-reviewed articles on this subject in the journal *Ecopsychology* and books on ecotherapy.
2. Devise an energy descent action plan for your college, town, or family. In the U.S. more than 20 federal agencies have published climate change adaptation plans (Newburger 2021). While some U.S. states and cities have proposed a plan, whole states in the U.S. have not, and no city in the state has proposed a plan. Be the first to propose something in your region. To check your state in the U.S. see georgetownclimate.org/adaptation/plans.html.

INTRODUCTORY PRINCIPLES

Anthropocene has been proposed as a new epoch in the geologic record, one we are now in. Put forth by Nobel Laureate Paul Crutzen, the name describes our situation in which humans have so much influence over nature that we have doubled the concentration of carbon dioxide in the atmosphere since the start of the Industrial Revolution. Current levels exceed all maxima for carbon dioxide over the last 800,000 years. This would seem to justify the naming of a new epoch.

In the past 50 years, the human population has doubled (quadrupled since 1900), the global economy has grown nearly fourfold, global trade has grown tenfold, all of this driving up the demand for energy and materials (IPBES 2019). Most of the observed increase in globally averaged temperature since the mid-20th century is the result of the observed increase in anthropogenic greenhouse

gas concentrations. Projections for the next 50 years place human population numbers at 9.1 billion by 2050 (Adam 2021).

Furthermore, global temperatures are expected to rise between 2°F and 9.7°F (1.1°C–5.4°C) warmer by 2100 (IPCC 2021, activesustainability.com). Sea levels could increase by up to 1.8 m. Intensification of the global water cycle could increase the frequency of torrential rains and droughts with a 6.8% increase in rainfall compared to 1995–2014. Heatwaves could be 39 times more common than in the 19th century. Cyclones, hurricanes, and droughts could no longer be seen as extreme because of their frequency. The Arctic could have 77% less ice than in pre-industrial times. Coral reefs could completely disappear. The trajectory is dire, grim, predictable, and still somewhat preventable. Even if we reduce emissions as much as possible, the best-case scenario is a temperature rise of 1.8°C by 2100, which will still cause major disruptions to life as we know it. How is one to reconcile this personally and collectively? Adaptation and mitigation are the simple words that express what must be done, although these require an introduction.

ESSENTIAL VOCABULARY

Wicked problem – a problem difficult or impossible to solve, "wicked" because of its complexity, often within the realm of economic, environmental, and political issues.

Wicked problems are without precedent, difficult even to formulate, with consequences difficult to imagine and considerable uncertainty and ambiguity. The term was introduced by Rittel and Webber in 1973. For wicked problems there may be great resistance to change. A solution may require that many people adjust their mindsets and behavior. Power structures and economic systems may need change. Stakeholders may have radically different views. The problems may change over time. Every problem may interact with other problems. Solutions may be contradictory. Arguments from the participants may be illogical, replaced by value-based thinking. The problem may not be understood or named until after the formulation of solutions.

THE SUPER WICKED

Climate change meets the definition of **super wicked**, which has the additional characteristics of time running out, no central authority, and the people trying to solve the problem are also causing it. Climate change is particularly complex because its effects are varied (www.epa.gov/climatechange). It can increase or decrease rainfall and crop yields. It can affect human health positively or negatively. It can cause changes to forests and other ecosystems and impact our energy supply. Access to energy and disruptions caused by human migration have been at the heart of all the world's wars for the last 20 years.

WHAT IS ADAPTATION?

Some governments, institutions, and individuals inside and outside the U.S. are preparing for the impacts of climate change through "adaptation." It is one of the ways to respond to climate change along with mitigation.

Adaptation – planning for the changes that are expected to occur through climate change.

Some communities have devised **energy descent action plans**, strategies to reduce a community's dependence on fossil fuels and reduce its carbon footprint. The community of Totnes and District, a market town, and its 15 encircling parishes in the U.K. was an early leader in writing a plan (Hodgson et al. 2010). The effort involved individuals, families, organizations, policy makers, and service providers with the underlying philosophy that if they waited for the government to do this, it would be too late. If they each tried it alone, it would be too little.

Unlike most councils, businesses, and governments that plan by assuming there will be more jobs, energy, cars, houses, businesses, and economic growth in their future, an energy descent action plan moves the community toward personal wellbeing. Some worldwide examples of local adaptation include (Dewan 2022):

- **Medellin, Colombia**: with a network of roadways and waterways shaded by 8,000 planted trees and 350,000 shrubs, shaded areas are 5.5°F (3°C) cooler than in the nearby sunshine.
- **Vienna, Austria**: offers parks with mist sprayers and water features for showering or sitting in the mist. The city also has abundant water fountains for drinking in a city with little air conditioning.
- **Abu Dhabi, United Arab Emirates**: uses an ancient form of architecture on modern buildings that consists of latticed screens diffusing sunlight without completely blocking light or breeze. In the modern version, light sensors fold out the lattices like an umbrella during heat, reducing the building's need for air conditioning by 50%.
- **Miami, Florida**: has planted trees around bus stops. For an added touch they etched haiku into the sidewalk of some of the stops, with the poems selected from a contest.
- **Athens, Greece**: plans to use ancient underground aqueducts for watering shade trees and misting in public places like Vienna.

RESILIENCE

Central to the idea of adaptation is resilience. For instance, the global economy is now highly networked. A shock or crisis in one part can pulse through the rest of the system. Resilience is about the ability to adapt to shock rather than disintegrate, like a surge protector in an electrical system.

Resilience – the ability of a system to withstand shock from the outside.

Ironically, it is an imperfect and redundant urban infrastructure (e.g., transportation or communication systems) that provide alternatives and ways out when one system fails. If the subway system floods, the bus system takes up the slack. If cell phone towers fail, a mechanical siren alerts the community to hurricanes, tornadoes, or forest fires. Innovation and creativity are important (Alberti 2013). When new frames of reference and new constraints emerge, people and planners see opportunities and risks they have never experienced. Instead of seeking to reduce uncertainty, future policies and management practices succeed or fail according to how well they can maneuver change, complexity, and uncertainty.

Five principles apply to planning under uncertainty in urban ecosystems (Alberti 2013):

- create and maintain diverse development patterns that support diverse human and ecosystem functions.
- maintain self-organization and increasing adaptation capacity instead of trying to control change.
- design strategies that are robust under the most divergent but plausible futures.
- create options for learning through experiments.
- expand the capacity for change through transformative learning. Challenge assumptions.

The key resource is individual people working together within a community. The creativity and pooled set of ideas solve problems in real time rather than depending completely on outside help. To summarize resilience, it could take the form of an economy that cycles more money locally, creates more local jobs, and is less at the mercy of major employers deciding to relocate elsewhere (Hodgson et al. 2010). It diversifies in terms of skills, livelihoods, land use, businesses, and housing. It produces most of its food locally, creates a range of livelihoods, retrofits old buildings rather than building new, buys energy locally, encourages locally owned businesses, uses food wastes to create bio-methane for powering vehicles, and has their banks reinvest in the community more so than in international markets.

To summarize this chapter, for a wicked problem, we cannot wait for outside entities to take leadership in finding solutions. Individuals working creatively and adroitly with other community members can find ways to adapt within the circumstances of an uncertain and changing situation.

THINKING QUESTIONS

1. What other issues qualify as wicked problems?
2. The chapter does not explain what permaculture is. Look up this concept and describe it.
3. Which do you think is the weaker system in your municipality, electricity generation, water filtration, or water production plant? Explain.
4. Our solar system will not exist forever. At what time is the sun expected to burn itself out or take another form?
5. The Earth is on several cycles that affect its temperature such as wobbling slightly on its axis, interacting with other planets, and moving around the sun. Use the internet to make a list of these cycles and compare their affect with the predicted changes due to greenhouse gas emissions.

REFERENCES

activesustainability.com. Retrieved August 2022. The IPPC answers the question: what will the world look like in 2100? Sustainability for all. Acciona, Business as Unusual.

Adam, D. 2021. How far will global population rise? Researchers can't agree. *Nature New Forum*, 21 September 2021 from *Nature* 597:462–465. doi: https://doi.org/10.1038/d41586-021-02522-6

Alberti, M. 2013. Planning under uncertainty: regime shifts, resilience, and innovation in urban ecosystems. https://www.thenatureofcities.com/2013/01/22/planning-under-uncertainty-regime-shifts-resilience-and-innovation-in-urban-ecosystems/

Dewan, A. 2022. These cities are better at enduring extreme heat. Here's what they're doing different. cnn.com/2022/08/04/world/cool-cities-heat-wave-climate-cmd-intl/index.html

Hodgson, J., R. Hopkins, and T.T. Totnes. 2010. Transition in action: Totnes and District 2030, an energy descent action plan. Transition Town Totnes. Green Books, Cambridge, U.K.

IPBES. 2019. Summary for policymakers of the global assessment report on biodiversity and ecosystem services of the Intergovernmental Science-Policy Platform on Biodiversity and Ecosystem Services. IPBES Plenary at its seventh session (IPBES 7, Paris, 2019). Zenodo. https://doi.org/10.5281/zenodo.3553579

IPCC. 2021: Summary for policymakers. In: [V. Masson-Delmotte, P. Zhai, A. Pirani, S.L. Connors, C. Péan, S. Berger, N. Caud, Y. Chen, L. Goldfarb, M.I. Gomis, M. Huang, K. Leitzell, E. Lonnoy, J.B.R. Matthews, T.K. Maycock, T. Waterfield, O. Yelekçi, R. Yu, and B. Zhou (eds.)], *Climate Change 2021: The Physical Science Basis. Contribution of Working Group I to the Sixth Assessment Report of the Intergovernmental Panel on Climate Change*. Cambridge University Press, Cambridge, UK and New York. doi:10.1017/9781009157896.001

Newburger, E. 2021. Here's how U.S. government agencies are planning to adapt to climate change. https://www.cnbc.com/2021/10/07/

Rittel, H.W., and M.M. Webber. 1973. Dilemmas in a general theory of planning. *Policy Sciences* 4:155–169. https://www.cc.gatech.edu/fac/ellendo/rittel/rittel-dilemma.pdf

33

Epilogue – the evolution of an idea

HISTORICAL OVERVIEW

What have we gained through our study of ecology and the scientific study of nature? We can start by concluding that:

- natural history was the great achievement of the 1700s,
- natural selection was the great achievement of the 1800s,
- return to nature was the great achievement of the 1900s.
- will life as we know it survive the 2000s?

Ecology became a household word by 1970, the nucleus of the environmental movement, and once and for all a part of the collective conscious (Worster 1994). Still, if our goal in ecology is to elucidate the principles that govern nature, what have we accomplished? If our goal is to conserve biodiversity and continue to make life livable for human beings and all living things, what have we gained?

EXISTING ORGANIZATIONAL IDEAS IN THEORETICAL CONCEPTS

Addressing principles first, the categorization of population, community, ecosystem, and landscape ecology has been rich and helpful. Specialization has occurred within each area. A profusion of journal articles, books, and society proceedings has made ecology more precise and predictive, allowing an understanding at different hierarchical levels as in the constraint hierarchy (Lomolino et al. 2016). However, the subdisciplines have diverged over time, developing different vocabularies, sometimes creating an obstacle to their mutual

enrichment. In reality, nature is neither a population, community, ecosystem, nor landscape. It is all at once, with subdisciplines identified merely for organization and simplicity. Despite attempts to remain holistic, population, community, and ecosystem ecology each became reductionistic in their quest to be quantitative.

Is reductionism the only way to elucidate principles that govern nature? Have emergent properties been missed? Is the mechanistic approach required to keep ecology rational, or will the great ecological achievement of the 21st century be a more synthetic and relevant understanding? Would reunification provide better solutions to environmental problems?

TOWARD A NEW UNDERSTANDING

The main question within theoretical ecology over the last 150 years has been what determines abundance and distribution of organisms. Once ecosystem ecology arrived, an additional question asked what regulates energy and matter in ecosystems. In the search for answers, helpful dichotomies emerged:

- is population growth controlled by density-dependent or density-independent factors,
- are communities better described by a holistic or an individualistic view,
- are energy and mater in ecosystems regulated by biotic or abiotic factors, top-down, or bottom-up determinants?

More than a century of ecological research has shown there are no simple answers. At the local level extrinsic abiotic factors (like fire and ice) often act together with intrinsic biotic interactions

DOI: 10.1201/9781003271833-33

(like facilitation, predation, and parasitism), affecting many species and individuals at once. Life history, chance, age, stage, and scale at which nature is examined make a difference in the answer we provide. It appears that ecological theory does not have the order of mathematics and physics. Too many complexities and exceptions occur when living beings are involved, or do they?

Proponents of the biogeography and macroecology approach argue that large spatial and temporal scales reveal simpler and clearer patterns than small-scale reductionist research characteristic of 20th-century ecology (Lomolino et al. 2016, Lawton 1999, Maurer 2009, Brown 1995). Specifically, macroecologists argue that species interactions of traditional community ecology are not the main drivers of species richness. Relationships among temperature, rainfall, and biomes/hotspots/ecoregions determine patterns more so than interspecific interactions.

A SIMPLIFIED ORGANIZATIONAL SCHEME

Perhaps we could bring order and synthesis to theoretical ecology by combining:

1. the historical context of when and by whom ecological theory was built,
2. the constraint hierarchy of Chapter 9,
3. a separation of the major ecological concepts into either descriptive patterns or explanations for patterns.

Laws, rules, and principles are descriptive patterns, and **hypotheses and theories** are explanations for these patterns. To qualify as a law, there must be no exceptions under ordinary circumstances. Concepts that meet the criteria of law when applied to living beings include (Dodds 2009):

- the physical laws of thermodynamics, Boyle's law of gases, Mendel's law of segregation and law of independent assortment, and other physical and chemical laws that apply to all entities including organisms,
- the physiological laws regarding the definition of life, such as there is no spontaneous generation, and all organisms require energy for metabolism, maintenance, and reproduction.

DOES ECOLOGY HAVE LAWS?

Most ecological patterns have exceptions and therefore do not qualify as laws. One true ecological law is limits to growth, the idea that no population can indefinitely increase in abundance while also maintaining a consistent individual body size. Beyond this, ecology may have only patterns, not laws (Lawton 1999). In Table 33.1 and the narrative to follow, "rules" and "principles" are stripped away because they become ambiguous in practice. Patterns are simply categorized as patterns and all explanations are "theories" (as long as they are widely accepted by ecology experts).

We could then attempt a summary of all the discoveries over the centuries in scientific ecology, while simultaneously working our way down the constraint hierarchy, recognizing patterns and explanatory theories. Table 33.1 adds more detail than the narrative by providing a column for major tenets of each theory. Many of these were identified and listed in a helpful book by Walter Dodds (2009).

A SUMMARY OF ECOLOGICAL UNDERSTANDING – NARRATIVE

Our current understanding of scientific ecological knowledge includes the Copernican view of the solar system, including the hypotheses about the formation of our star-planet system (Kricher 2009). We recognize the atomic theory and quantum mechanics, Einstein's theories of relativity, the concept of space-time, the big bang and subsequent expansion of the universe, plate tectonics, and continental drift.

IDEAS SPECIFIC TO ECOLOGY – GLOBAL SCALE

Even before Darwin's time we recognized a pattern of vegetative and animal life forms categorized into what we now call biomes or ecoregions. Biomes fall into four moisture and temperature bands from the equator to the poles: humid tropical, dry, humid temperate, dry polar areas.

In working toward an explanation for the pattern, we can place our biomes in a modified Whittaker diagram with axes labeled as temperature and moisture (see Figure 8.1). The biome theory of life form explains that plant life forms largely depend on climate. Animal life forms

Table 33.1 Highlights of an entire ecology course

Pattern	Theory name	Some major tenets
Biome pattern of life forms	Biome theory of life forms	Plant life forms are largely a function of evapotranspiration. Animal life forms are largely a function of plant life forms.
Ecographic patterns for animals – many exceptions exist.	Ecographic theory	For related forms of endothermic animals, darker colors occur in more humid environments (Gloger's). For warm-blooded animals, those in hotter environments have longer limbs and appendages (Allen's). For mammals, turtles, and some other animal groups, there is a tendency for average body mass of the geographic population of an animal species to increase with latitude (Bergmann's). Groups tend in one direction, e.g., larger body size with evolutionary time (Cope's). Environmentally similar, but isolated regions have distinct assemblages of mammals and birds (Buffon's). Antipredator defenses (such as thicker, more sculpted shells) are developed to a greater degree in low latitudes (Vermeij's). Invertebrates with direct development are more likely to become isolated and have higher speciation rates (Thorson's). With distance from the equator, number of vertebrae for marine fishes increases (Jordan's). Geographic range size tends to increase with latitude (Rapoport's). There is an increase in litter sizes of mammals and clutch sizes of birds in colder climates (Rensch's). Smaller-bodied animals become larger and larger-bodied animals become smaller on islands compared with mainland areas (Forster's).
Hot spot maps for global species distribution, and individual range maps.	Theory of limiting factors and tolerances	Organisms exhibit a temperature optimum. For terrestrial plant species, nitrogen is often the limiting factor. For aquatic plant species, phosphorus is often the limiting factor. Interactions among more than one factor can limit populations.
Scaling – numerous allometric patterns occur, "power laws." Many exceptions exist within these patterns.	Macroecology theory	Most species exhibit small ranges, and a small number of species exhibit large ranges. Within related species groups, range size appears to be related to average local abundance, such that species with large ranges tend to be abundant throughout those ranges, whereas species with small ranges tend to be less abundant. A positive relationship exists between range size and animal body mass, dispersal capabilities of marine mollusks, and germination niche breadth of weedy plant species. Geographic range size decreases with decreasing latitude and decreasing elevation. The number of fragments and discontinuities tends to increase near the range periphery. Within a range, abundance typically decreases from center to periphery. A maximum size exists for organisms with an exoskeleton, and size of mobile animals, and size of flying animals.

(Continued)

Table 33.1 (*Continued*) Highlights of an entire ecology course

Pattern	Theory name	Some major tenets
Three-domain tree of life. Species specialize and become adapted. No two species are alike.	Theory of biological evolution Theory of natural selection Theory of plate tectonics Endosymbiont theory describing the origin of mitochondria and chloroplasts.	Spontaneous generation does not occur. Both biotic and abiotic processes can drive evolution. All organisms die. All species reproduce. Organisms that move have sensory capabilities. All organisms require energy for metabolism, maintenance, and reproduction as they grow, move, and survive. Enzymes operate under Henri-Michaelis-Menten kinetics. Genes sort under Hardy-Weinberg principles. Organisms need liquid water. All life on Earth has a common biochemical basis with mechanisms of homeostasis and heredity. All organisms interact with the environment. They influence it and are influenced by it. Species extinction is irreversible.
The law of segregation and the law of independent assortment.	Mendel's laws of inheritance. Chromosome theory of inheritance. Neutral theory of molecular biology, and nearly neutral theory.	Transposable genes, horizontal gene transfer – by any of these names some genetic elements are prone to frequently change their position in the genome or can transfer to another genome.
No population can increase indefinitely while maintaining a consistent body size.	Growth limit theory Equilibrium theory Non-equilibrium theory	A population of organisms will increase in size exponentially (geometrically) given abundant nutrition and no disease, predation, emigration, or immigration. For populations, the mathematical properties of density-dependent growth can give rise to one stable population value, stable oscillations, chaotic behavior, or instability. Populations often do not grow as in a logistic model when in a natural situation because abiotic or biotic factors limit their growth first.
More than one species is present in all systems and all species interact with some other species.	Community ecology theory	Heterotrophic organisms need autotrophic organisms. All species have symbionts and/or facilitators. No community in which competition does not occur has been described yet. Probably all organisms have diseases (viral or otherwise). Humans interact with more species more strongly than any other species.
Larger areas will have more species. Extinction of a population is more probable the smaller the population.	Theory of island biogeography	Heterogeneity increases biodiversity. Spatial variation within an area leads to an increase in diversity relative to a spatially homogeneous area.

(Continued)

Table 33.1 (*Continued*) Highlights of an entire ecology course

Pattern	Theory name	Some major tenets
Areas with an intermediate amount of disturbance will have the most species compared to areas with little disturbance or a great deal of disturbance.	Intermediate disturbance theory	Many ecologists find this idea too vague to be helpful in practice.
No two species require the same resources and conditions.	Niche theory	
Various null hypotheses can be proposed under the idea that all species or individuals are equivalent. It begins with the premise that no pattern except randomness would be expected, then explanations are sought in the absence of randomness.	Neutral theory of ecology	For instance, at Barro Island Colorado, Hubbell found that the key to maintaining high biodiversity was that individual organisms only interacted with their local environment, and only with the individual competitors and predators nearby.
Populations and resources are distributed heterogeneously over space and time.	Metapopulation theory Patch dynamics theory Landscape ecology theory Shifting-mosaic steady-state theory	
When foraging, animals and animal-like organisms tend to maximize energy gain and minimize energy expended.	Optimal foraging theory	Giving up time theory

(Continued)

Table 33.1 (*Continued*) Highlights of an entire ecology course

Pattern	Theory name	Some major tenets
Some species have an inordinate influence on all others in the community.	Keystone species theory	
Ecosystems tend toward maximum energy storage, maximum power, and maximum ascendancy as an emergent property.	Succession theory	
1. Energy cannot be created or destroyed under ordinary circumstances, but it can change forms. 2. All energy pathways tend toward entropy.	The laws of thermodynamics	Ecological systems are all open with regard to energy. At less than a global scale, ecological systems are open with regard to nutrients. Biomass pyramids will occur. Trophic cascades can occur, especially in systems in which organisms can see each other easily, as in the open water area of lakes. Regulating these processes are a combination of stoichiometry rising from the bottom of the food chain, and predatory effects descending from the top.
Characteristics of a river change downstream and laterally.	River continuum theory Flood pulse theory	
Occam's razor – the principle of parsimony holds with many exceptions.		If an organism regularly displays a characteristic that costs it energy, it must have some adaptive value.

Theory is defined as a well-accepted explanation for a pattern. Many of the tenets come from a compilation by Dodds (2009).

largely depend on the type of plant that occurs. Because plants are greatly limited by their need for water, life forms are mainly a function of evapotranspiration. More specifically, at the global scale, latitudinal effects combined with Earth's tilt on its axis produce air circulation patterns. This accounts for the major bands of moisture and dryness from the equator to the poles. Much has been learned about biogeography since the idea of biomes was conceived in the early 20th century. Ecoregions provide a ready-made numbering system at three spatial scales, made more useful for ecologists at the global level by the World Wildlife Fund categorizations that combine ideas about biomes with bioregions.

In explaining why species are where they are and why they are clustered in hotspots, the scheme presented by biogeographers Lomolino et al. (2016) can basically be summarized in the constraint hierarchy schematic presented in Chapter 9. Lomolino et al. (2016) further make the case that geographic range of species is a more evolutionarily meaningful unit of study than arbitrary assignments in community ecology and ecosystem ecology. In particular, geographic range provides a more standard spatial scale than the variety of often tiny scales used by community ecologists. Geographic range provides other insights, for instance, in asking questions about gaps where species diversity is low as one moves away from the center of a range. With the technology now provided by landscape ecology, macroecologists and biogeographers implore ecologists to investigate the world's ecology at larger spatial scales, for good reason. While theorists have struggled to come to unified theories, especially in community ecology, generalized patterns based largely on climate come to light easily at large scales (e.g., Rapoport's rule).

A further summary message from contemporary ecology research is that within the constraint hierarchy at the global level, chemical interactions make Liebig's Law of the Minimum too simple when concluding there is but one limiting nutrient for any situation. A combination of abiotic, physiological, behavioral, and ecological capacities of the species determines its range. Likewise, Shelford's Law of Tolerance should be expanded to include a host of tolerance variables when defining a niche. Together Liebig's Law of the Minimum and Shelford's Law of Tolerances might be better labeled simply as the theory of limiting factors and tolerances.

In contrast to life form, more specificity is possible when describing species distribution and abundance. Overarching patterns of species richness can be depicted in hotspot maps, with ever finer detail through spatially explicit models in landscape ecology. The same factors that determine life form largely determine species richness. In other words, species richness is a function of evapotranspiration. Using the constraint hierarchy, physical and chemical tolerances (especially evapotranspiration, freezing, and fire) determine where species are found.

With increasing latitude, many species are pushed to their most poleward limits, suggesting limitation by temperature, but undoubtedly this conclusion has the same issue with multiple tolerance variables as when defining a niche. At the least, populations may be found only at low elevations and equator-facing mountainsides at high latitudes. In these cold places, it might be competition with another species that limits a population rather than cold itself if the competitor is superior

at tolerating cold. Perhaps species are limited on the poleward side of their range by abiotic influences, and on the equator-facing side by a range of biotic interactions.

Over long temporal scales, the speciation machine has been driven by tectonic plate movements, astronomic effects (meteors), and variations in the Earth's orbit and tilt. All this is probably best labeled as the biological evolution theory, with the inclusion of what began with the theory of natural selection, but now includes the modern synthesis, emerging ideas in postmodern thought, and contributions from plate tectonic theory.

IDEAS SPECIFIC TO ECOLOGY – REGIONAL SCALE

Regional climate processes operate in conjunction with land forms such as altitude, aspect, and air circulation to create situations like rain shadows and maritime influences, which affect biome presence. Altitude follows the same pattern in life forms as latitude to some extent. To explain the patterns of species richness at regional scales, the concepts of dispersal, speciation, and vicariance are central as outlined in the constraint hierarchy. Our understanding of speciation has benefited from 150 years of study on evolution. In the new understanding through molecular evolutionary biology, natural selection is but one of the processes that shape evolving genomes. Neutral processes such as genetic drift and draft are what mainly shape evolution.

We have learned there is no life without viruses. The tree of life might be better conceived as a tree with tangled roots at its base. One of the greatest questions in biology still unanswered is how this tangle resulted in a single common language of genetic coding for protein synthesis. Horizontal gene transfer is widespread in prokaryotes and makes pivotal contributions to the evolution of eukaryotes. Especially important is endosymbiosis along with various forms of genome fusion and exchange of genetic material between hosts and parasites. In developmental biology the ability of cells to self-organize into complex structures can lead to major evolutionary innovations such as the origin of the vertebrate limb with perhaps little or no genetic change. We now understand that switches in the genome turn on and off genes, and natural selection acts in conjunction with phenotypic plasticity, facilitated variation, and epigenic inheritance.

Dispersal and colonization niche variables alone do not explain all patterns of distribution and abundance. As landscape ecologists have shown, habitat conditions are not equal at all patches across the range. A gradient of habitats may be present. Some areas may be source habitats, producing surplus individuals that disperse to other areas. Other areas may be sink habitats, making it necessary to receive immigrants to maintain population abundance.

Consistent with island biogeography theory, suitability for colonization may be dependent on distance, size of the area, and, importantly, age of the island. Adequate habitat may exist without residents at all, or with intermittent occupation as predicted by metapopulation theory. Sink areas may absorb excess individuals during boom years or serve as a link between two subpopulations in a corridor. Very small subpopulations, even in favorable patches, may go extinct by chance if caught in an extinction vortex. The smallest subpopulations are likely to occur on the periphery of the range and contribute to dynamic shifts in range boundaries.

Furthermore, individuals are often clumped (aggregated) in patches where niche requirements best match the habitat offered within the geographic range. The abundance of a population may vary substantially over time. Populations at least for vertebrates tend to be boom and busty including those that show irruptions. It is typical for abundance to be high near the center of the geographic range, and near zero at the boundaries.

Many habitats are dependent on disturbance for their continued survival, but a large area is required to experience large-scale disturbance, for example, grasslands dependent on fire. In other cases, disturbance removes dominant plants or sessile animals, allowing fast-growing but competitively inferior species to colonize. This results in a patchwork of micro-successional stages. In forests, single tree falls may create this effect, and a shifting-mosaic steady-state model most accurately describes the situation.

IDEAS SPECIFIC TO ECOLOGY – LOCAL SCALE

At the local level, geologic features create patchiness in life form patterns. From here, it is only at the third-deepest and local level in the constraint hierarchy that inter- and intraspecific interactions are most important, such as competition, predation, parasitism, facilitation, and the ability to withstand disturbance and disease. More than 100 years of study in population and community ecology steered us away from an old paradigm in which ecological systems were considered to be closed, self-regulating, and existing at or close to an equilibrium. In the old days a balance of nature was assumed, a point of stability, as long as humans did not disturb it. This is still the thinking of most people outside of academic ecology, and some ecologists. According to Pickett et al. (2007), the idea persists because it is a cultural metaphor more so than a scientific one.

On the one hand we can be grateful; the investigation of the equilibrium concept was central in a successful effort especially in the 1950s and 1960s to launch ecology as a bone fide discipline of biology, one based on mathematical theory (Pickett et al. 2007). Experiences like those described by E.O. Wilson made the focus on theory necessary. Wilson and James Watson of DNA fame began teaching at Harvard in 1956. Ecologists were considered by their fellow biologists to be nothing more than stamp collectors, counseled to avoid using the dirty word "ecology" in faculty meetings (Wilson 2000, Kingsland 1995). To counter, Wilson turned to experiments on islands, and became the co-author of one of the most famous theories to be proposed in ecology, the island biogeography theory (MacArthur and Wilson 1967).

So too, G.E. Hutchinson and R. MacArthur worked to turn ecology into something more than natural history (Kingsland 1995, McIntosh 1985). They built on the efforts of earlier ecologists who worked in laboratories and gardens. Others were able to make the science into an environmental movement. Conservation biology came into existence in the 1970s, largely transforming the mission of ecology from seeking theory as its main goal, to conserving biodiversity.

Robert MacArthur, a charismatic person, knew how to make use of the opportunities available to him, in the opinion of Sharon Kingsland (1995). MacArthur and his small group demonstrated that it is possible for a handful of people to inordinately affect a science, or at least this science. (Even more remarkable was the influence of Aristotle and his group, influencing the preeminent belief about nature in Western thought for most of two millennia.)

For the MacArthur/Hutchinson group, at stake was their surety that populations and communities held a steady equilibrium, and it was competition that regulated this. The niche was crucial because there had to be a unit structure over which species fought for possession (Kingsland 1995). To make it mathematical, they had to draw attention to species in defined dimensions and make it at least theoretically measurable.

The accumulation of long-term study brought about a new ecological paradigm that might be best termed, the flux of nature (Pickett et al. 2007). At some large spatial scales (the whole Earth) ecological systems may be closed and self-regulating. At others,

- populations, communities, ecosystems, and landscapes are permeable to fluxes of energy, materials, and information from the outside,
- their regulation can result from frequent disturbances, competitors, mutualists, or consumers,
- equilibrium, if it exists at all, may be a case of patch dynamic stability, requiring an overall matrix at a large spatial scale to operate.

Ecological systems may change by accidents of history from human influences and other organisms, and changes in climate and other environmental effects. Coexistence among species may occur because of niche partitioning, character displacement, mutualisms, and spatial and temporal patchiness. In these patches, intraspecific exploitation competition can limit the growth of a population before interspecific competition has a chance to take hold. When this is combined with low dispersal abilities or strong competitors, highly biodiverse situations result, with many similar species coexisting. In summary, nature is patchy with some degree of randomness, which is why the metapopulation idea and patch models are helpful. Advances in animal behavior allow predictions regarding food choice and when it will be profitable to move to a different patch. We recognize the dominating influence of certain species as keystones.

Stunting sometimes takes place when conditions are in short supply, which can allow exponential growth for an extended time, but individual body size declines. Some populations are self-thinned. Others have doomed surplus built into population growth, which may look like repeated boom and bust periods when plotted.

Failing to understand the flux of nature metaphor, we make ourselves vulnerable to the myths that (a) no human management is justified because it will upset the balance, or (b) any sort of human impact is justified because nature can rebound. More right-minded approaches are to follow the rich strategies provided by conservation biology, Aldo Leopold's principles of wildlife management, and the success of practices in restoration ecology. Basically, this entails protecting and restoring habitat for indigenous species, while removing non-natives.

THE PRESENT AND FUTURE

Threads now trending in research ecology have the potential to bring insight and synthesis to the subdisciplines of ecology. Landscape ecology and biogeography are providing prospective from large spatial scales and constraints working down toward smaller scales rather than a reductionist philosophy. Such a prospective makes principles easier to understand and links population, community, and ecosystem ecology in a hierarchy.

A second trend is the emphasis on mutualisms, which is bringing further enlightenment to every aspect of ecology and provides an exciting and different way to view nature. We have moved from species considered mainly as competitors and predators, to species often as cooperators and facilitators. Further discoveries in genetics and virology will inventively move ideas forward in evolution. Recognition of anthropogenic biomes and a change in mindset from humans outside of nature to humans as part of it has profound implications. Interestingly, ecology has now benefited from a more worldwide diversity of ecologists who contribute to theory and practices rather than leadership among a small group of individuals in certain graduate universities. Scientific ecological knowledge is beginning to recognize traditional ecological knowledge.

The task still ahead is restoring human relationships with the Earth. In the past 50 years, the human population has doubled, the global economy has grown nearly fourfold, and global trade has grown tenfold, driving up the demand for energy and materials. Projections for the next 50 years place human population numbers at 9.1 billion by 2050, and we have doubled the concentration of carbon dioxide in the atmosphere since the start of the Industrial Revolution. Global temperatures are expected to rise between 2°F and 9.7°F (1.1°C–5.4°C) warmer by 2100. Sea levels could increase by up to 1.8 m. The

Arctic could have 77% less ice than in pre-industrial times. Coral reefs could completely disappear. The best-case scenario is a temperature rise of 1.8°C by 2100, which will still cause major disruptions to life as we know it. Adaptation and mitigation are the simple words that express what must be done.

Beyond these environmental problems a cultural and spiritual collision course is in effect too. This is reflected in the disconnect and lack of familiarity many people have with nature. There is hope. Forming a reciprocal relationship with the Earth – healing self by healing Earth – is a powerful and motivating idea. With allies beyond the borders of ecology in medicine and psychology this effort may have surprising traction, led by efforts in ecopsychology and ecotherapy (Tamasso and Chen 2022).

WHAT CAN WE DO?

If human society in general had a closer relationship with nature, we might be better caretakers. Toward this goal ecopsychologists Kolan and Poleman (2009) argue that we need a different approach to education and learning toward nature. Evolutionarily, our brains are patterned to learn about nature. Those who were not attuned to cycles and patterns or who could not interact and move with the land did not survive to reproduce. It is only in very recent times that our lives have become sedate and virtual. Kolan and Poleman argue that the practice of natural history is a doorway to the study of wholeness, writing:

> We must "recognize our place in the fabric of life, to witness birth and life, joy and suffering, to be attuned to the beauty, rhythms, cycles, and preciousness of life, to heighten our sensitivity of what needs attention in our lives via reflection when in nature. These are the true gifts and wisdom of nature."

Likewise, Steve Trombulak and Tom Fleischner (2018) have called for "a renaissance in natural history education." Fleischner has in mind not just the study of nature, but the practice of natural history, which he describes as "the intentional focused attentiveness and receptivity to the more-than-human world, guided by honesty and accuracy."

Natural history has been separated from ecology to distance theoretical ecology from amateur nature study, a strategy attributed to Robert MacArthur (Kingsland 1995). Sagarin and Pauchard (2010) argue that it is time to reunite. Understanding nature as story is comforting and familiar. Nature as story is as old as humans, recorded in pictograms on the walls of caves, eventually written in words, and now available via movies and every form on the internet. Sagarin and Pauchard ask, which is more effective in gaining instant supporters, the latest technical report on climate change, or a photo of polar bears hopelessly searching for icebergs on which to fish? Emotions play a monumental role in our experience of nature. Sagarin and Pauchard believe it is time to use it advantageously.

In what Sagarin and Pauchard (2010) call "observation ecology" the strictly theoretical and experimental approach taken in ecology ignores the goldmine of information available through citizen science and existing historical documents for collecting information. Old photographs, church records, logs from old whaling ships describing ice pack, Alaskan gambling records on when the ice would go out – these can be used in conservation efforts by embracing observation methods. Modern technology can be used too. Critter-cams can put animals to work watching other animals. TikTok reels are a treasure trove.

WHAT ECOLOGY IS AND IS NOT

Ecology at best has been a body of knowledge, methods, and vocabulary. The body of practices for solving some of humanities' most serious problems and to restore the more spiritual relationship with the land has been outside its parameters. We need a change. Our fate is not yet sealed on climate change and biodiversity loss, but the window is closing. Immediate action is vital in a way that many in the human population are terrified to embrace. For enlightenment we must realize the ways of thinking about nature that are less mechanistic. It is time to embrace natural history, traditional ecological knowledge, and ecopsychology as portals that are not as scary. It could work.

For our task at hand here and now, the evolution of an idea has been reviewed to this point. Test yourself comprehensively on the next page by answering several overarching questions in ecology. An ecology student should be able to answer most of them by the end of the course.

LAST WORDS

"It is impossible to gain pure knowledge of nature. We can only reflect on our limited human experience of nature acknowledging that it is always partial, evolving and in need of application." *Salzman and Lawler (2008)*

AFTER READING THIS BOOK TEST YOURSELF WITH THESE QUESTIONS

1. What is the difference between natural history and ecology?
2. What was the reigning paradigm in understanding nature before the 1800s?
3. Do individuals within a species generally sacrifice themselves for the good of the species? Explain.
4. If natural selection can only eliminate unfavorable traits, what produces favorable traits?
5. What are the three most common limiting factors for plants?
6. Why are organisms where they are?
7. What regulates population growth?
8. Competition can regulate population growth, but only in some situations. In what cases would competition tend to occur?
9. What are the greatest controlling factors in ecological communities?
10. Why is it that predators do not eat all the prey then go extinct?
11. Globally, where would you expect the greatest species diversity?
12. Why are there so many plants? Why is it that herbivores have not proliferated to the point where they eat all the plants then go extinct?
13. What regulates ecosystems?
14. Is there a balance of nature?
15. Make a list of at least ten laws, principles, or theories that are prominent in ecology.
16. Of the main types of ecology, which do you think will be most productive in the 21st century? How will the 21st century be known in ecology?
17. Now that you have taken an entire course in ecology, what definition would you give to "ecology?" Explain your answer.
18. How many years ago was the big bang? When was the formation of Earth?
19. How many years ago was the beginning of life on Earth?

20. How many years ago was the beginning of the age of reptiles? When was the beginning of the age of mammals and birds?
21. When did *Homo sapiens* first evolve? When did *H. sapiens* reach the Americas?
22. What has been left out of the explanations in the summary? What needs more or less emphasis?

REFERENCES

Brown, J. 1995. *Macroecology*. University of Chicago Press, Chicago, IL.

Dodds, W. 2009. *Laws, Theories, and Patterns in Ecology*. University of California Press, Berkeley, CA.

Kingsland, S.E. 1995. *Modeling Nature: Episodes in the History of Population Ecology*. 2e. The University of Chicago Press, Chicago, IL.

Kolan, M., and W. Poleman. 2009. Revitalizing natural history education by design. *Journal of Natural History Education* 3:30–40.

Kricher, J. 2009. *The Balance of Nature: Ecology's Enduring Myth*. Princeton University Press, Princeton, NJ.

Lawton, J.H. 1999. Are there general laws in ecology? *Oikos* 84:177–192. https://doi.org/10.2307/3546712

Lomolino, M.V., B.R. Riddle, R.J. Whittaker, and J.H. Brown. 2016. *Biogeography*. Sinauer, Sunderland, MA.

MacArthur, R., and E.O. Wilson. 1967. *The Theory of Island Biogeography*. Princeton University Press, Princeton, NJ.

Maurer, B.A. 2009. Spatial patterns of species diversity in terrestrial environments. In: S.A. Levin (ed.), *The Princeton Guide to Ecology*. Princeton University Press, Princeton, NJ.

McIntosh. R. 1985. *The Background of Ecology*. Cambridge University Press, Cambridge, UK.

Pickett, S.T.A., J. Kolasa, and C.G. Jones. 2007. *Ecological Understanding: The Theory of Nature and the Nature of Theory*. 2e. Academic Press, San Diego, CA.

Sagarin, R., and A. Pauchard. 2010. *Observation and Ecology: Broadening the Scope of Science to Understand a Complex World*. Island Press, Washington, DC.

Salzman, T.A., and M.G. Lawler. 2008. *The Sexual Person: Toward a Renewed Catholic Anthropology*. Georgetown University Press, Washington, DC.

Tamasso, L.P., and J.T. Chen. 2022. Toward a theory of nature experience and health. *Ecopsychology* 14:282–297. http://doi.org/10.1089/eco.2022.0005

Trombulak, S.C., and T.L. Fleischner. 2018. The natural history renaissance continues. *Journal of Natural History Education and Experience* 12:1–4.

Wilson, E.O. 2000. *Naturalist*. Island Press, Washington, DC.

Worster, D. 1994. *Nature's Economy*. Cambridge University Press, New York.

Index

Note: **Bold** page numbers refer to tables; *italic* page numbers refer to figures.

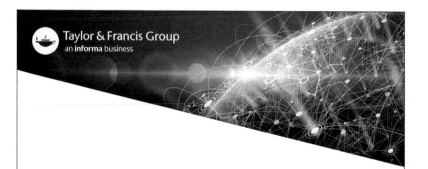